JONES AND BARTLETT SERIES IN BIOMEDICAL INFORMATICS
SERIES EDITOR JULES J. BERMAN

Python
for Bioinformatics

Jason Kinser
George Mason University

JONES AND BARTLETT PUBLISHERS
Sudbury, Massachusetts
BOSTON   TORONTO   LONDON   SINGAPORE

World Headquarters
Jones and Bartlett Publishers
40 Tall Pine Drive
Sudbury, MA 01776
978-443-5000
info@jbpub.com
www.jbpub.com

Jones and Bartlett Publishers
Canada
6339 Ormindale Way
Mississauga, Ontario L5V 1J2
Canada

Jones and Bartlett Publishers
International
Barb House, Barb Mews
London W6 7PA
United Kingdom

Jones and Bartlett's books and products are available through most bookstores and online booksellers. To contact Jones and Bartlett Publishers directly, call 800-832-0034, fax 978-443-8000, or visit our website www.jbpub.com.

Substantial discounts on bulk quantities of Jones and Bartlett's publications are available to corporations, professional associations, and other qualified organizations. For details and specific discount information, contact the special sales department at Jones and Bartlett via the above contact information or send an email to specialsales@jbpub.com.

Copyright © 2009 by Jones and Bartlett Publishers, LLC

All rights reserved. No part of the material protected by this copyright may be reproduced or utilized in any form, electronic or mechanical, including photocopying, recording, or by any information storage and retrieval system, without written permission from the copyright owner.
This publication is designed to provide accurate and authoritative information in regard to the Subject Matter covered. It is sold with the understanding that the publisher is not engaged in rendering legal, accounting, or other professional service. If legal advice or other expert assistance is required, the service of a competent professional person should be sought.

Production Credits
Publisher, Higher Education: Cathleen Sether
Acquisitions Editor: Shoshanna Goldberg
Managing Editor: Dean W. DeChambeau
Associate Editor: Molly Steinbach
Production Director: Amy Rose
Senior Marketing Manager: Andrea DeFronzo
Manufacturing Buyer: Therese Connell
Composition: Northeast Compositors
Cover Design: Kristin E. Ohlin
Cover Image: © Ozger Aybike Sarikaya/ShutterStock, Inc.
Printing and Binding: Malloy, Inc
Cover Printing: Malloy, Inc

Library of Congress Cataloging-in-Publication Data
Kinser, Jason M., 1962-
  Python for bioinformatics / Jason M. Kinser. — 1st ed.
    p. cm.
  Includes bibliographical references and index.
  ISBN-13: 978-0-7637-5186-9
  ISBN-10: 0-7637-5186-3
  1. Bioinformatics. I. Title.
  QH324.2.K55 2009
  572.80285'5133—dc22
                          2008003030
6048

Printed in the United States of America
13 12 11 10 09   10 9 8 7 6 5 4 3 2

To Suzy, Johann, and Jacob.

Preface

The content of this text comes from several semesters of an advanced programming course in the Bioinformatics and Computational Biology department at George Mason University. The intent of this course was to learn an advanced programming language and apply it to current publications in the field of bioinformatics. Along the way a toolbox was created to assist in the replication of published algorithms. The chapters in this book describe the construction of the tools or, more importantly, their use in bioinformatics applications. All of the codes and examples in this text are also available at **www.jbpub.com/biology/bioinformatics**. The reader will be able to replicate all of the examples in this book using the Python codes and data files found at this website.

The Python language was selected because it was the easiest for the students to learn. Attempts at C++ and Java spent more time on the language and very little time on the applications. So, the latter classes were taught exclusively in Python.

When documents were read by the students they believed that they understood the content of the documents. However, when the simple question "How would we put this in code?" was presented to the students they realized that they didn't understand enough of the document. So, a greater amount of learning was accomplished by "doing" rather than "reading." Students that returned after graduation indicated that this approach was highly beneficial once they were in the working environment.

This text is designed from these courses and is intended for readers with knowledge in sequential programming and some biology. It is assumed that the reader has very little Python experience but does have access to the instructional documents that accompany a Python installation or that can be found at http://python.org. This text is not designed to demonstrate the most efficient Python code possible or to explore the very depths of particular bioinformatics algorithms. Rather it is intended as a "good place to start."

Brief Contents

Preface   v

Chapter 1   Introduction   1

Chapter 2   NumPy and SciPy   13

Chapter 3   Image Manipulation   29

Chapter 4   The Akando and Dancer Modules   37

Chapter 5   Statistics   53

Chapter 6   Parsing DNA Data Files   71

Chapter 7   Sequence Alignment   89

Chapter 8   Dynamic Programming   101

Chapter 9   Tandem Repeats   119

Chapter 10   Hidden Markov Models   125

Chapter 11   Genetic Algorithms   139

Chapter 12   Multiple Sequence Alignment   157

Chapter 13   Gapped Alignments   179

Chapter 14   Trees   197

Chapter 15   Text Mining   227

Chapter 16   Measuring Complexity   241

Chapter 17   Clustering   255

Chapter 18   Self-Organizing Maps   275

Chapter 19   Principal Component Analysis   289

Chapter 20   Species Identification   309

Chapter 21   Fourier Transforms   319

Chapter 22   Correlations   333

Chapter 23   Numerical Sequence Alignment   345

Chapter 24   Gene Expression Array Files   365

Chapter 25 Spot Finding and Measurement 375

Chapter 26 Spreadsheet Arrays and Data Displays 389

Chapter 27 Applications with Expression Arrays 401

Index 413

Contents

Preface v

Chapter 1 Introduction 1
    1.1   The Purpose of This Book   1
    1.2   Use of Third-Party Software   2
    1.3   Required Background of Readers   2
    1.4   Object-Oriented Programming   3
    1.5   Presentation Convention   3
    1.6   Conversion from C/C++ to Python   3
            1.6.1   Similarities   4
            1.6.2   Fundamental Python Commands that Differ from C/C++   7
    1.7   The Environment   11
    1.8   Biopython   12
            Bibliography   12

Chapter 2 NumPy and SciPy 13
    2.1   Introduction to NumPy and SciPy   13
    2.2   Basic Array Manipulations   13
    2.3   Basic Math   14
    2.4   More on Multiplication   16
    2.5   More Math   17
            2.5.1   Equals or Copy   17
            2.5.2   Comparisons   18
            2.5.3   More on Slicing   20
            2.5.4   Sorting and Shaping   21
            2.5.5   Random Numbers   23
            2.5.6   Statistical Methods   24
    2.6   Thinking About Problems   24
    2.7   Array Conversions   25
    2.8   SciPy   27
    2.9   Summary   27
            Bibliography   27
            Problems   27

Chapter 3   Image Manipulation   29
- 3.1   The Image Module   29
- 3.2   Colors and Conversions   30
- 3.3   Digital Image Formats   31
- 3.4   Simple Image Manipulations   33
- 3.5   Conversions to and from Arrays   34
- 3.6   Summary   36
-       Bibliography   36
-       Problems   36

Chapter 4   The Akando and Dancer Modules   37
- 4.1   The Akando Module   37
  - 4.1.1   Plotting Routines   37
  - 4.1.2   Algebraic and Geometric Functions   41
  - 4.1.3   Correlation   47
  - 4.1.4   Image Conversions   48
- 4.2   The Dancer Module   48
- 4.3   Summary   52
-       Problems   52

Chapter 5   Statistics   53
- 5.1   Simple Statistics   53
- 5.2   Distributions   55
- 5.3   Normalization   56
- 5.4   Multivariate Statistics   63
- 5.5   Probabilities   66
- 5.6   Odds   68
- 5.7   Decisions from Distributions   68
- 5.8   Summary   69
-       Problems   69

Chapter 6   Parsing DNA Data Files   71
- 6.1   FASTA Files   71
- 6.2   Genbank Files   72
  - 6.2.1   File Overview   73
  - 6.2.2   Parsing the DNA   73
  - 6.2.3   Gene and Protein Information   75
  - 6.2.4   Gene Locations   76
  - 6.2.5   Normal and Complement   77
  - 6.2.6   Splices   78
  - 6.2.7   Extracting All Gene Locations   79
  - 6.2.8   Coding DNA   80
  - 6.2.9   Proteins   82
  - 6.2.10   Extracting Translations   83
- 6.3   ASN.1 File Format   84

	6.4	Summary 87
		Bibliography 87
		Problems 87

Chapter 7 Sequence Alignment   89
- 7.1 Alphabets   89
- 7.2 Matching Sequences   90
  - 7.2.1 Perfect Matches   90
  - 7.2.2 Insertions and Deletions   90
  - 7.2.3 Rearrangements   90
  - 7.2.4 Global Versus Local Alignments   91
  - 7.2.5 Sequence Length   91
- 7.3 Simple Alignments   91
  - 7.3.1 Direct Alignment   91
  - 7.3.2 Statistical Alignment   92
  - 7.3.3 Brute Force Alignment   96
- 7.4 Summary   98
- Bibliography   98
- Problems   99

Chapter 8 Dynamic Programming   101
- 8.1 The Problem with the Brute Force Approach   101
- 8.2 The Dynamic Programming Algorithm   101
  - 8.2.1 The Scoring Matrix   102
  - 8.2.2 The Arrow Matrix   103
  - 8.2.3 Extracting the Aligned Sequences   105
- 8.3 Efficient Programming   107
  - 8.3.1 Flowing along the Diagonals   107
  - 8.3.2 Slicing Matrices   108
  - 8.3.3 Extracting Diagonal Element Locations   108
  - 8.3.4 Extracting Values from the Substitution Matrix   109
  - 8.3.5 Computing the Scoring Matrix Values for a Single Diagonal   110
  - 8.3.6 An Efficient Computation of the Scoring Matrix   110
- 8.4 Global Versus Local Alignments   112
- 8.5 Gap Penalties   114
- 8.6 Does Dynamic Programming Find the Best Alignments?   114
- 8.7 Summary   116
- Problems   117

Chapter 9 Tandem Repeats   119
- 9.1 Tandem Repeats   119
- 9.2 Hauth's Solution   119
  - 9.2.1 Foundation   119
  - 9.2.2 Multiple Words   122

		9.2.3 Tandem Repeats 123
	9.3	Summary 123
		Bibliography 123
		Problems 124

Chapter 10   Hidden Markov Models   125

	10.1	The Emission HMM 125
	10.2	The Transition HMM 128
	10.3	The Recurrent HMM 130
	10.4	Constructing a Transition HMM 132
	10.5	Considerations 136
		10.5.1 Assuming Data 136
		10.5.2 Spurious Strings 136
		10.5.3 Recurrent Probabilities 137
	10.6	Summary 137
		Problems 137

Chapter 11   Genetic Algorithms   139

	11.1	Simulated Annealing 139
	11.2	The Genetic Algorithm 143
		11.2.1 Energy Surfaces 143
		11.2.2 The Genetic Algorithm Approach 144
		11.2.3 Checking the Solution 149
	11.3	Nonnumerical Genetic Algorithms 149
		11.3.1 Notes on Copying 149
		11.3.2 Creating Random Arrangements 151
		11.3.3 The Genetic Algorithm 152
	11.4	Summary 155
		Problems 155

Chapter 12   Multiple Sequence Alignment   157

	12.1	The Greedy Approach 157
		12.1.1 Sequence Comparison 158
		12.1.2 Assembly 160
	12.2	Nongreedy Approach 169
		12.2.1 Creating Genes 170
		12.2.2 Steps in the Genetic Algorithm 174
		12.2.3 The Test Run 176
		12.2.4 Improvements 177
	12.3	Summary 178
		Problems 178

Chapter 13   Gapped Alignments   179

	13.1	Theory of Gapped Alignments 179
	13.2	Chopping the Data 180

13.3 Pairwise Alignments 182
13.4 Building the Assembly 185
  13.4.1 Creating New Contigs 186
  13.4.2 Adding to a Contig 187
  13.4.3 Joining Contigs 191
  13.4.4 Performing the Assembly 193
13.5 Summary 194
  Bibliography 194
  Problems 194

Chapter 14 Trees 197

14.1 Basic Tree Theory 197
14.2 Python and Trees 198
14.3 An Example Using UPGMA 199
14.4 Examples of Trees 203
  14.4.1 Sorting Trees 203
  14.4.2 Dictionary Trees 207
  14.4.3 Percolation Trees 209
  14.4.4 Suffix Trees 217
14.5 Decision Trees and Random Forests 220
14.6 Summary 224
  Problems 225

Chapter 15 Text Mining 227

15.1 An Introduction to Text Mining 227
15.2 Collecting Bioinformatic Textual Data 227
15.3 Creating Dictionaries 228
15.4 Methods of Finding Root Words 229
  15.4.1 Porter Stemming 230
  15.4.2 Suffix Trees 230
  15.4.3 Combining Simplified Porter Stemming with Slicing 231
15.5 Document Analysis 232
  15.5.1 Text Mining Ten Documents 232
  15.5.2 Word Frequency 232
  15.5.3 Indicative Words 236
  15.5.4 Document Classification 237
15.6 Summary 238
  Bibliography 238
  Problems 239

Chapter 16 Measuring Complexity 241

16.1 Linguistic Complexity 241
16.2 Suffix Trees 244
16.3 Superstrings 246

16.4 Summary 254
Bibliography 254
Problems 254

Chapter 17 Clustering 255
17.1 The Purpose of Clustering 255
17.2 k-Means Clustering 259
17.3 Solving More Difficult Problems 262
17.3.1 Preprocessing Data 264
17.3.2 Modifications of k-Means 266
17.4 Dynamic k-Means 268
17.5 Comments on k-Means 272
17.6 Summary 273
Bibliography 273
Problems 273

Chapter 18 Self-Organizing Maps 275
18.1 SOM Theory 275
18.2 An SOM Example 276
18.2.1 Reading an Image 276
18.2.2 Initializing the SOM 277
18.2.3 The Best Matching Unit (BMU) 278
18.2.4 Updating the SOM 280
18.2.5 SOM Iterations 281
18.2.6 Interpreting the SOM 282
18.3 Summary 286
Bibliography 287
Problems 287

Chapter 19 Principal Component Analysis 289
19.1 The Purpose of PCA 289
19.2 Eigenvectors 290
19.3 The PCA Process 291
19.3.1 Case 1: More Dimensions than Vectors 292
19.3.2 Case 2: Linear Combinations in the Data 294
19.3.3 Case 3: Imperfect Dimensionality Reductions 295
19.3.4 Coordinate Selection 296
19.4 Using SVD to Compute PCA 297
19.5 Describing Systems with Eigenvectors 298
19.6 Eigenimages 302
19.7 Summary 306
Bibliography 306
Problems 306

Chapter 20 **Species Identification** 309
    20.1   Data Collection   309
    20.2   The First Clustering   311
    20.3   Using Principal Component Analysis   312
    20.4   The Second Clustering   313
    20.5   Using a Self-Organizing Map   314
    20.6   Summary   317
           Bibliography   317
           Problems   317

Chapter 21 **Fourier Transforms** 319
    21.1   Fourier Theory   319
    21.2   Digital Fourier Transform   320
           21.2.1   DFT Theory   320
           21.2.2   Example with a Simple Sawtooth Signal   320
           21.2.3   Features of the DFT   321
           21.2.4   Power Spectrum   322
    21.3   Fast Fourier Transform   322
           21.3.1   Duplicate Computations   322
           21.3.2   The FFT Method   323
           21.3.3   FFTs in SciPy   324
           21.3.4   The Swap Function   325
    21.4   Frequency Analysis   325
           21.4.1   Simple Signals   325
           21.4.2   DNA Coding Regions   327
    21.5   Summary   331
           Bibliography   331
           Problems   331

Chapter 22 **Correlations** 333
    22.1   Correlation Theory   333
    22.2   Random Signal Correlation   334
    22.3   Structured Signal Correlation   335
    22.4   Correlation of DNA Strings   337
    22.5   Higher Dimensions   338
           22.5.1   Two-Dimensional FFTs in SciPy   338
           22.5.2   Image Frequencies   339
    22.6   The Onset of Image Processing   341
    22.7   Two-Dimensional Correlations   342
           Summary   343
           Bibliography   343
           Problems   343

Chapter 23	**Numerical Sequence Alignment** 345	
	23.1 Alternate Encodings 345	
	23.1.1 Hydrophobicity 345	
	23.1.2 GC Content 347	
	23.1.3 Numerical Methods 348	
	23.2 Numerical Alignments 350	
	23.3 Measuring the Hurst Exponent 351	
	23.4 Chaos Representation 354	
	23.4.1 Representing the Data 354	
	23.4.2 A Simpler Method 356	
	23.4.3 Comparing Chaos Images of Different Species 357	
	23.4.4 Organizing the Data 358	
	23.5 Summary 362	
	Bibliography 362	
	Problems 363	
Chapter 24	**Gene Expression Array Files** 365	
	24.1 Raw Data 365	
	24.1.1 Reading Raw Data in Python 365	
	24.1.2 Dealing with 16-Bit Data 367	
	24.2 GEL Files 369	
	24.2.1 TIFF Headers 370	
	24.2.2 The Image File Directory 370	
	24.2.3 Reading the Data 372	
	24.3 Summary 373	
	Bibliography 374	
	Problems 374	
Chapter 25	**Spot Finding and Measurement** 375	
	25.1 Spot Finding 375	
	25.1.1 Intensity Variations 376	
	25.1.2 Block Location 376	
	25.1.3 The Coarse Grid 380	
	25.1.4 Fine-Tuning the Spot Locations 381	
	25.2 Spot Measurements 383	
	25.3 Summary 386	
	Bibliography 386	
	Problems 386	
Chapter 26	**Spreadsheet Arrays and Data Displays** 389	
	26.1 Reading Spreadsheets 389	
	26.1.1 The Platform File 389	
	26.1.2 The Z-Ratio File 390	
	26.1.3 Reading Two Channel Files 391	

26.2 Displaying the Data   393
  26.2.1 The Heat Map   393
  26.2.2 The R Versus G Graph   395
  26.2.3 The R/G Versus I Graph   396
  26.2.4 M Versus A Graph   397
26.3 Summary   398
Bibliography   399
Problems   399

Chapter 27 Applications with Expression Arrays   401
27.1 LOESS Normalization   401
27.2 Expressed Genes   404
27.3 Multiple Slides   405
  27.3.1 Normalization   405
  27.3.2 Extracting Outliers   409
27.4 Summary   410
Bibliography   411
Problems   411

Index   413

1 Introduction

Bioinformatics is a burgeoning field that has generated a considerable amount of research, as well as a variety of products and associated revenues. The solutions of problems in bioinformatics have historically relied on numerous algorithms developed for other applications. Thus, to understand research results, it is important to understand the algorithms that are used. A good manner in which to investigate algorithms is to create working programs that implement them.

Of course, there are many options available to implement algorithms. One of them is to use previously written computer programs. For a product facility, this may be an appropriate option; for an individual wishing to understand the mechanics of bioinformatics research, this is a less attractive approach. Users of canned software may not understand what the algorithm is doing, its limitations, or even the results that they return.

For individuals wishing to learn the tools of the trade, it is therefore important to consider basic versions of the algorithm and to understand how they are implemented in code. The next concern is deciding which computer language should be used. Again, there are myriad choices—from C++, Java, and Fortran to Python, the language used in this book. Python was chosen because it has several attributes that other languages may be missing: low cost, easily written code, large supporting community, and fast execution, to name a few.

1.1 The Purpose of This Book

The purpose of this book is to demonstrate the construction of algorithms commonly used in the bioinformatics field. Although these demonstrations will be written in Python, it is not the goal here to create the ultimate Python code. In all languages it is possible to generate code that is more efficient (either in code length or execution time), but at the expense of making the code more difficult to read. Since the goal of this book is educational, the flashy scripting will be avoided to keep the instruction set clearly readable.

The algorithms used here are common ones that have been studied and altered over the past few decades. Again, it is possible to spend hundreds of pages on each algorithm, its variations, and efficiency issues. Because the purpose is to demonstrate the use of the algorithms, such details will not be discussed. The goal is to show the base algorithm and how it is applied in bioinformatics.

1.2 Use of Third-Party Software

Some of the programs presented in this book can be found in other forms written by other individuals. It is certainly plausible to replace programs in this text with other programs. There are, however, some concerns in doing so. The first is that the other programs may not be performing exactly the same tasks as described in the text. An example is shown in Code 1-1. Often the covariance matrix is described as a dot product of a matrix with its transpose. In this example the matrix b is computed as such and this is compared to the covariance matrix as computed by a program provided by NumPy. The final statement should produce a 0 if both b and c were equivalent. (As yet these Python commands have not been discussed and are shown merely as a demonstration.)

A second concern is that some third-party programs attempt to catch user errors or accept many forms of data. While this is commendable it makes for a much larger program that is more difficult to read. The programs in this text are intentionally short.

A third concern is that programs written by different individuals may not be directly compatible. Program A written by person 1 will create an output in a certain format. This needs to be fed into program B written by person 2 who is expecting a different format for the input. It would then be up to the user to write a script to convert the data from one format to another. The use of programs written by many individuals can be a bit tedious to manage.

```
>>> from numpy import cov, dot, random
>>> a = random.ranf( (5,5) ) # create a random 5x5 matrix
>>> b = dot( a, a.transpose() ) # covariance via dot product
>>> c = cov( a ) # covariance matrix from NumPy
>>> abs(b-c).sum() # should produce 0.00 if b = c
48.5811449902
```

Code 1-1

The philosophy adopted here is that these programs in their base form are simple to write and to understand. It is probably just as easy to create code to run the algorithm than it is to understand what each individual programmer requires to use their programs.

1.3 Required Background of Readers

Readers of this book will need to have some background in order to benefit from the content. It is assumed that they will have some knowledge of bioinformatics (such as sentencing and alignment) and that they will have some mathematical skills, including sector/matrix math, complex numbers, and a small amount of calculus. In the case of the latter, they should not be frightened by simple integrals or derivatives.

The reader should also have a little programming experience and some familiarity with fundamental instructions such as a *for* loop or an *if* statement even if the experience is with a different language. Readers completely unfamiliar with Python should read the Python tutorial that accompanies the software and presents fundamental commands with simple examples. Readers familiar with languages such as C/C++ may wish to read Section 1.6.

1.4 Object-Oriented Programming

Python has the ability of defining object classes but this is not required. Certainly object-oriented programming can keep large programs well organized and more easily maintained, but this is not a serious consideration for this book.

One of the advantages of a scripting language is that it is possible to try different versions of a function without rerunning an entire program. For example, if a function needs to be altered in a compiled language such as C, then the code is changed and compiled, and the data presented to the new function must either be reloaded or regenerated. In a scripting environment, the original data can be maintained and fed into function without requiring it to be reloaded or regenerated.

Since this book focuses on understanding basic algorithms, the use of object-oriented programming will not be employed. However, it should be understood that all native data types are objects and that some data types used by other modules are also objects.

1.5 Presentation Convention

This book presents several functions and Python commands. Coding regions are captioned similar to figures and referred to in the text by this caption. The executed Python commands will be shown with the typical prompt ">>>". Functions that are stored in files in the accompanying website will be presented with the file name at the top of the code and without the prompt. The user can import the files and run the commands given at the prompts to replicate the results printed here.

In many cases the algorithm will use a random number generator. In order to replicate the results in this text, it would be necessary to use the same set of random numbers. This is accomplished by seeding the random number generator. In all cases the random number seed command can be removed and the results will be similar but not exactly the same.

Within the text, function names will be presented in boldface, and variable names will be italicized.

1.6 Conversion from C/C++ to Python

This section reviews some of the similarities and dissimilarities between Python and C/C++. Readers familiar with other languages such as Java and Fortran may find this section useful in quickly understanding the basic syntax used in Python. Readers just beginning their adventure into programming should read the Python tutorial that accompanies the installation.

1.6.1 Similarities

Python and C/C++ have several commands that are quite similar. Assignment and simple math functions are identical. Code 1-2 displays an assignment, addition, subtraction, multiplication, division, and modulus as well as a combination of an integer division and multiplication. A word of warning: A division result will be an integer if both the numerator and denominator are integers, as shown in Code 1-3.

```
>>> abc = 2.5
>>> abc + 5
7.5
>>> abc - 6
-3.5
>>> abc * 7
17.5
>>> abc /4.5
0.55555555555555558
>>> abc % 6
2.5
>>> divmod( abc, 3 )
(0.0, 2.5)
```

Code 1-2

```
>>> 9/4
2
```

Code 1-3

Python allows many data types similar to C—for example, *int*, *float*, *long*, and *double*. It uses a string class but does not use a char type, and it also offers a complex number data type.

The *if* statement in Python is quite similar to C, with some exceptions. Python does not require the use of parentheses, and it uses words such as *and* and *or* instead of symbols such as && and ||. Finally, Python does not use curly braces to enclose commands within an *if* statement and relies instead on a colon and indentations. Some examples are shown in Code 1-4. The use of parentheses is allowed when creating complicated arguments to the *if* statement.

```
>>> if abc > 2 and abc < 3:
    print 'Yes it is'
Yes it is
>>> if 2 < abc < 3:
    print 'Yes it is'
else:
    print 'No it is not'
Yes it is
```

Code 1-4

Python offers a *while* loop that is also quite similar to the one used in C/C++. Code 1-5 shows a simple example that contains two commands within the *while* loop. Similar to the *for* loop, Python relies on the use of a colon and proper indentation to denote the commands that are within the *while* loop.

```
>>> while i < 4:
        print i,
        i+=1
0 1 2 3
```

Code 1-5

Python has a string class but does not use a char data type. Strings can be denoted by either a set of single or double quotes. Code 1-6 shows a few examples of string usage.

```
>>> st = 'I am a string.'
>>> st + st
'I am a string.I am a string.'
>>> st * 3
'I am a string.I am a string.I am a string.'
>>> st[0]
'I'
>>> st[1]
' '
>>> st[2]
'a'
```

Code 1-6

The reading and writing of files in Python use the same philosophy as in C. The file is opened, it is read, and then it is closed. By default, Python reads and writes text files, but it can be triggered to read and write binary data files. Code 1-7 shows simple reading and writing operations. Similar to C, the files can be opened as read, write, append, etc. Python also offers the **seek** and **tell** commands to manage movement within a file.

```
# read a file
>>> filehandle = file( 'myfile.txt' )
>>> data = filehandle.read( )
>>> filehandle.close( )
# write to a new file
>>> fp = file( 'my_other_file.txt', 'w' )
>>> fp.write( data )
>>> fp.close( )
```

Code 1-7

Functions in Python follow a format similar to the one used in C. The function name is defined, and arguments are passed into a function. The differences are that the arguments do not require type declaration and the function is not required to have a **return** statement. The presence of the **return** command is all that is required to return data from a function. Code 1-8 defines **Myfun**, which receives three arguments. The last argument *c* has a default value of 6. If only two arguments are passed to the function then *c*=6; if a third argument is passed, then *c* is assigned its value. As seen in these examples, *a* and *b* are used as integers or strings. This function will work either way, but the command *print a+b* is not valid if they are different types. Thus, the final example creates an error statement at that line.

```
>>> def Myfun( a, b, c=6 ):
    print a, b, c
    print a+b
    return c + 8

>>> Myfun( 1,5 )
1 5 6
6
14
>>> Myfun( 1,5, -1 )
1 5 -1
6
7
>>> Myfun( 'abc', 'xyz' )
abc xyz 6
abcxyz
14
>>> Myfun( 'abc', 7 )
abc 7 6

Traceback (most recent call last):
  File "<pyshell#41>", line 1, in <module>
    Myfun( 'abc', 7 )
  File "<pyshell#37>", line 3, in Myfun
    print a+b
TypeError: cannot concatenate 'str' and 'int' objects
```

Code 1-8

Objects in Python are somewhat similar to C in that an object contains class variables and functions. In Python each function requires the first argument to refer to itself (commonly with the word *self*, which behaves similar to the **this* in C++). All class variables are public in Python. Code 1-9 shows a simple example. Classes in Python can inherit other classes and can define constructors, destructors, and operators. Python does not allow overloading.

```
>>> class Myclass.
    g = 1  # class variable
    def fun( self, a ):
```

```
        self.g += a    # a is a local variable
        print a

>>> d = Myclass( )
>>> d.g
1
>>> d.fun( 2 )
2
>>> d.g
3
>>> d.g = 'pyt'
>>> d.fun( 'hon' )
hon
>>> d.g
'python'
```

Code 1-9

A Python module is a file that contains Python commands, definitions, and functions. These are included in the Python environment by using the **import** command. This is similar to the *#include* from C. Code 1-10 shows two import statements. The first requires the user to include the module name for each call to a function or variable defined therein. The second uses the **from** *module* **import** *function* format, which relinquishes the requirement that the module name be used in each call. However, it does offer the chance of overwriting a previously defined variable or function. In this example, if a function was named **atof**, then the **from string import atof** statement would have eliminated the previous definition.

```
>>> import string
>>> string.atoi( '57' )
57
>>> from string import atof
>>> print atof( '6.523' )
6.523
```

Code 1-10

One advantage of using **import** *module* is that the user can modify the code within the module and then reload it using the **reload** command. This is not possible with the **from module import function** command.

A third method of including code from a file is the **execfile** command. Equivalent to typing in all of the commands in the file into the interpreter, this command is useful when the user needs to reload the functions several times during development.

1.6.2 Fundamental Python Commands That Differ from C/C++

One major difference between Python and C/C++ commands is the use of indentation instead of curly braces to delineate code that is contained within a *for* loop, *if* statement, *while* loop, function, or class. Python requires that lines within one of these commands be aligned vertically. The leading spaces can be space characters or tabs, but they cannot be mixed. Users that change editors may find that one editor uses spaces and the other uses tabs. While the code may appear to be aligned, errors will still occur. Code 1-11 shows examples demonstrating nested *if* statements and the use of indentations. The number of spaces used in the indentation is not critical as long as statements are aligned. All the statements—*if*, *while*, function, etc.—end with a colon, which is equivalent to the opening curly brace in C.

```
>>> a = 6
>>> if a > 3:
    print 'inside'
    if a > 4:
        print 'further inside'
        print 'second line'
    print 'finished'

inside
further inside
second line
finished

>>> a = 3.5
>>> if a > 3:
    print 'inside'
    if a > 4:
        print 'further inside'
        print 'second line'
    print 'finished'

inside
finished
```

Code 1-11

Python offers a more advanced method of slicing—the process of accessing elements within a string, list, tuple, or array. Code 1-12 shows several examples of slicing a string. A slice is denoted by two indices with the first being inclusive and the second being exclusive. In the first case the characters 0, 1, 2, and 3 but not 4 are printed. It is not necessary to explicitly denote the beginning or end of the string. The second example also prints the first four characters but does not explicitly denote the first. In the third example, the negative index indicates a position from the end of the string, so this prints out the last five characters of the string. The fourth example uses a third number to indicate the step, so this command prints out every second character starting from the first and ending with the tenth. The final command uses the entire string and steps by -1, which prints out the string in reverse order.

```
>>> st = 'I am a string.'
>>> st[0:4]   # first four characters
'I am'
>>> st[:4]    # first four characters
'I am'
```

```
>>> st[4:]    # from the fourth to the end of the string
' a string.'
>>> st[-5:]   # from the fifth element from the end to the end of the string
'ring.'
>>> st[0:10:2] # from the first to the tenth stepping every second position
'Ia  t'
>>> st[::-1]  # string reversal
'.gnirts a ma I'
```

Code 1-12

Two very useful tools in Python are tuples and lists. A *tuple* is a collection of data of any type enclosed by parentheses. This is somewhat similar to a *struct* in C in that it keeps a collection of data together. Tuples however cannot be altered. A *list* is also a collection of data enclosed by brackets, but it can be altered. Code 1-13 shows a simple example of a tuple and a list. The list *b* is altered by adding a new element at the end (**append**), inserting new data (**insert**), removing data denoted by the data itself (**remove**), or removing data by index (**pop**). Both tuples and lists can contain other tuples and lists. Lists are quite useful in cases where the amount of data being generated by a routine is unknown. For example, a DNA string can be searched for a particular pattern, but the number of occurrences is not known before the search begins. In this case, when the results are found, they can be appended to a list, and the list will grow to whatever size is needed.

```
>>> a = ( 'tuple', 1, 5.6, 7+1j )
>>> a[0]
'tuple'
>>> a[1:4]
(1, 5.6, (7+1j))

>>> b = [1, 'list', 5.6]
>>> b[0] = 4
>>> b.append( a )
>>> b
[4, 'list', 5.6, ('tuple', 1, 5.6, (7+1j))]
>>> b.insert( 2, 'new' )
>>> b
[4, 'list', 'new', 5.6, ('tuple', 1, 5.6, (7+1j))]
>>> b.remove( 'list' )
>>> b
[4, 'new', 5.6, ('tuple', 1, 5.6, (7+1j))]
>>> b.pop( 0 )
4
```

Code 1-13

Python offers a dictionary that is similar to a hash table in other languages. A dictionary contains a *key:value* pair in which the index of the dictionary entry is the *key* and the data is the *value*. A dictionary is denoted by curly braces, and any type of data can be used as the key. Code 1-14 shows some examples of a dictionary. Three entries are created using different data types as the keys. The **keys** command returns the dictionaries' keys and the **has_key** command indicates if the dictionary has a specified key. Dictionaries offer advantages in situations where a data set will be searched using only a single key. Searching a dictionary is much faster than searching a collection of lists for a particular entry.

```
>>> dct = { }
>>> dct[1] = 'first entry'
>>> dct['j'] = 'second entry'
>>> dct[9.6] = 'third entry'
>>> dct.keys()
[1, 'j', 9.6]
>>> dct.values()
['first entry', 'second entry', 'third entry']
>>> dct[1]
'first entry'
>>> dct['j']
'second entry'
>>> dct.has_key( 'k' )
False
```
Code 1-14

The final difference between Python and C/C++ to be discussed involves the *for* loop, which employs a list. The index in the *for* loop will become each element of a list. The example in Code 1-15 shows that *i* becomes each element of the list in Code 1-13. The **range** command can be used to generate a list of sequential integers, and the *for* loop can be used in a manner that is traditional in C, as shown in Code 1-16. The *for* loop can also assign multiple indices, as shown in Code 1-17.

```
>>> for i in b:
    print i

new
5.6
('tuple', 1, 5.6, (7+1j))
```
Code 1-15

```
>>> range( 4 )
[0, 1, 2, 3]
>>> for k in range( 4 ):
    print k

0
1
2
3
```
Code 1-16

```
>>> t = [(4,5), (7,6), (5,4)]
>>> for k,l in t:
    print k,l

4 5
7 6
5 4
```

Code 1-17

1.7 The Environment

The Microsoft Windows installation of Python comes with the Idle interface, which contains a text editor, a debugger, and hot keys. However, when Idle is started, the working directory is the same as that of the Python installation. A standard installation would set the working directory as C:/Python25. This is a dangerous setting as users that save files take a chance of overwriting important Python files. Thus, the first command in a session should be to move to a better working directory. Commands to manipulate files and directories are in the *os* module, thus requiring an **import** statement, as shown in Code 1-18.

```
>>> import os
>>> os.chdir( 'c:/mydir' )
```

Code 1-18

Python has a system path variable that indicates the directories that will be searched when the **import** statement is used. This variable is defined in the *sys* module and is merely a list of strings. Developers who wish to collect modules in their own directories will need to append the directory string to this list. Code 1-19 displays the current list of paths (which may differ for each computer) and two additions to this list.

```
>>> import sys
>>> sys.path
['C:\\Python25\\Lib\\idlelib', 'C:\\WINDOWS\\system32\\python25.zip',
'C:\\Python25\\DLLs', 'C:\\Python25\\lib', 'C:\\Python25\\lib\\plat-win',
'C:\\Python25\\lib\\lib-tk', 'C:\\Python25', 'C:\\Python25\\lib\\site-
packages', 'C:\\Python25\\lib\\site-packages\\PIL']
>>> sys.path.append( '.' )# append the current working directory
>>> sys.path.append( 'c:/pysrc' )  # my personal directory
```

Code 1-19

The first two steps that users should invoke when starting an Idle shell in Windows is to change the working directory and to establish the *sys.path* directories for their personal use.

1.8 Biopython

Biopython is an extremely powerful tool that offers the ability to read many different types of files used in bioinformatics. It also provides the ability to access commonly used databases either locally or over the Internet. Biopython also offers commonly used alignment algorithms, which are extremely useful tools in several bioinformatics applications. The reader is encouraged to examine these tools. This book will not attempt to replicate or match the capability of Biopython, but will instead pursue a completely different goal: to review and convert to Python code published bioinformatics applications—from tools such as genetic algorithms and self-organizing maps to principal component analysis and text mining.

Certainly, Biopython and the tools presented here are complementary. A research project could use Biopython to extract large amounts of data from databases, read standard bioinformatics files, and perform alignments. Tools within this book can then be used for further analysis.

Therefore, this text will not attempt to review Biopython or replicate its functions, but will present programs to perform sequence alignment through dynamic programming. The reason for this duplication is that the dynamic programming algorithm demonstrates the necessity of creating efficient Python script.

Bibliography

Biopython. Retrieved from http://biopython.org.
Python. Retrieved from http://python.org.

2 NumPy and SciPy

Numerical arrays are vectors, matrices, and tensors that are important to the speedy manipulation of data. There are two reasons that arrays are so useful: (1) they provide dramatic savings in computational time and (2) they can drastically reduce the amount of coding that is required to accomplish a task. The NumPy and SciPy packages offer a wide range of array functions and data types that will be used in almost all of the applications in this book.

2.1 Introduction to NumPy and SciPy

NumPy and SciPy are packages designed for the manipulation of arrays and scientific data. The NumPy package contains the array manipulation routines and the SciPy package contains a variety of scientific packages that are not part of the standard Python installation. Currently, the packages can be downloaded from http://scipy.org. This website also contains documentation for the myriad functions included in these libraries.

The predecessor to NumPy was a package named Numeric. Users who are switching from Numeric to NumPy should note that there are several differences between the two packages, including variable name changes, function name changes, and function operation changes. One of the major changes that affect work in this book is the removal of type names (such as Float, Integer, and Complex). Functions such as **nonzero** now return a tuple instead of an array, and functions such as **eig** (eigenfunctions) return vectors as columns instead of rows in a matrix. There are many more changes, and users who are switching will need to be on the alert for them.

2.2 Basic Array Manipulations

An array is a vector, matrix, or tensor. There are several methods in which arrays can be created, as shown in Code 2-1. The functions **zeros** and **ones** create arrays filled with either 0's or 1's. Some versions of NumPy will create these arrays as integers while newer versions create them as floats. The type of data contained in an array can be controlled with data types *int, float, complex, double, byte, long*, etc.

```
>>> from numpy import *
>>> a = array( [1,2,3,4,5] ) # creates a vector from a list
>>> b = zeros( 4 )            # creates a vector of length 4
>>> c = ones( 4 )             # vector of length 4 filled with 1
```

```
>>> d = array( [[1,2],[3,4]] )    # matrix
>>> f = zeros( (4,3) )            # matrix of size 4x3
>>> g = zeros( (4,3), int )       # matrix of integers
>>> h = zeros( (4,3), complex )   # complex valued matrix
```

Code 2-1

It should be noted that the argument to the **zeros** function in the creation of a matrix is a tuple that contains two integers. These integers indicate the vertical and horizontal dimension of the matrix. In Code 2-1, f is thus a matrix with 4 rows and 3 columns. Creating a tensor merely requires a tuple with more integers, as illustrated in the example shown in Code 2-2.

```
>>> tensor = zeros( (3,4,5), float )
```

Code 2-2

Python can print out an entire array in a single command, as shown in Code 2-3. Larger arrays are treated differently in various versions of NumPy. Older versions would attempt to print out the entire array, which can lock the Python shell for some time. Newer versions of NumPy print out just a few elements of the array and uses ellipses (…) to indicate that only a small fraction of the elements has been printed.

```
>>> f = zeros( (4,3) )
>>> f
array([[0, 0, 0],
       [0, 0, 0],
       [0, 0, 0],
       [0, 0, 0]])
```

Code 2-3

2.3 Basic Math

One of the many advantages of arrays is that a single command is used to perform basic math operations. In other languages such as C and Java these operations usually require a *for* loop or nested loops. Code 2-4 demonstrates the simple arithmetic of vectors. For these operations the arrays must have compatible lengths or an error occurs. Access to the elements in an array is achieved through slicing.

```
>>> f = zeros( (4,3) )
>>> a = array( [1,2,4.3] )
>>> b = array( [-0.9, 3, 4] )
# add two vectors
>>> a + b
array([ 0.1, 5. , 8.3])
# subtract two vectors
>>> a - b
array([ 1.9, -1. , 0.3])
```

```
# multiple two vectors
>>> a * b
array([ -0.9,    6. ,   17.2])
# divide two vectors
>>> a / b
array([-1.11111111,  0.66666667,  1.075     ])
```
Code 2-4

```
#slicing
>>> c = array( [[1,2,3,4],[5,6,7,8],[9,10,11,12]] )
>>> c[0] # first row
array([1, 2, 3, 4])
>>> c[-1] # last row
array([ 9, 10, 11, 12])
>>> c[0,0] # first row, first column
1
>>> c[1,2]
7
>>> c[:2] # first two rows
array([[1, 2, 3, 4],
       [5, 6, 7, 8]])
```
Code 2-5

Code 2-5 demonstrates a few slices. Extracting an element in a matrix requires a vertical and horizontal location. As seen with *c[1,2]* the slices receive the vertical dimension first and the horizontal dimension second. Extracting a row from a matrix is performed by commands such as *c[0]*. This makes sense since the first argument in the slice is the vertical dimension. Extracting a column in a matrix requires that all of the elements in the vertical dimension be extracted for a single horizontal dimension, as demonstrated in Code 2-6. The colon indicates that all of the first dimension is used, and the 1 indicates that only a single index of the second dimension is used.

```
>>> c[:,1]
array([ 2,  6, 10])
```
Code 2-6

The dimensions of an array are returned by *shape*, as shown in Code 2-7. Note that this is not calling a function (since there are no parentheses). The *dtype* variable returns a code to indicate the type of data contained in an array. In the case that follows, the code 'i4' indicates an integer that uses 4 bytes.

```
>>> c.shape
(3, 4)
>>> c.dtype
dtype('<i4')
```
Code 2-7

2.4 More on Multiplication

There are several ways to multiply arrays. The command $(a*b)$ shown in Code 2-4 performs an element-by-element multiply, which is

$$c_i = a_i b_i; \quad \forall i. \tag{2-1}$$

The *inner product* (or *dot product*) returns a scalar value,

$$c = \sum_i a_i b_i. \tag{2-2}$$

The *outer product* creates a matrix from two vectors, as in

$$c_{i,j} = a_i b_j; \quad \forall i, j. \tag{2-3}$$

Code 2-8 demonstrates these multiplications. The first command performs the element-by-element multiplication in Equation (2-1). The **dot** command performs the inner product, and the **outer** command performs Equation (2-3). An alternative is to use the function **multiply.outer** shown in Code 2-8, which uses a function named **outer** from the module named *multiply*.

```
>>> c = a * b
>>> c
array([ -0.9,   6. ,  17.2])
>>> c = dot( a,b )
>>> c
22.3
>>> c = outer(a,b)
>>> c
array([[ -0.9 ,   3.  ,   4.  ],
       [ -1.8 ,   6.  ,   8.  ],
       [ -3.87,  12.9 ,  17.2 ]])
>>> c = multiply.outer(a,b)
```

Code 2-8

The NumPy package offers other forms of matrix math such as outer addition and subtraction. One example is the outer addition shown in Code 2-9, which performs

$$c_{i,j} = a_i + b_j; \quad \forall i, j \tag{2-4}$$

```
>>> add.outer( a,b )
array([[ 0.1,  4. ,  5. ],
       [ 1.1,  5. ,  6. ],
       [ 3.4,  7.3,  0.3]])
```

Code 2-9

There are many more functions than are listed here, and the reader is encouraged to consult NumPy documentation.

2.5 More Math

The NumPy package offers math functions that apply to an entire array. Code 2-10 begins with a random array and then computes the square root, square, exponential, and logarithm; the last function performs the $\log_2$. Each of these functions performs the mathematical operation on each element in the array. There are other functions as well, including trigonometric ones (sin, cos, tan, tan2, cosh, etc.).

```
>>> from numpy import random, sqrt, exp, log, log2
>>> a = random.ranf( (4,3) )
>>> a
array([[ 0.26978006,  0.63229368,  0.79582955],
       [ 0.96121207,  0.29647525,  0.5989877 ],
       [ 0.45980855,  0.76866697,  0.5310724 ],
       [ 0.58531764,  0.28331364,  0.59143308]])
>>> sqrt( a )
array([[ 0.51940356,  0.79516896,  0.89209279],
       [ 0.98041423,  0.5444954 ,  0.77394296],
       [ 0.67809185,  0.87673654,  0.72874715],
       [ 0.76506054,  0.53227215,  0.76904687]])
>>> a**2
array([[ 0.07278128,  0.3997953 ,  0.63334468],
       [ 0.92392865,  0.08789757,  0.35878627],
       [ 0.21142391,  0.5908489 ,  0.2820379 ],
       [ 0.34259673,  0.08026662,  0.34979309]])
>>> exp( a )
array([[ 1.30967637,  1.88192216,  2.21627876],
       [ 2.61486395,  1.34510926,  1.82027521],
       [ 1.58377075,  2.15688913,  1.70075522],
       [ 1.79556123,  1.32752146,  1.80657553]])
>>> log( a )
array([[-1.31014824, -0.45840131, -0.22837024],
       [-0.03956022, -1.21579155, -0.51251421],
       [-0.77694506, -0.26309748, -0.63285692],
       [-0.53560061, -1.26120073, -0.52520673]])
>>> log2( a )
array([[-1.89014437, -0.66133329, -0.32946862],
       [-0.05707333, -1.75401644, -0.73940171],
       [-1.12089479, -0.37956943, -0.91301954],
       [-0.77270835, -1.81952804, -0.75771315]])
```

Code 2-10

2.5.1 Equals or Copy

The one math operator that does behave as expected at times is the equals sign. The equals sign does not copy one array into another but rather copies the address. Basically, this means that both arrays are pointing to the same location in memory and there is only one version of the array. Changes to one will cause changes to the other variable. The code in Code 2-11 shows the creation of an array b, and the c is set equal to b. The first element of c is changed, but it is b that is printed and, as can be seen, the first element of b was changed. In this case b and c are using the same memory in the computer, and there is no difference between the two.

```
>>> b = array( [5.6, 4.2, 3.8] )
>>> c = b
>>> c[0] = -1
>>> b
array([-1. , 4.2, 3.8])
```

Code 2-11

For applications in this book, this feature is a bit unfortunate as it is desired that b and c occupy different locations in memory, changes in one array will thus not change the other. There are, however, two methods to easily accomplish this. The first is to use the **copy.copy** function, and the second is to simply add 0. These are shown in Code 2-12.

```
>>> import copy
>>> b = array( [5.6, 4.2, 3.8] )
>>> c = copy.copy( b )
>>> c[0] = -1
>>> b
array([ 5.6, 4.2, 3.8])
>>> c
array([-1. , 4.2, 3.8])
>>>
>>> c = b + 0
>>> c[0] = -1
>>> b
array([ 5.6, 4.2, 3.8])
>>> c
array([-1. , 4.2, 3.8])
```

Code 2-12

The addition of a 0 may seem a bit odd at first. The use of the addition sign will actually create a copy in memory of the array. The $b+0$ will create a new array, and with the use of 0 it has the same values as b. The equal sign still shares the memory with the data on the right-hand side of the equation, but this time it is the $b+0$ array, not the original b.

2.5.2 Comparisons

The elements in two arrays can be compared to each other as long as the arrays have axes (or dimensions) of the same length. For example, Code 2-13 shows the comparison of two vectors to determine if the values in one vector are greater than the values in the other vector. The **greater** function returns an array of Boolean values that represent the element by element comparisons.

2.5 More Math

```
>>> from numpy import greater
>>> a = random.rand( 5 )
>>> b = random.rand( 5 )
>>> a
array([ 0.46302608,  0.77595201,  0.9810886 ,  0.71411958,  0.14652436])
>>> b
array([ 0.32756491,  0.37556837,  0.55483111,  0.12703247,  0.72902361])
>>> greater( a, b )
array([True, True, True, True, False], dtype=bool)
```

Code 2-13

Other comparison functions are **less**, **less_equal**, **greater_equal**, **equal**, and **not_equal**. The functions return an array of Boolean values that can be converted to integers (or other types) using **astype**, as shown in Code 2-14. The sign of an array is returned by the **sign** function, as shown in Code 2-15.

```
>>> greater( a, b ).astype(int)
array([1, 1, 1, 1, 0])
```

Code 2-14

```
>>> a = random.rand( 5 )-0.5
>>> a
array([-0.49158702,  0.3527218 , -0.01372066, -0.4705638 , -0.48136565])
>>> sign( a )
array([-1.,  1., -1., -1., -1.])
```

Code 2-15

Finally, it is possible to combine results from the logical operators using functions such as **logical_and**, as shown in Code 2-16. These logical functions perform the functions on the values in the array. To perform functions on the bit patterns in the data, functions such as **bitwise_and** are used.

```
>>> a = random.rand( 5 )
>>> a
array([ 0.07869375,  0.94517074,  0.62396456,  0.11409584,  0.47865114])
>>> a1 = greater( a, 0.5 )
>>> a1
array([False, True, True, False, False], dtype=bool)
>>> a2 = less( a, 0.9 )
>>> a2
array([True, False, True, True, True], dtype=bool)
>>> logical_and( a1, a2 )
array([False, False, True, False, False], dtype=bool)
>>> logical_or( a1, a2 )
array([True, True, True, True, True], dtype=bool)
>>> logical_xor( a1, a2 )
```

```
array([True, True, False, True, True], dtype=bool)
>>> logical_not( a1)
array([True, False, False, True, True], dtype=bool)
```

Code 2-16

2.5.3 More on Slicing

Code 2-5 showed some of the ways to extract information from an array. Of course, the same slicing techniques can be used to alter the data in an array such as the one shown in Code 2-17.

```
>>> c = array( [[1,2,3,4],[5,6,7,8],[9,10,11,12]] )
>>> c[0,2] = -1
>>> c[0]
array([ 1, 2, -1, 4])
>>> c[1] = ones( 4 )
>>> c[:,2] = zeros( 3 ) -2
>>> c
array([[ 1, 2, -2, 4],
       [ 1, 1, -2, 1],
       [ 9, 10, -2, 12]])
```

Code 2-17

A caution needs to be exhibited here. Code 2-17 illustrates an array of integers, and attempting to convert an element to a float will not change the data type. This is shown in the first example in Code 2-18. However, an entire array can be changed to a new data type by adding a scalar of the higher-level data type. The last example uses the **astype** command to change the data type of the entire array. The **dtype** command returns the type of data stored in an array.

```
>>> c[0,2] = 5.5
>>> c
array([[ 1, 2, 5, 4],
       [ 1, 1, -1, 1],
       [ 9, 10, -1, 12]])
>>> c = c + 0.0 # adding a float to everything
>>> c
array([[ 1., 2., 5., 4.],
       [ 1., 1., -1., 1.],
       [ 9., 10., -1., 12.]])
>>> c = c.astype(int)
>>> c
array([[ 1, 2, 5, 4],
       [ 1, 1, -1, 1],
       [ 9, 10, -1, 12]])
>>> c.dtype
dtype('int32')
```

Code 2-18

The methods shown so far can access rows, columns, or individual elements in a regular pattern in an array. Accessing a regular pattern of elements can be achieved by the third argument in slicing. As demonstrated in Chapter 1, the following will extract every other element *a[::2]*. If *a* is a matrix, then this will extract every other row.

Accessing an irregular pattern of elements is also accomplished through list slicing. In Code 2-19 a vector is created using the **arange** function. This function is quite similar to the **range** function except that it returns an array rather than a list. A list *b* is then created, which contains the indices of the array that are to be accessed. In the command *a[b]*, the elements of *a* are extracted according to the list *b*. In the following command those same elements are changed. It should be noted that they are changed according to the order of *b*. Thus, *a[b[0]]* becomes the first element in the tuple and so on. This is an excellent method of accessing irregular patterns of data, which does occur often in the applications.

```
>>> a = arange( 15 )/4.
>>> a
array([ 0.  ,  0.25,  0.5 ,  0.75,  1.  ,  1.25,  1.5 ,  1.75,  2.  ,  2.25,
        2.5 ,  2.75,  3.  ,  3.25,  3.5 ])
>>> b = [6,9,14,7,3]
>>> a[b]
array([ 1.5 ,  2.25,  3.5 ,  1.75,  0.75])
>>> a[b] = -1,-2,-3,-4,-5
>>> a
array([ 0.  ,  0.25,  0.5 , -5.  ,  1.  ,  1.25, -1.  , -4.  ,  2.  , -2.  ,
        2.5 ,  2.75,  3.  ,  3.25, -3. ])
```

Code 2-19

2.5.4 Sorting and Shaping

There are several other useful functions that are used throughout this book. An array can be sorted using the **sort** command. The indices of the sort order can also be returned using the **argsort** command. These are shown in Code 2-20. The **sort** command changes the arrangement of the elements in the array, whereas the **argsort** command indicates the order of the elements but does not change the arrangement of the data.

```
>>> a = random.rand( 5 )
>>> a
array([ 0.09541768,  0.02161618,  0.46247117,  0.83177906,  0.95871478])
>>> a.argsort()
array([1, 0, 2, 3, 4])
>>> a.sort()
>>> a
array([ 0.02161618,  0.09541768,  0.46247117,  0.83177906,  0.95871478])
```

Code 2-20

The **take** command will extract from an array the elements according to a list. In Code 2-21 an array *a* and a list *b* are created. The **take** command gathers the data from *a* according to the prescription defined in *b*. An error will occur if a number in *b* is outside of the range of *a*.

```
>>> a = random.rand( 5 )
>>> a
array([ 0.99325464,  0.37953944,  0.72033516,  0.58302183,  0.65205877])
>>> b = [3,1,2,1,1]
>>> take( a, b )
array([ 0.58302183,  0.37953944,  0.72033516,  0.37953944,  0.37953944])
```

Code 2-21

Arrays can be reshaped as long as the number of elements does not change. The **ravel** function will convert a matrix to a long vector, as shown in Code 2-22. The **reshape** command can convert the array to any other shape as long as the number of elements is the same, as shown in Code 2-23. The **concatenate** function joins two arrays. A second argument can be used to indicate which axes are being joined, as shown in Code 2-24.

```
>>> a = random.ranf( (3,2) )
>>> a
array([[ 0.04897374,  0.35842065],
       [ 0.60357182,  0.98938948],
       [ 0.39334855,  0.44884645]])
>>> ravel( a )
array([ 0.04897374,  0.35842065,  0.60357182,  0.98938948,  0.39334855,
0.44884645])
```

Code 2-22

```
>>> a.reshape( (2,3) )
array([[ 0.04897374,  0.35842065,  0.60357182],
       [ 0.98938948,  0.39334855,  0.44884645]])
```

Code 2-23

```
>>> a = random.ranf( (3,2) )
>>> b = random.ranf( (2,2) )
>>> a
array([[ 0.20894554,  0.72381952],
       [ 0.08718544,  0.9803592 ],
       [ 0.51509276,  0.72995728]])
>>> b
array([[ 0.33436572,  0.23557135],
       [ 0.78765853,  0.41989476]])
>>> concatenate( (a,b) )
array([[ 0.20894554,  0.72381952],
       [ 0.08718544,  0.9803592 ],
       [ 0.51509276,  0.72995728],
       [ 0.33436572,  0.23557135],
       [ 0.78765853,  0.41989476]])
>>> b = random.ranf( (3,3) )
```

```
>>> concatenate( (a,b),1 )
array([[ 0.20894554,  0.72381952,  0.44493734,  0.52981483,  0.17825647],
       [ 0.08718544,  0.9803592 ,  0.57896141,  0.26381222,  0.13927053],
       [ 0.51509276,  0.72995728,  0.15169618,  0.51958984,  0.69107468]])
```

Code 2-24

Finally, a useful function is **indices** that will return two (or more) arrays with the indices indicated as rows or columns. The example in Code 2-25 considers arrays that are 4×3. The indices returns two matrices (in the form of a tensor) with the rows or columns incrementing.

```
>>> indices( (4,3) )
array([[[0, 0, 0],
        [1, 1, 1],
        [2, 2, 2],
        [3, 3, 3]],

       [[0, 1, 2],
        [0, 1, 2],
        [0, 1, 2],
        [0, 1, 2]]])
```

Code 2-25

2.5.5 Random Numbers

NumPy contains a random module that can be imported with *from numpy import random*. This module contains many routines for generating random numbers. A vector of N random numbers with values between 0 and 1 is obtained by *random.rand(N)*. A matrix of random numbers uses a different function *random.ranf((V,H))* where V and H are the vertical and horizontal dimensions of the array. One case that is used later in the book is a set of random numbers based on a Gaussian distribution. Code 2-26 shows the **random.normal** function with no argument, which returns a number from a standard distribution. By adding two arguments, the user defines the mean and standard deviation of the distribution. The third argument returns an array of values along the specified distribution. The *random* module has several other types of random number generators.

```
>>> random.normal()
0.9299042168312539
>>> random.normal(0.5, 0.1)
0.54425343582683983
>>> random.normal(0.5, 0.1, 10)
array([ 0.41583017,  0.47527431,  0.62681263,  0.52641006,  0.55507713,
        0.64093938,  0.46411351,  0.61719065,  0.55209002,  0.5232665 ])
```

Code 2-26

Another useful function is **random.shuffle**, which rearranges items in an array in a random order. In Code 2-27 an array that can consist of any type of items is created and shuffled.

```
>>> a = [1,2,3,4,5]
>>> random.shuffle( a )
>>> a
[4, 3, 2, 5, 1]
>>> a = ['t', (3,4), 45, 6.7]
>>> random.shuffle( a )
>>> a
[(3, 4), 45, 6.7, 't']
```

Code 2-27

2.5.6 Statistical Methods

Arrays are *objects* and therefore contain functions. Code 2-28 shows how to extract information about an array. Without any argument, these functions will compute the values for the entire array. An integer argument determines over which axis the functions are computed. In a two-dimensional array the axes are vertical and horizontal. If the argument to a function is 0, then the function is performed over the vertical axis; if the argument is 1, it is performed over the horizontal axis. Arrays with higher dimensions can receive arguments with larger integer values.

```
>>> d = random.ranf( (4,3) )
>>> d.max( )    # max value in entire array
>>> d.max( 0 )  # max values in each column
>>> d.max( 1 )  # max values in each row
>>> d.min( )    # min value in entire array
>>> d.sum( )    # sum of entire array
>>> d.sum( 0 )  # sum of columns
>>> d.mean( 1 ) # average of rows
>>> d.std( )    # standard deviation of entire array
```

Code 2-28

2.6 Thinking About Problems

It is possible to use *for* loops to perform the addition of two arrays, as shown in Code 2-29. Note that the use of a direct command (>>> d = a + b) is much faster than using loops.

```
>>> d = zeros( 3, float )
>>> for i in range( 3 ):
d[i] = a[i] + b[i]

>>> d
array([ 0.1, 5. , 8.3])
```

Code 2-29

Python is an interpreted language, which means that each step of the program is first converted to machine language and then executed. Thus, a command inside of a *for* loop will be interpreted several times, which is time-consuming. When a command such as $c=a+b$ can be used to perform the same function, a significant amount of time will be saved. In actuality there is still a *for* loop that is performing all of the additions necessary to compute $c=a+b$, but these are performed "beneath the Python" in the compiled language in which Python was written. Functions such as **dot** and **add.outer** are much faster than writing the steps out, as shown in Code 2-29.

Programmers who have experience in C, Fortran, and similar languages are used to thinking in terms of *for* loops in order to solve problems. In order to make Python efficient, this thinking needs to change a little. Problems with data arrays need to be conceived in a more parallel fashion in which operations are performed over the entire array rather than for individual elements. The basic rule is to avoid *for* loops. If a program is being developed that contains nested *for* loops, then there is a good chance that it can be rewritten using faster and more efficient array commands.

2.7 Array Conversions

Arrays often need to be converted to data types such as lists and strings. The **array** command has already been used to convert a list to an array (see the first line in Code 2-1). Code 2-30 shows the **tolist** command, which converts an array to a list.

```
>>> a = array([ 0., 0.25, 0.5, -5., 1., 1.25, -1., -4., 2., -2., 2.5, 2.75,3., 3.25,-3.])
>>> a.tolist()
[0.0, 0.25, 0.5, -5.0, 1.0, 1.25, -1.0, -4.0, 2.0, -2.0, 2.5, 2.75, 3.0, 3.25, -3.0]
```

Code 2-30

Conversion from an array to a string is a little more complicated. The command **str** will convert the characters from a **print** statement into a string, as shown in Code 2-31. However, this includes newline characters. To convert the elements of an array into a string, each element must be sent to the **str** function, which can easily be accomplished by employing the **map** function.

```
>>> str( a )
'[ 0.    0.25  0.5  -5.    1.    1.25 -1.   -4.    2.   -2.    2.5   2.75\n  3.
  3.25 -3.  ]'
>>> map( str, a )
['0.0', '0.25', '0.5', '-5.0', '1.0', '1.25', '-1.0', '-4.0', '2.0', '-2.0',
'2.5', '2.75', '3.0', '3.25', '-3.0']
```

Code 2-31

Another type of string that can be used is a binary string in which the elements of the array are converted to their binary equivalents. Consider an array of integers. A 32-bit processor will encode an integer in four bytes. When applied to an array of integers, the command **tostring** will convert each integer to its four-byte representation. Code 2-32 displays this conversion. The first four bytes (shown as \x00 \x00 \x00 \x00) are the bytes that encode the first integer (which is 0) in the array *a*.

```
>>> a = arange( 4 )
>>> a
array([0, 1, 2, 3])
>>> b = a.tostring()
>>> b
'\x00\x00\x00\x00\x01\x00\x00\x00\x02\x00\x00\x00\x03\x00\x00\x00'
```

Code 2-32

Conversion from a byte string is performed by **fromstring**, which may need an argument to indicate what type of conversion is taking place. The first command in Code 2-33 converts the byte-string to integers (which is the default conversion). The second command converts the same byte string to short integers that use only two bytes instead of four. The third conversion produces two floats (eight bytes apiece).

```
>>> fromstring( b )
array([0, 1, 2, 3])
>>> fromstring( b, Int8 )
array([0, 0, 0, 0, 1, 0, 0, 0, 2, 0, 0, 0, 3, 0, 0, 0], dtype=int8)
>>> fromstring( b, float )
array([  2.12199579e-314,   6.36598737e-314])
```

Code 2-33

These conversions can be very useful when reading and writing to binary files. Some bioinformatics applications use files produced by machines. The data files may contain data stored in binary format that needs to be converted to arrays for processing in Python. The **fromstring** command will convert data into an array in a single command. In the early days of DNA sequencing, many of the sequencing machines were attached to MACs rather than PCs. The problem for PC users was that MACs stored information differently than PCs and so a conversion was necessary to read the data properly.

A MAC stores information in a *big endian* format whereas a PC uses the *little endian* format. Consider a two-byte (short) integer. The least significant byte can store numbers from -128 to $+127$. Any integer bigger than that uses the most significant byte. The big endian format will store the most significant byte first, whereas the little endian format will store the least significant byte first.

Code 2-34 shows an example that begins with array a, which contains four elements that are two bytes each (shorts). The byte string for this is printed for a PC. Now consider a case in which the same data is stored on a MAC. String c is the byte string in this example. When this data is read in on a PC, the conversion will be incorrect (the first **fromstring** command). The **byteswap** command will reverse the bytes and produce the correct data.

```
>>> a = array([0, 1, 2, 3]).astype(short)
>>> a.tostring()
'\x00\x00\x01\x00\x02\x00\x03\x00'

>>> c = '\x00\x00\x00\x01\x00\x02\x00\x03'
```

```
>>> fromstring( c )
array([16777216, 50332160])
>>> fromstring(c,short).byteswap()
array([0, 1, 2, 3], dtype=int16)
```

Code 2-34

2.8 SciPy

SciPy is a large library of scientific routines for Fourier transforms. Solving equations, integrations, differential equations, wavelets, and so on are functions available in this package. Certainly, before creating a scientific program it would be prudent to see if SciPy contains the needed routines. In this book only the Fourier transform module will be used.

2.9 Summary

There are many third-party modules available for Python. This chapter reviewed the NumPy and SciPy packages, which provide a plethora of vector and matrix tools. SciPy is a large package that provides many scientific functions. This text uses only a small fraction of these tools, and readers are encouraged to explore their offerings.

Bibliography

SciPy. Retrieved from http://scipy.org.

Problems

1. What is the difference between the **range** and **arange** functions?
2. Using the **numpy.random** module, create a vector of random numbers.
3. Without using the **for** or **while** commands, compute the average of a vector of 1 million random numbers.
4. Compute the sum of the columns of a large matrix of random numbers.
5. Create a vector of random numbers having a range of −1 to +1.
6. Create a matrix of random numbers. Multiply this matrix by scalar in a single Python command.
7. Create a 5 × 3 matrix of random numbers. Create a vector of three random numbers. What happens when you run *matrix * vector*?
8. Compute the transpose of a matrix.
9. Create a random DNA string and write a Python script to convert the letters to an array of numbers (A=1, C=2, G=3, and T=4). Do not use loops (*for, while*) to accomplish this conversion. (This can be accomplished in five commands, including the creation of the DNA string.)

3 Image Manipulation

Several bioinformatics applications use images to represent the data. Raw data coming off of a gel sequencer is an image of vertical lanes. Expression arrays originate as an image of thousands of circles. Newer applications associate phenotypes extracted from images with genotype information.

The Python Image Library (PIL) is a third-party module that provides a decent amount of image reading, writing, and manipulation software. This chapter will review the basics of this package and explore some of the basics of digital image processing.

3.1 The Image Module

To employ the PIL, the module is imported in one of two ways. For Versions 1.1.5 or earlier, the command **import Image** is used; for Version 1.1.6, the **from PIL import Image** command is used. These are shown as the first two lines in Code 3-1. Note that unlike other modules this one uses a capital letter.

The variable *mg* is an *Image* and has several internal variables that provide descriptive data. The second command in Code 3-1 indicates the dimensions of this image. Images are treated differently than arrays. In the case of an array, the dimensions are returned with the vertical dimension first and the horizontal dimension second. Images are just the opposite, with the horizontal dimension being the first number returned. The (0,0) pixel is in the upper-left corner of the image. The other commands return information about the filename and the compression of the image. The two commands that will be used most in this text are *size* and *mode*.

```
>>> import Image   # versions 1.1.5 or earlier
>>> from PIL import Image  # version 1.1.6
>>> mg = Image.open( 'data/bird.jpg')
>>> mg.size
(1280, 960)
>>> mg.filename
'c7/bird.jpg'
>>> mg.format
'JPEG'
>>> mg.format_description
'JPEG (ISO 10918)'
>>> mg.info
{'jfif_version': (1, 1), 'jfif': 257, 'jfif_unit': 1, 'jfif_density': (300, 300), 'dpi': (300, 300)}
```

Code 3-1

There are three main modes for an image. In this case the 'RGB' indicates that this is a color image with three different color planes. The mode 'L' indicates that an image is grayscale, and the mode 'P' indicates that the image contains a palette. These are used in the next section.

Writing an image to the hard drive is also performed in a single command. The filename extension indicates the type of compression used to store the file. Even though this file was read in as a JPEG, it can easily be stored as a PNG, as shown in the first line of Code 3-2. Many computer environments have a default image viewer, and the **show** command will send an image to this default viewer, which will then be opened and display the image. If a reader is using a computer in which the **show** command does not work, then the alternative is to use the first command to save the image and then to use any image viewer to see it.

```
>>> mg.save( 'mypict.png' )
>>> mg.show()
```

Code 3-2

3.2 Colors and Conversions

In an RGB image there are three values for every pixel, which correspond to the amount of red, green, and blue contained in the pixel. This information is extracted by using **getpixel** and modified by using **putpixel**. Code 3-3 extracts the color values for a pixel that is 10 pixels from the top and 20 over from the left. Note that the pixel location and the returned RGB values are both tuples.

```
>>> mg.getpixel( (10,20) )
(202, 224, 201)
```

Code 3-3

A grayscale image contains one byte per pixel with values ranging from 0 to 255, where a value of 0 indicates a black pixel and a value of 255 indicates a white pixel. An RGB image has three such planes that describe the amount of red, green, and blue. A palette image contains a look-up table and can contain up to 256 different colors. Each pixel contains one byte, which references an entry in the look-up table that then retrieves the RGB values. These are discussed more in the next section.

It is possible to convert from one format to the other using the **convert** command. To change a color image to a grayscale image, the switch 'L' is used, as shown in Code 3-4. Converting a grayscale image to an RGB image would use the 'RGB' switch and all three color channels would obtain the same intensity values.

```
>>> newmg = mg.convert( 'L' )
```

Code 3-4

The three channels of an RGB image can be separated using the **split** command. This will create three new images, each being a grayscale image that contains the values for the red channel, green channel, and blue channel. Code 3-5 shows the splitting of an RGB image. Each of the grayscale images can be viewed using the **show** command.

```
>>> r, g, b = mg.split()
>>> r.mode
'L'
>>> r.size
(600, 450)
```

Code 3-5

The **merge** command will take three grayscale images (of the same dimensions) and create a new color image. Since it is possible to create an RGB image or a palette image, the first argument indicates the type of image. The second argument is a tuple that contains the grayscale images. The first line in Code 3-6 recreates the image. It is possible to change the colors by just rearranging the grayscale images. The second command creates a new image in which the colors are mixed up.

```
>>> mg2 = Image.merge( 'RGB', (r,g,b) )
>>> mg2 = Image.merge( 'RGB', (g,b,r) )
>>> mg2.show()
```

Code 3-6

3.3 Digital Image Formats

There are dozens of methods in which an image is stored on the hard drive. The simplest versions are bitmaps that do not compress the image. A color image is stored using three bytes for every pixel. This method is not very efficient, and a photograph from a 3.3 megapixel camera would consume 10 megabytes of storage. Bitmap images have filenames that have extensions such as *.bmp, .ppm, .pgm,* and *.tga*. Generally, these are used for very small images such as icons.

Commonly, larger images are stored in a compressed format of which there are many. The most common are TIFF, JPEG, GIF, and PNG. The TIFF image format has many variants, and it is recommended that this format should be avoided. It does, however, offer the advantage of supporting 16 bit data planes, which allows a grayscale value to range from 0 to 65,536. Such precision has been used by some microarray software (Chapter 23).

The JPEG image format is commonly used for photographs. It has a very good compression ratio and tends to replicate photographs without too much error. However, it is not perfect. JPEG compression is sensitive to sharp edges in an image. Consider Code 3-7, in which a new image (**Image.new**) and a broad vertical stripe are created. This stripe is 20 pixels wide and 170 pixels tall, and all values in this stripe are set to a bright 250. The **save** command is used to store the image as a bitmap ('line.bmp'). This is a grayscale image of size 200 × 200, and it consumes 41,078 bytes. (The data consumes 40,000 bytes, and the other 1,078 bytes are used to store information about the image.) The fourth command saves the image as a JPEG, and the fifth command reads it back in as *mg2*. The final two *for* loops print out pixels across the top of the stripe (11th row) and across a row more toward the middle (the 50th row). Recall that originally all values inside of the stripe were 250, but as can be seen, the JPEG compression has altered them.

```
>>> mg = Image.new( 'L', (200,200) )
>>> for i in range( 10, 180):
        for j in range( 90,110):
            mg.putpixel( (j,i), 250 )
>>> mg.save('line.bmp')

>>> mg.save('line.jpg')
>>> mg2 = Image.open( 'line.jpg')

>>> for j in range( 90,110):
        print mg2.getpixel( (j, 11) ),
255 244 235 241 240 255 248 248 248 248 248 248 248 248 255 240 241 235 244 255

>>> for j in range( 90,110):
        print mg2.getpixel( (j, 50) ),
248 253 249 249 254 249 250 250 250 250 250 250 250 250 249 254 249 249 253 248
```

Code 3-7

JPEG compression will create ghost edges next to the location of sharp edges in an image. Since photographs generally do not have sharp edges, this effect is much less noticeable. Naturally, an argument occurs that many photographs have sharp edges, but in reality it is the mind's interpretation of the image information. By zooming in on an image, quite often the seemingly sharp edge is shown to be actually a few pixels wide. While JPEG is good for photographs, it does terrible things to cartoons and banners. In some bioinformatics applications (such as those in the field of forensics), it is necessary to keep track of the flow of all information. Alterations to an image can have unfortunate legal effects (such as disqualifying the image for a patent application or a courtroom application). It is important to note that compression such as JPEG does infuse some error into the image.

The GIF format is one that uses a palette. Thus, it is capable of storing only 256 colors, which means that storing a photograph as a GIF can have devastating effects on the colors. However, GIF can store grayscale images without a flaw. This format is also good for color images with a small number of colors (such as cartoons). The GIF format also has the ability to store a sequence of images and display them as a movie. Commonly, web pages that have "dancing flames" or other moving images have stored the information as a sequence of images in a single GIF file.

The PNG format compresses images without alteration. However, it is usually slower to compress and uncompress images using PNG and the compression ratio is not nearly as good as the other formats.

3.4 Simple Image Manipulations

The PIL provides functions to manipulate images. Some of the basic ones are reviewed here. To extract a rectangular section of the image, the **crop** command is used. This command receives a single argument that is a tuple of four numbers. These are the locations of the box in a west-north-east-south format. The first number is the west coordinate or the location of the left portion of the bounding box. Code 3-8 loads an image and cuts out a section of that image. The **resize** command will create a new image of a different size. It will squeeze or stretch the image to fit the new coordinates. Code 3-9 forces the image to have 100 pixels in the horizontal direction and 300 pixels in the vertical dimension. Code 3-10 rotates an image and the argument is in degrees. A positive angle rotates the image counterclockwise while a negative argument rotates the image in the clockwise direction.

```
>>> mg = Image.open( 'data/bird.jpg')
>>> mg2 = mg.crop( (10,10,30,50) )
>>> mg2.show()
```

Code 3-8

```
>>> mg2 = mg.resize( (100,300) )
>>> mg2.save('test.jpg')   # to save it to disk
```

Code 3-9

```
>>> mg2 = mg.rotate( 5 )
>>> mg2.show()
```

Code 3-10

A rotation will not be perfect. Code 3-11 is an example of an image being rotated just 5 degrees in a counterclockwise direction and then rotated again 5 degrees in the clockwise direction. This should reproduce the original image with *mg* and *mg3* being exactly the same as each other. However, the **getpixel** statements reveal that they are not the same.

```
>>> mg2 = mg.rotate( 5 )
>>> mg3 = mg2.rotate( -5 )
>>> mg.getpixel( (50,50) )
(80, 121, 81)
>>> mg3.getpixel( (50,50) )
(75, 116, 84)
```

Code 3-11

Consider an image of a cell on a microscopic slide. This cell can be at any orientation, and so it is conceivable that a program designed to analyze the cell would have to consider all orientations of the cell. A brute force solution would be to consider the cell at every 5 degrees of rotation. The incorrect approach is to rotate the image 5 degrees, analyze it, rotate it another 5 degrees, analyze it, and so on. Each rotation infuses a little error, and by performing the task in this manner each subsequent rotation will add on even more error. Code 3-12 rotates an image every 5 degrees until it is back to the original orientation. The display of this image shows the destructive influences of consecutive rotations.

```
>>> mg2 = mg.rotate( 5 )
>>> for i in range( 5,360, 5 ):
        mg2 = mg2.rotate( 5 )
>>> mg2.show()
```

Code 3-12

The better approach is to rotate the original image 5 degrees and analyze it, rotate the original image 10 degrees and analyze it, etc. In this fashion each image used has been rotated only once keeping the rotation errors to a minimum.

The **transpose** command can rotate or flip images. It receives a single argument that performs one of these functions. The rotation using this method is exactly the same as in the **rotate** function. Code 3-13 shows the five options.

```
mg2 = mg.transpose(Image.FLIP_LEFT_RIGHT)
mg2 = mg.transpose(Image.FLIP_TOP_BOTTOM)
mg2 = mg.transpose(Image.ROTATE_90)
mg2 = mg.transpose(Image.ROTATE_180)
mg2 = mg.transpose(Image.ROTATE_270)
```

Code 3-13

The *Image.FLIP_LEFT_RIGHT* is merely a definition in the Image module. As shown in Code 3-14, this one is defined as an integer 0 and another is defined as an integer 2. The commands can also be called by using the integer equivalents.

```
>>> print( Image.FLIP_LEFT_RIGHT )
0
>>> print( Image.ROTATE_90 )
2
>>> mg2 = mg.transpose( 0 )   # flip left to right
```

Code 3-14

The PIL offers many other transformations as well that will not be included in this book. The reader is encouraged to read PIL documentation to become familiar with the variety of operations that are offered.

3.5 Conversions to and from Arrays

The PIL does not offer a single command to convert an image into an array. However, through a short combination of commands this is possible. Code 3-15 shows the steps that can convert a grayscale image to an array. The first command performs two transformations (*rotate_90* and *flip_top_bottom*). This is to change the orientation of the data for the next command. The **getdata** command extracts the information from a grayscale image (or from a single band of an RGB image) into the variable *f*, which is of type ImagingCore. This can be converted to an array using the **array** command. The variable *z* is a vector array, and the final command converts it to an array of the appropriate size and orientation.

```
>>> mgt = mg.transpose(2).transpose(1)
>>> f = mgt.getdata()
# a structure
>>> z = array(f)
>>> zz = transpose(reshape(z,mg.size))
```

Code 3-15

The conversion of an array into an image requires that the user define a new image as a grayscale. A scaling operation is also needed. A pixel in an image can range in value from 0 to 255, but the data in the array is not at all restricted. Thus, the maximum value in the data is set to 255, the minimum value is set to 0, and all intermediate values are scaled accordingly. Code 3-16 outlines the set of steps used to convert an array to a grayscale image.

The **Image.new** command creates a new grayscale image of the appropriate size. Recall that *data* is an array, and its dimensions are (vertical, horizontal). In contrast, an image uses dimensions (horizontal, vertical), so the **transpose** function is used to rearrange the dimensions. The **min** and **max** functions are used in conjunction with simple algebra to scale the data so that all values are between 0 and 255. The final command uses **ravel** to convert the matrix to a long vector, and it uses the **putdata** command to place this data into the image.

```
>>> mg = Image.new( 'L', transpose(data).shape)
>>> mn = data.min()
>>> a = data - mn
>>> mx = a.max()
>>> a = a*256./mx
>>> mg.putdata( ravel(a))
```

Code 3-16

Code 3-15 and Code 3-16 work only for grayscale images. A color image has three color planes, and so it will need to be converted to three different matrices. This is performed by splitting the color bands of the image (using the **split** function) and then using Code 3-15 on each of the three grayscale images.

Converting three matrices to a color image does require a bit of forethought. If the data scaled in Code 3-16 is used for three different matrices, then the scaling factors for each could very well be different. Thus, all three matrices need to be scaled by the global maximum (the maximum of all three channels) and the global minimum. Each channel can then be converted to a grayscale image using **putdata**, and the three images can be combined into a single color image using **Image.merge**.

3.6 Summary

The Python Image Library (PIL), an easy-to-use third-party module, offers the ability to read and write images in a large number of formats. The PIL also offers functions for the manipulation of images. However, in this book the PIL is mostly used to read and write images. Codes are then provided to convert the images to arrays that can be used in the applications. Tools are also provided to convert arrays back to images so that results can be displayed. Many of the images in this book were created using these tools.

Bibliography

PIL. Retrieved from http:/www.pythonware.com.

Problems

1. Read an image and return its mode and size.

2. Read a color image and exchange the red and green channels.

3. Convert a grayscale image to an array. Double the values of pixels in the top half of the array (without using a *for* or *while* loop). Convert back to an image.

4. Read a color JPG image. Save it as a GIF image. Describe the changes.

5. Repeat Problem 3, but this time double the values of all of the pixels. Does the image change? Why or why not?

6. Repeat Problem 3, but this time compute the square root of all of the values (using **numpy.sqrt**). What does this do to the image?

7. Read in a color image and **split** the color channels. Convert each to an array. Using the **numpy.random** module, add random numbers (ranging from 0 to 25) to each pixel. Convert the three channels back to a single color image.

8. Read in an image and rotate it 20 degrees. Save the image.

4 The Akando and Dancer Modules

The Akando and Dancer modules are unique to this book. The Akando module contains many different but generic routines. The Dancer module is used for display and simple interaction with images.

4.1 The Akando Module

Akando is a Native American word for "ambush" and is used to name the file that contains a variety of generic functions useful in many different types of applications. For example, two routines are used to convert a matrix into an image and an image into a matrix. Because such functions are not unique to a specific application, they are stored in the Akando module.

Like all modules, Akando needs first to be imported. However, since the Akando module is not part of a Python installation, the directory that contains it needs to be added to the search path, as shown in Code 4-1.

```
>>> import sys
>>> sys.path.append( '/mydir/pysrc' )
>>> import akando
```

Code 4-1

Akando is divided into four sections: (1) a section that contains routines for plotting, (2) a section dedicated to algebra and geometry, (3) a section dedicated to correlations, and a section that performs image conversions.

4.1.1 Plotting Routines

Throughout the text, functions will create vectors and matrices that need to be plotted. There are several options for plotting data in Python, but this book will simply save data as text files and use programs such as GnuPlot or a spreadsheet to plot the functions. GnuPlot commands are given in cases that go beyond simple plotting.

To plot a vector in GnuPlot, the data is stored in a single column in a text file, and there is a newline character after each data entry. The function **PlotSave** in Code 4-2 receives the filename and the vector. A file is opened using the "w" option to indicate that it is writable. The *for* loop is used to write each element of the vector. The data is converted to a string using the **str** function, and a newline character is appended. An example of generating a function and saving it as a file is shown. The reader can verify this function by viewing the file "plot.txt" with any text viewer.

CHAPTER 4 THE AKANDO AND DANCER MODULES

```
# akando.py
def PlotSave( fname, data ):
    L = len( data )
    fp = file( fname, 'w' )
    for i in range( L ):
        fp.write( str(data[i]) + '\n' )
    fp.close()

>>> from numpy import cos, arange
>>> a = cos( arange( 100 )*0.3 )
>>> akando.PlotSave('plot.txt', a )
```

Code 4-2

GnuPlot can perform many different functions and the reader is encouraged to view the example files that accompany the program. This section will quickly review the commands that are used within this book. The examples used here apply to GnuPlot v4.0, and the reader should note that there are some differences from previous versions.

Code 4-3 displays commands to be used in the GnuPlot window, not in the Python shell. The first command moves to the directory where the data file is stored, and this, of course, is unique to each user. The second command reads the file and creates a plot in a new window. The third command replots the file and replaces the data points with a connecting line. Figure 4-1 shows the results. Code 4-4 removes the legend and sets the plotting style so that the default is a line plot. The file is then plotted again.

```
gnuplot> cd 'c:/mydir'
gnuplot> plot 'file.txt'
gnuplot> plot 'file.txt' with lines
```

Code 4-3

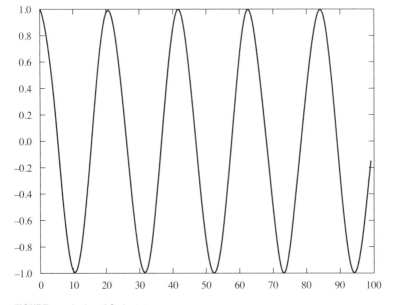

FIGURE 4-1 A plot of Code 4-4.

4.1 The Akando Module

```
gnuplot> unset key
gnuplot> set style data lines
gnuplot> plot 'file.txt'
```

Code 4-4

GnuPlot uses "terminals" to direct the display, and the terminal is set to a default that displays the plot on the user's computer screen. The terminals are different for the various operating systems, and it can also be a file output. Code 4-5 first assigns the filename of "file.gif" to be used when writing the plot to an image. The terminal is then set to a GIF file with small font and a screen size of 400×300. The file is plotted, and this time the information is written to a file named "file.gif" as a GIF image. Any display that is previously showing on the computer monitor will not be updated. The final command in Code 4-5 is to reset the terminal to a Microsoft Windows environment. Users of other operating systems will need to reset GnuPlot to their specific system, which will be listed when GnuPlot starts. Commands will be given as necessary within the text. Again the user is encouraged to view example files that demonstrate the power of GnuPlot; this section merely provides a set of commands commonly used here.

```
gnuplot> set output 'file.gif'
gnuplot> set terminal gif small size 400,300
Terminal type set to 'gif'
Options are 'small size 400,300 '
gnuplot> plot 'file.txt'
gnuplot> set terminal windows
Terminal type set to 'windows'
Options are 'color "Arial" 10'
```

Code 4-5

The previous commands create a dependent variable that is plotted as the y values in the plot. The x values are assumed to be equally spaced integers starting at 0. To create a plot in which the user specifies both the x and y data values, a text file with two columns is needed: the first column for the x values and the second for the y values. The data points will be plotted in the order in which they are listed in the file. If the user is connecting these data points with lines, then it will be necessary for some cases to sort the data according to the x values. The sorting order of the X column can be obtained by using the **argsort** command on this column, and then by using ordered slicing both columns can be rearranged by this sort order. Code 4-6 uses Python to create values for both the x and y values. The final command for GnuPlot changes the style to lines with data points, and it plots the data. The result is shown in Figure 4-2.

```
>>> from numpy import zeros
>>> mat = zeros( (100,2), float )
>>> mat[:,1] = cos(arange(100)*0.3)
>>> mat[:,0] = cos(arange(100)*0.2+5 )
>>> akando.PlotMultiple('file.txt', mat )

gnuplot> plot 'file.txt' with linespoints
```

Code 4-6

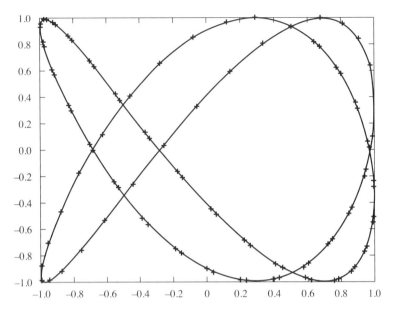

FIGURE 4-2 A plot of Code 4-6.

A surface plot requires a two-dimensional array of *y* values. GnuPlot will read a file with multiple columns as a surface plot when using the **splot** command. Code 4-7 creates a surface to be plotted. The first GnuPlot command changes the viewing angle of the surface, and the second command creates the surface plot. The viewing angle is defined as two angles and a *z* scale. The default for these three values is 60, 30, 1, and the **set view** command changes them. By using this command, the user can rotate the view to find the optimal presentation. This result is shown in Figure 4-3.

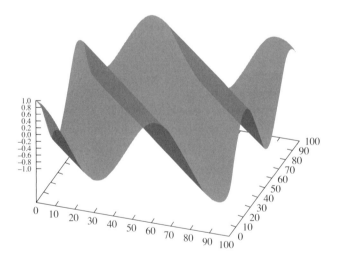

FIGURE 4-3 A plot of Code 4-7.

```
>>> mat = zeros( (100,100))
>>> for i in range( 100 ):
        for j in range( 100):
            mat[i,j] = cos( (i+j)*0.1)
>>> akando.Plot2D('file.txt', mat )

gnuplot> set view 45,20,1
gnuplot> splot 'file.txt'
```

Code 4-7

4.1.2 Algebraic and Geometric Functions

This section will focus on Akando functions that are based on geometry or algebra. The simplest is to measure the Euclidean distance between two vectors, as shown in Code 4-8. The *dvec* is the subtraction of two vectors of any length. The **sqrt** function comes from *numpy* and computes the square root of all elements of the input vector.

```
from numpy import sqrt

# akando.py
def Distance( vec1, vec2 ):
    a = vec1 - vec2
    dist = sqrt( (a*a).sum() )
    return dist
```

Code 4-8

A vector or matrix can be smoothed by performing a local averaging function, as in

$$y_i = \frac{1}{2k + 1} \sum_{n=-k}^{k} x_{i+k} \qquad (4\text{-}1)$$

The k represents window size, and making it larger will smooth the data more. While this equation is the correct operation, it is also repetitive. Consider two cases ($i = 4$ and $i = 5$) with $k = 2$:

$$y_4 = \frac{1}{5}(x_2 + x_3 + x_4 + x_5 + x_6) \qquad (4\text{-}2)$$

and

$$y_5 = \frac{1}{5}(x_3 + x_4 + x_5 + x_6 + x_7) \qquad (4\text{-}3)$$

As can be seen, four of the five elements in these two computations are the same. A more efficient method of computing y_5 is

$$y_5 = y_4 - \frac{1}{5}(x_7 - x_3) \qquad (4\text{-}4)$$

However, for values of i near the beginning or end of the vector, a little more care is required. The function **Smooth** in Code 4-9 has two main sections for the case of a vector or matrix. The *if ndim == 1* will be used if the number of dimensions is 1 and therefore applies to a vector. There are three parts inside this section. The first considers values of *i < k*, the second considers intermediate values of *i*, and the third considers values of *i* near the end of the vector.

The second half of the major *if* statement considers a matrix. This section will smooth each row by sending it as a vector to the same function, after which each column will be smoothed.

```
# akando.py
def Smooth(data,window):
    # data is the input data
    # window is the linear dimension of the smoothing kernel
    dim = data.shape
    ndim = len( dim )   # the number of dimensions
    # for a 1D vector smooth it
    if ndim == 1 :
        ans = zeros( dim, float )
        K = sum(data[0:window+1])
        ans[0] = K/(window+1)
    # ramp up
        for i in range(1,window+1):
            K = K + data[i+window]
            ans[i] = K / (i+window+1)
    # steady as she goes
        for i in range(window+1,dim[0]-window) :
            K = K + data[window+i] - data[i-window-1]
            ans[i] = K / (2*window+1)
    # Ramp down
        j = 0
        if dim<window+window : j =window+window-dim[0]
        for i in range(dim[0]-window,dim[0]):
            K = K - data[i-window-1]
            ans[i] = K / (2*window-j)
            j = j+1
    # end of vector smooth
    else :  # you have more than 1 dimension.
        # smooth the columns and then the rows
        t = data + 0
        for i in range(0,dim[0]):
            t[i,:] = Smooth(t[i,:],window)
        for j in range(0,dim[1]):
            z = (Smooth(t[:,j],window))[0:dim[0]]
            t[:,j] = z
        ans = t
    # end the 2D
    return ans
```

Code 4-9

An example in Code 4-10 creates a cosine function with some noise. These points are plotted as *points* in Figure 4-4. Two smoothed versions of this function are created using different smoothing windows. They are plotted as lines in the same figure. A certain amount of smoothing will reduce the small additive noise. However, too much smoothing will decrease the true maximums and minimums of the function. The amount of smoothing is unique to each application.

4.1 The Akando Module

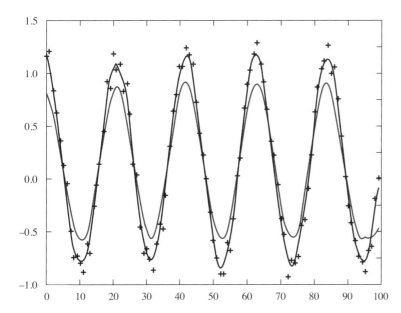

FIGURE 4-4 A cosine function with noise (dots) and two smoothed versions, based on the example shown in Code 4-10.

```
>>> a = cos( arange( 100 )*0.3 )+0.3*random.rand( 100)
>>> b = akando.Smooth( a, 1 )
>>> c = akando.Smooth( a, 4 )
>>> akando.PlotSave('plot.txt', a )
>>> akando.PlotSave('plot0.txt', b )
>>> akando.PlotSave('plot1.txt', c )

gnuplot> plot 'plot.txt' with points , 'plot0.txt', 'plot1.txt'
```

Code 4-10

The **Baseline** function in Code 4-11 removes the baseline from a function. In many experiments there will be a floating bias that distorts the data. Consider the case in Code 4-12 where the original signal is *a* but a bias of *b* is added to it. The **Baseline** function will separate the function into equally spaced sections with a width *WN*. Within each group, the minimum *y* value is identified. Straight lines connect these minimal points as shown in Figure 4-5. The values at the straight lines are subtracted from the original function to create a function that is mostly based at $y = 0$. In the function **Baseline**, the first *for* loop finds these minimum points. If the window size is too big, then this function will not accomplish much. If the window size is too small, then data peaks may be destroyed. The size of the window should be bigger than the main wavelength inherent in the data.

The second *for* loop creates these connecting lines by computing the slope (*m*) and intercept (*b*) for each line segment. The last line in this *for* loop performs the subtraction. The *mask* is used to find the elements that are less than 0, and these are set to 0.

```
# akando.py
def Baseline(data, WN=100):
    L = len(data)
    pts=[]
    for i in range(0, L, WN):
        a=data[i:i+WN]
        mn = a.min()
        x=nonzero(equal(a, mn))[0][0]
        pts.append([i+x, mn])
    nd=zeros(len(data), float)
    nsegs=len(pts)-1
    for i in range(nsegs):
        x1, y1 = pts[i]
        x2, y2 = pts[i+1]
        m = (y2-y1)/(x2-x1)
        b = y1 - m*x1
        w = arange(x1,x2)
        y=m*w + b
        nd[x1:x2]=data[x1:x2]-y
    # create a vector containing 0 if nd is less than zero
    mask=greater_equal(nd, 0)
    # new new data
    nnd = mask*nd
    return nnd
```

Code 4-11

```
>>> a = cos( arange( 100 ) )+2
>>> b = cos( arange( 100 )*0.1 )
>>> c = a + b
>>> nnd = akando.Baseline( c, 10 )
```

Code 4-12

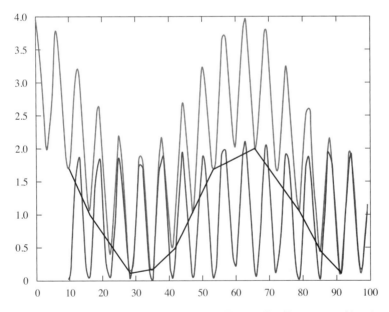

FIGURE 4-5 An example of baseline subtraction. The baseline (line segments) is subtracted from the original but biased signal.

The **RangeHistogram** function shown in Code 4-13 receives a vector (*indata*) and the number of bins (*nbins*), and the histogram is computed. The user has the option of defining the upper and lower limits of the data. The histogram of a single vector can automatically determine the size of the bins from the *max* and *min* values. In cases when multiple histograms need to be compared, it is necessary to ensure that the bins encapsulate the same ranges of data. For these cases, it is necessary to specify the upper and lower ranges of the histogram and thus to specify *mn* and *mx*.

```
# akando.py
def RangeHistogram( indata, nbins, mn=-1, mx=-1 ):
    # indata is the input data
    # nbins is the number of bins
    ans = zeros(nbins)
    L = len( indata )
    data = indata + 0
    fix = 0
    if mn==-1 and mx==-1:
        # no limits were given so - create an autoscale
        mx = indata.max()*1.01
        mn = indata.min()
    else:
        mx *= 1.01
    data = (indata-mn)/(mx-mn)*nbins
    print data.max()
    hst = zeros( nbins, int )
    for i in range( L ):
        k = int( data[i] )
        hst[k] += 1
    return hst
```

Code 4-13

Code 4-14 presents the function **linearRegression**. Given a set of *x* and *y* data points, this function will estimate the line that best fits the data. In Code 4-15 data is generated and plotted as points in Figure 4-6. The slope and intercept of these data points are estimated and plotted as a line.

```
# akando.py
def linearRegression( x,y ):
    """Returns m,b. x and Y are vectors"""
    sxy = ( x * y +0.0).sum() # also ensures at least a float type
    sx = ( x +0.0).sum()
    sy = ( y +0.0).sum()
    sx2 = ( x*x +0.0).sum()
    n = len( x )
    m = ( n * sxy - sx * sy ) / ( n * sx2 - sx*sx)
    b = ( sy - m * sx ) / n
    return m,b
```

Code 4-14

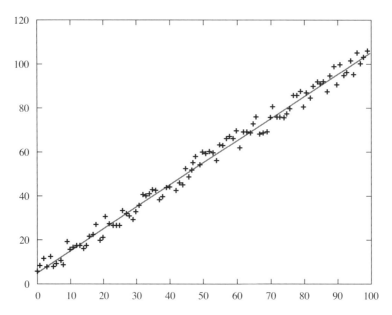

FIGURE 4-6 Results from Code 4-15.

```
>>> x = arange( 100 )
>>> y = arange( 100 )+10*random.rand(100)
>>> mat = zeros( (100,2) )
>>> mat[:,0] = x; mat[:,1] = y
>>> akando.PlotMultiple('plot.txt', mat )
>>> m,b = akando.linearRegression( x, y )
>>> m,b
(1.00369238725, 4.8955591252)

gnuplot> plot 'plot.txt' with points, 1.004*x + 4.896
```

Code 4-15

The final function in this section creates a circle in a two-dimensional frame. The function **Circle** in Code 4-16 creates a matrix with the dimension *size*, a tuple with a vertical and horizontal dimension. Another tuple, *loc*, identifies the location of the center of the circle and *rad* represents the radius of the circle. This function will return a matrix in which all elements are 1 if they are inside the specified circle and 0 outside.

```
# akando.py
def Circle( size, loc, rad):
    b1,b2 = indices( size )
    b1,b2 = b1-loc[0], b2-loc[1]
    mask = b1*b1 + b2*b2
    mask = less_equal( mask, rad*rad ).astype(int)
    return mask
```

Code 4-16

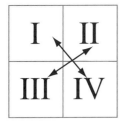

FIGURE 4-7 Image of a matrix Swap.

4.1.3 Correlation

There are two functions in Akando that support correlation, a topic discussed in Chapter 9 and therefore are not discussed in any detail here. Code 4-17 shows the **Correlate** and **Swap** functions, two functions that will receive data that is either a vector or a matrix and perform the necessary operations. The **Correlate** function performs the correlation and the **Swap** function rearranges the data. For a vector, the **Swap** function will cut the vector in half and swap locations of the halves.

Figure 4-7 shows the swap for a matrix. The first and fourth quadrants exchange positions, as do the second and third quadrants.

```
# akando.py
def Correlate( a, b ):
    # performs Fourier space correlation
    n = len( a.shape )
    if n==1:
        A = fftpack.fft(a)
        B = fftpack.fft(b)
        C = A * B.conjugate( )
        d = fftpack.ifft( C )
        d = Swap(d);
    if n==2:
        A = fftpack.fft2(a)
        B = fftpack.fft2(b)
        C = A * B.conjugate( )
        d = fftpack.ifft2( C )
        d = Swap(d)
    return d

def Swap( A ):
    #performs a quadrant swap
    if len(A.shape) == 2:
        (v,h) = A.shape
        ans = zeros(A.shape,A.dtype)
        ans[0:v/2,0:h/2] = A[v/2:v,h/2:h]
        ans[0:v/2,h/2:h] = A[v/2:v,0:h/2]
        ans[v/2:v,h/2:h] = A[0:v/2,0:h/2]
        ans[v/2:v,0:h/2] = A[0:v/2,h/2:h]
    # perform a vector swap
    if len(A.shape) ==1 :
        v = A.shape[0]
        ans = zeros(A.shape,A.dtype)
        if v%2==0:      # even number of elements
            ans[0:v/2] = A[v/2:v]
            ans[v/2:v] = A[0:v/2]
```

```
        else:              # odd number of elements
            ans[0:v/2] = A[v-v/2:v]
            ans[v/2:v] = A[0:v/2+1]
    return ans
```

Code 4-17

4.1.4 Image Conversions

Data may come in the form of an image but mathematic operations need to be performed on a matrix. Thus, it is necessary to convert an image to a matrix. Likewise, a computational result in the form of a matrix may need to be displayed as an image. Thus, it is also necessary to convert a matrix to an image. Code 4-18 presents **a2i**, which converts a matrix to a gray scale image, and **i2a**, which converts a gray scale image to a matrix. In the function **a2i**, the max and min of the matrix are determined, and these are set to 255 and 0 respectively. The rest of the data is scaled accordingly. In some cases this automatic scaling is not desired. The function **a2if** does not perform the scaling. However, it is up to the user to make sure that the data is in the range of 0 to 255 before being sent to this function. This function is used in cases where multiple matrices need to be converted into images using the same scale.

```
# akando.py
def a2i( data ):
    mg = Image.new( 'L', transpose(data).shape)
    mn = data.min()
    a = data - mn
    mx = a.max()
    a = a*256./mx
    mg.putdata( ravel(a))
    return mg

def a2if( data ):
    mg = Image.new( 'L', transpose(data).shape)
    mg.putdata( ravel(data))
    return mg

def i2a( mg ):
    mgt = mg.transpose(2).transpose(1)
    f = mgt.getdata()    # a structure
    z = array(f)
    zz = transpose(reshape(z,mg.size))
    return zz
```

Code 4-18

4.2 The Dancer Module

The purpose of the Dancer program is to interactively display images and to extract RGB (red, green, blue) values at specific locations. Thus, if an application generated a series of images, they could be displayed as soon as the calculations for the individual images were completed. Some applications require that the user

4.2 The Dancer Module

select specific points in an image before further processing commences. For example, if the user needed to manually locate specific points on a microscopic slide image, then it would be useful to have a fast interface that collected locations of the mouse clicks.

Because Python does not have built-in graphical commands, it instead uses the graphical capabilities of the TCL/TK language that comes with Python. The string-based TCL/TK language contains several simple tools for creating graphics. In fact, the IDLE interface uses TCL/TK to create the command window and text editing files.

It is possible to create Python programs that use the TCL/TK interface to generate GUIs. In versions of Python that use version 0.8 (or earlier) of IDLE, the user could interact with the GUI development directly through the main IDLE window. Such an interaction was well suited for a program like Dancer, which allows the user to interact with the display windows. Unfortunately, the IDLE interface version 1.0 and later made changes that usually disallows the IDLE window to interact with active TCL/TK windows. In short, this prevents IDLE from running programs such as Dancer. Windows users will find that Dancer works perfectly well outside of the IDLE environment. This chapter will thus consider the use of Python directly from a DOS shell.

A command-line interface of Python is accessed in Windows by Start:Python 2.5:Python (command line). Although this window can be used to run any Python command, it does not have the editing conveniences offered by IDLE. This new window has a more tedious method of copy-and-paste and does not have the fast editing hot keys that IDLE offers. There are, however, two conveniences to this more archaic interface: (1) it tends to respect the escape (control-C) command under almost all circumstances and (2) it does not inherently use the TCL/TK interface. The first can be used to stop programs stuck in infinite loops and the second can be used to run interactive TCL/TK programs. It is the latter that is useful for the Dancer program.

The Dancer program requires a startup image that is named "startup.jpg." Users can certainly replace the current image with their favorites. Code 4-19 displays the environment setup and the start of the Dancer program. Users should change the working directory and path names to suit their individual environments. When the last command is executed a new window should appear that contains the image stored as "pysrc/startup.jpg." The argument to the **dancer.Dancer** command is the path in which "startup.jpg" is stored. Figure 4-8 shows the initial window that appears when an instance of Dancer is created. The somewhat lengthy Dancer program code is shown in Code 4-21.

```
########## RUN IN A DOS SHELL ##  DO NOT RUN IN IDLE ##########
>>> import os, sys
>>> os.chdir('mydir')   # your working directory
>>> sys.path.append('pysrc')
>>> import dancer
>>> q = dancer.Dancer( 'pysrc')
```

Code 4-19

FIGURE 4-8 The Dancer startup window

This window has the ability to interact with the mouse. By placing the mouse over the image and clicking on the *right* mouse button, the **dancer.PointShoot** function is called and the location of the mouse and the RGB color values will be printed. For an application where it is necessary to locate the features manually, this mouse click feature is a fast way of collecting the points of interest. By simply performing a copy-paste from the DOS window to an editor, the information can be saved for future use.

The Dancer program also allows users to change the image that is in the display window through the **Load** command. Additional windows can be created through the **Add** command. The **Destroy(n)** command will eliminate a window. The variable n corresponds to the Image number that is located in the title bar of each display window. This variable can also be used to specify which window the **Load** command is going to use.

The final command of this simple program is the **Paste** command, which is useful for cases in which an image has been created through computation. Code 4-20 displays a simple case. As each computation is completed, the image of the results will be displayed in the specified window. In this fashion, the user can watch the results change as the program is running.

```
# assume that Dancer has been loaded
>>> for i in range( 10 ):
        data = Your Computation for this array
        mg = akando.a2i( data ) # convert to an image
        q.Paste( mg, 1 ) # display image in window 1
```

Code 4-20

Dancer is an extremely simple program. However, it provides two features that are difficult to find in the other programs. The first is the ability to click on several features creating a list of the mouse click locations. The second is the ability for Python to refresh an image while computations are being performed.

```
import Tkinter
import Image
import ImageTk
import ImageEnhance

class Dancer:
    "The dancer image machine"
    def __init__( self,JPY ):
        self.rt = Tkinter.Tk()
        self.rt.withdraw()
        self.tp = [ Tkinter.Toplevel(self.rt,visual="best")]
        self.tp[0].title('Image #1')
        self.mg = [ Image.open( JPY + '/startup.jpg')]
        self.ph = [ ImageTk.PhotoImage( self.mg[0]) ]
        self.lb = [ Tkinter.Label( self.tp[0], image = self.ph[0] ) ]
        # use Button-2 for point and shoot
        self.lb[0].bind( "<Button-3>", self.PointShoot )
        self.lb[0].pack( anchor='nw', side='left')
    def Show( self, N=0 ):
        # used by many of the routines here.
        self.ph[N] = ImageTk.PhotoImage( self.mg[N])
        self.lb[N].destroy()
        self.lb[N] = Tkinter.Label( self.tp[N], image=self.ph[N])
        self.lb[N].bind( "<Button-3>", self.PointShoot )
        self.lb[N].pack(anchor='nw', side='left')
    def Load( self, filename, N=0):
        self.mg[N] = Image.open( filename )
        self.Show( N )
    def Add( self, filename):
        self.mg.append( Image.open( filename) )
        N = len(self.mg)-1
        self.tp.append( Tkinter.Toplevel(self.rt, visual="best"))
        self.tp[N].title( 'Image #' + str(N+1))
        self.ph.append( ImageTk.PhotoImage( self.mg[N]))
        self.lb.append( Tkinter.Label(self.tp[N], image=self.ph[N]))
        self.lb[N].bind( "<Button-3>", self.PointShoot )
        self.lb[N].pack( anchor='nw', side='left')
    def Paste( self, mg, N=0):
        self.mg[N] = mg
        self.Show( N )
    def Destroy(self, N1 ):
        # This will change the numbers of all subsequent images
        N = N-1   # coordinates the window title (+1) with the list index
        del self.mg[N]
        del self.ph[N]
        self.lb[N].destroy()
        self.tp[N].destroy()
        del self.lb[N]
        del self.tp[N]
      # rename
        for i in range( 0,len(self.mg)):
            self.tp[i].title( 'Image #'+str(i+1))
    def PointShoot(self, event ):
        # who has focus
        #a = self.rt.winfo_children()
        print (event.x, event.y), self.ph[0]._PhotoImage__photo.get( event.x, event.y)
```

Code 4-21

4.3 Summary

The NumPy, SciPy, and PIL tools were written by third parties and contain a vast quantity of useful functions. However, there are many functions required for this book that are still generic in nature. For example, the functions that convert images to and from arrays are not unique to bioinformatics applications. These functions are collected in the Akando module, a varied collection of commonly used generic functions that will communicate with plotting programs, algebraic and geometric functions, correlation functions, and image conversions.

This book also provides the Dancer module, which is designed to display images interactively with the Python program. If a user creates a program that generates several images, the Dancer module can be used to display the images while the program is still running. Unfortunately, the Idle interface and programs that use Tkinter such as Dancer do not get along well and so Dancer needs to be run in command line interface.

Problems

1. Create a vector with 300 elements containing the values for $sin(2* \pi / 300 * i)$. The value of π can be imported from **numpy.pi**. Plot this data.

2. Add 10% random noise to the column *mat[:,1]* in Code 4-6. Create a plot similar in nature to Figure 4-2.

3. Smooth the column *mat[:,1]* from Problem 2. Create a plot similar in format to that of Figure 4-2.

4. Using **RangeHistogram**, show that **random.rand** creates a set of evenly distributed random numbers.

5. Using the **RangeHistogram** function, show that **random.normal** produces a Gaussian distribution of random numbers. Compute the mean and standard deviation of these numbers. Concurrently plot the histogram and the Gaussian function

$$y = A\exp[-(x - m)/(2*\sigma^2)],$$

where A is the height of the histogram, m is the mean, and σ is the standard deviation.

6. Create a new version of **linearRegression** that also computes the error (the sum of the distances from each point to the line).

5 Statistics

Many types of analyses in bioinformatics involve the use of statistical measures. Quite often these are simple algorithms. This chapter will review basic statistics that are common in the field. This will include analysis of distributions, probabilities, multivariant statistics, and analysis of multiple distributions.

Statistics, of course, is an extensive field, and this chapter covers only some of the basic topics that are common in bioinformatics algorithms. More in depth analysis techniques are certainly available for readers who are interested in delving deeper into this topic.

5.1 Simple Statistics

Given a vector of numbers, the most common statistical measure is to compute the average of the numbers. Code 5-1 demonstrates the single *numpy* command that accomplishes this task. A vector of random numbers is generated, and the average (or mean) is computed in two different manners.

```
>>> from numpy import random, set_printoptions
# use only to replicate the following results
>>> random.seed( 520987 )
>>> set_printoptions( precision=3 )   # set print precision
>>> vec = random.rand( 10 )
>>> vec
array([ 0.319,  0.269,  0.778,  0.972,  0.543,  0.416,  0.109,  0.355,
        0.226,  0.41 ])
>>> vec.mean()
0.439735631122
```

Code 5-1

The average is simply the sum of the elements divided by the number of elements, as in

$$\mu = \frac{1}{N}\sum_{i=1}^{N} d_i \qquad (5\text{-}1)$$

where d is the data, N is the total number of elements, and μ is the average. The average does not tell everything about the data. For example, the average of (2, 3, 4) is 3 and the average of (1, 3, 5) is also 3, but the second set of numbers has a wider variation among the elements.

The *standard deviation* measures this variation in values. It is computed by

$$\sigma = \sqrt{\frac{1}{N}\sum_{i=1}^{N}(x_i - \mu)^2} \qquad (5\text{-}2)$$

The standard deviation measures the square of the distance between each data point and the average. Obviously, elements that are close to the average will contribute little to the standard deviation and those that are far away will contribute greatly. Code 5-2 displays the simple one-line call in Python.

```
>>> vec.std()
0.24822049
```

Code 5-2

The *variance* is the square of the standard deviation and is denoted by σ^2

$$\sigma^2 = \frac{1}{N}\sum_{i=1}^{N}(x_i - \mu)^2 \qquad (5\text{-}3)$$

The average as computed in Equation (5.1) determines the average of the entire data stream. In many applications, the goal is to compute for small segments of the data stream. This is called a *running average* or equivalently a *smoothing operation*. It is defined as

$$p_i = \frac{1}{2K+1}\sum_{j=i-K}^{i+k} d_j \qquad (5\text{-}4)$$

Each value of p is computed for a small segment of the data. In reality this equation works well for elements in the interior of the data stream. However, the elements at the beginning and end of the stream are not quite defined. Consider the case of $i = 0$ in which case $i - K$ is meaningless. Thus, there are modifications required to compute the running average for the beginning and ending elements.

The computation in Equation (5-4) is also inefficient, as discussed in Section 4.1.2. A more efficient process is given in **akando.Smooth**, as presented in Code 4.9. Code 5-3 smoothes a vector with a window size of 1.

```
>>> import akando
>>> akando.Smooth( vec, 1 )
array([ 0.39151046,  0.55783005,  0.68315583,  0.71286299,  0.60174151,
        0.32187529,  0.52461403,  0.51524529,  0.80902141,  0.73289845])
```

Code 5-3

5.2 Distributions

Measurements from biological systems are quite difficult to achieve, and thus the extracted data is usually noisy. Scientific measurements therefore rely on collections of measurements that usually provide a distribution of results.

The most common type of distribution is described as *normal* or *Gaussian*. A pure normal distribution is computed as

$$y_i = A \exp\left(-\frac{(x_i - \mu)^2}{2\sigma^2}\right) \tag{5-5}$$

Here the μ is the average of the data and the σ is the standard deviation. The A is the amplitude of the distribution; for a distribution in which the area under the curve is 1.0,

$$A = \frac{1}{\sigma\sqrt{2\pi}} \tag{5-6}$$

Figure 5-1 displays two normal distributions and indicates the three variables of Equation (5-5). In this example, the distribution with the solid line has a mean of 3.2 and a standard deviation of 0.2. The height of the distribution is computed by Equation (5-6). The location of the peak along the *x*-axis is the value of the mean, and the

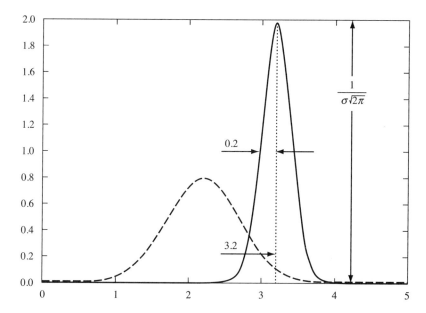

FIGURE 5-1 Two normal distributions with one marked.

half-width at half of the maximum (HWHM) is governed by the standard deviation. The curve with the dotted line has a mean of 2.2 and a standard deviation of 0.3. So, it is located to the left, wider, and shorter than the solid curve.

The standard deviation is also used to identify outliers. Given a set of data points that produce a normal distribution, a little more than 68% will be within one standard deviation of the mean. That means that two-thirds of the data points lay between $\mu - \sigma$ and $\mu + \sigma$. Outliers are usually considered as those points that are outside of the range $\mu - 3\sigma$ to $\mu + 3\sigma$. For normal data, the number of outliers should be about 0.3% of the total data samples.

The mean and standard deviation are related to the first and second *moments* of the data. Higher moment measurements can be made of distributions that are not perfect. For example, the third moment is related to *skewness*, and the fourth moment is related to *kurtosis*.

Skewness measures the imbalance of the distribution in a left-right sense. If the distribution has more data points above the mean, it will then have a positive measure of skewness. Kurtosis measures the vertical imbalance of the distribution. A set of data that creates a pointed distribution will have a positive kurtosis, and a set that creates more of a flat-top distribution will have a negative kurtosis.

Equation (5-2) describes the standard deviation as the square of the difference between the data and the mean. Skewness is related to the cube of this difference divided by the cube of the standard deviation. Kurtosis is related to the fourth power of this difference also divided by the fourth power of the standard deviation. There are, however, many variants of skewness and kurtosis for the normalizing factors.

5.3 Normalization

Data received from biological experiments will commonly have experimental biases that make comparisons of two separate experiments difficult to accomplish. These biases stem from different concentrations of chemicals used in experiments or biases infused by the detection equipment. In order to compare separate experiments, it is necessary to remove these biases, a process usually accomplished through normalization.

There are several degrees of normalization. In most cases, normalization is accomplished by equalizing the mean and also the standard deviation. Methods such as LOESS rely on local normalizations. These methods will be reviewed here, but the reader should be aware that significant work is ongoing for optimizing normalization methods.

Before these methods are discussed, a simple data set is generated. Code 5-4 contains the function **MakeNormData**, which creates random data points from a normal distribution. This and other functions of this section are stored in the file **pysrc/-normalizer.py**. This particular function uses the **random.normal** function from the *numpy* module to generate random data points.

5.3 Normalization

```
from numpy import array, random, zeros

# normalizer.py
# generate a normal distribution of data
def MakeNormData( N=100, mu=3.2, dev=2.1 ):
    v = random.normal( mu, dev, N )
    return v
```

Code 5-4

Even in cases in which all of the parameters are constant, biological experiments create data with different biases and variances. Thus, **MakeNormData** is used to simulate the results from several experiments. In this example, it is run ten different times with different biases, as shown in Code 5-5. The variable *exps* contains ten experiments each with 1,000 data points. Each experiment had a mean somewhere between 2 and 4 and a standard deviation between 0.5 and 1.5.

```
>>> random.seed( 91057 )
>>> exps = zeros( (10,1000), float )
>>> for i in range( 10 ):
       mu = 2 * random.rand()+2
       sigma = random.rand()+0.5
       exps[i] = MakeNormData( 1000, mu, sigma )
```

Code 5-5

The **akando.RangeHistogram** function will generate a histogram from a data vector. The arguments to this function are the data vector and the number of bins in the histogram. There are two additional optional arguments for the minimum and maximum value of the bins. If these are not used, then the function will create a histogram in which the minimum data value is in the first bin and the maximum value is in the last bin. This works well for single experiments. However, multiple experiments need to have histograms generated on the same scale. So, if the *mn* and *mx* arguments are used, then the histogram will have a set scale for the bins.

The Akando module also has a function that will save data in a format suitable for plotting with a spreadsheet program or plotters such as GnuPlot. The function **PlotSave** receives two arguments: (1) the filename to store the data and (2) the data vector. The function **PlotMultiple** will save a matrix, or as in this case, a set of vectors.

Code 5-6 generates the histograms for all of the data vectors and stores the information for plotting. Figure 5-2 displays the distributions for each of the vectors. Anomalies in the distributions (such as outliers or odd shapes to the distribution) are of interest to the investigator but in the present form the curves are difficult to compare.

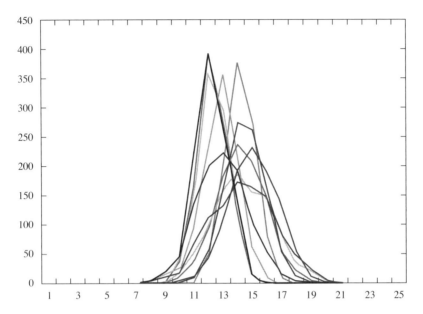

FIGURE 5-2 Distributions for each of the ten vectors.

```
>>> hsts = zeros( (10,25) )
>>> for i in range( 10 ):
        hsts[i]= akando.RangeHistogram( exps[i], 25, -5., 10.)
>>> akando.PlotMultiple( 'plot.txt', hsts.transpose() )
```

Code 5-6

Figure 5-2 presents a view of the data that is difficult to understand. Another presentation is shown in Figure 5-3, where there are ten columns—one for each data vector. The middle of the box is the average of the data for that particular vector. The extent of the box is the average plus or minus one standard deviation. The extent of the bars represents the maximum and minimum value of the data. This representation is much better at presenting the simple statistics of each sample.

Figure 5-3 was generated using GnuPlot, as demonstrated in Code 5-7. The function **PlotCandle** creates a matrix that has five columns. The first column is the average of the different vectors. The second column is the bottom extent of the box, which in this case is one standard deviation below the mean. The third and fourth columns indicate the extents of the bar, and the fifth column indicates the upper limit of the box. This data is saved using **akando.PlotMultiple** and is then plotted in GnuPlot. The *set bar 2* command widens the boxes from the default presentation.

```
# normalizer.py
# plot using GnuPlot candelsticks
def PlotCandle( filename, data ):
    N, L = data.shape # number of vectors, length of vectors
    avgs = zeros( N, float ) # will contain the averages
    devs = zeros( N, float )
    for i in range( N ):
```

```
        avgs[i] = data[i].mean()
        devs[i] = data[i].std()
    # save the data
    M = zeros( (10,5), float )
    for i in range( 10 ):
        mx = data[i].max()
        mn = data[i].min( )
        M[i] = i+1, avgs[i]-devs[i], mn, mx,avgs[i]+devs[i]
    akando.PlotMultiple( filename, M )

>>> PlotCandle( 'candle.txt', exps )
gnuplot> set bar 2
gnuplot> plot [0:11] 'candle.txt' with candlesticks
```

Code 5-7

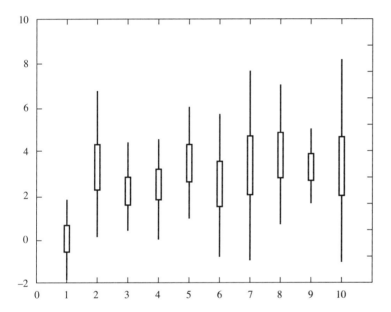

FIGURE 5-3 Representation of the average and deviations of ten samples.

Normalization by the average simply subtracts the average of each vector from itself. In other words, each vector becomes a *zero sum vector*. Code 5-8 accomplishes this task, and the result is shown in Figure 5-4. The middle boxes are now aligned, and the data has been normalized by the mean.

```
# normalizer.py
# normalize via the mean. Create data that has a 0-mean
def MeanNorm( vec ):
    sm = vec.sum()
    answ = vec - sm/len(vec)
    return answ
```

```
>>> for i in range( 10 ):
        exps[i] = MeanNorm( exps[i] )
>>> PlotCandle( 'candle.txt', exps )
```

Code 5-8

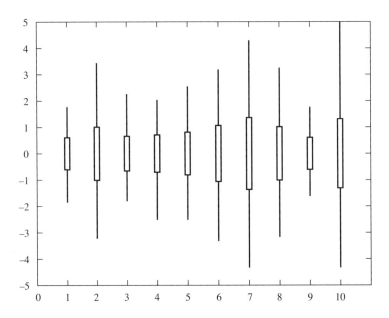

FIGURE 5-4 The data is now normalized by the mean.

However, the variance among the different samples is still quite different. This is indicated by the extent of the boxes. Normalization via the standard deviation requires that the data in a vector be divided by the standard deviation of that data. This will convert the vector so that its standard deviation becomes 1. Code 5-9 and Figure 5-5 depict the code and results.

```
# normalizer.py
# normalize via the std dev
def StdNorm( vec ):
    b = vec.std()
    return vec/b

>>> for i in range( 10 ):
        exps[i] = StdNorm( exps[i] )

>>> PlotCandle( 'candle.txt', exps )
```

Code 5-9

The distributions can now be compared. Figure 5-6 displays the ten distributions after mean normalization and standard deviation normalization. As can be seen, the distributions are now quite similar. Since they were created from a normal distribution program, this is to be expected. However, in those cases in which one of the experiments had anomalies in the data, this type of normalization would reveal the

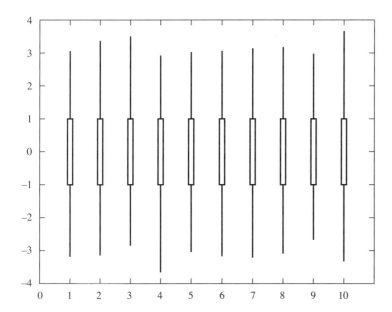

FIGURE 5-5 The data after normalization using the standard deviation.

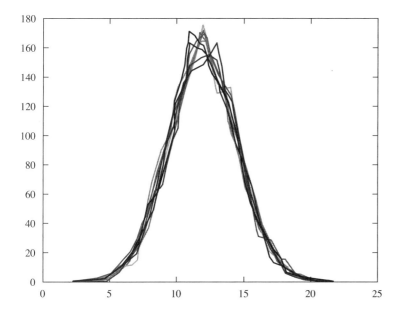

FIGURE 5-6 Ten distributions after normalization.

difference between the anomalous vector and the other vectors collected from the experiment.

The final normalization to be reviewed is LOESS (locally weighted scatterplot smoothing) which is reviewed in more detail in Section 26.1. Biological experiments such as gene expression arrays provided two data points for every measurement. In this case, one of the measurements is an overall intensity and the second is a ratio of two reactions. The data points of interest are those that have a ratio that exceeds a threshold. The problem is that the threshold is dependent on the average of all of the data points and that average is sensitive to the intensity.

Consider the data shown in Figure 5-7 in which the two measurements for each data point are plotted on the x- and y-axes. As can be seen, the average in the y-axis changes with increasing x.

A simple version of the LOESS algorithm groups data into subsets. The first subset in this example contains the first 50 data points (as sorted by their x-values). The second subset contains the next 50 data points and so on. The average for each subset is computed and removed from the subset of data. This is repeated for each subset of data, and the results are shown in Figure 5-8. The drifting of the average is removed from the data.

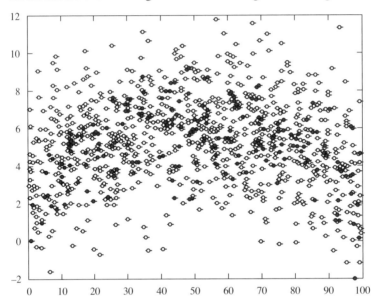

FIGURE 5-7 Data with an average sensitive to the x-axis.

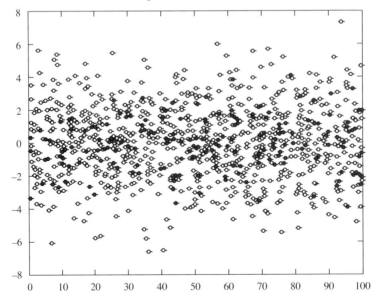

FIGURE 5-8 Data after LOESS normalization.

5.4 Multivariate Statistics

The data in Figure 5-1 shows the distribution of experimental values based on a single independent variable (the x variable). Because many experiments have more than a single independent variable, it is necessary to review multivariate distributions.

The first is the *covariance* matrix. In the case of a single independent variable, the variations were measured by the standard deviation of the variable. When there are multiple independent variables, there are two types of deviations that can be measured: (1) the standard deviation of each variable, and (2) how one variable changes with respect to another. For the case of N variables, there are N^2 measures of deviation. These measures are contained in the covariance matrix, which is computed by

$$c_{i,j} = \frac{1}{N-1} \sum_k v_{i,k} v_{j,k} \qquad (5\text{-}7)$$

The $v_{i,k}$ is the k-th element of the i-th data vector after the normalization that occurs with respect to each element. This can be better demonstrated in Code 5-10. The data is stored in a matrix a in which each row is a data vector. The variable $a0$ is a copy of this data, and in the *for* loop each column is normalized to a zero sum. The *covmat* is the covariance matrix.

```
>>> V,H = a.shape
>>> a0 = a + 0
>>> for i in range( H ):
        a0[:,i<XPressMath Error: You have not added a terminator to the end of this equation string.>] -= a0[:,i<XPressMath Error: You have not added a terminator to the end of this equation string.>].mean()
>>> covmat = dot( transpose(a0), a0)
>>> covmat /= (V-1)
```

Code 5-10

The *numpy* package provides a covariance function **cov**, and its use is shown in Code 5-11.

```
>>> c = cov( a )
```

Code 5-11

The diagonal elements of the covariance matrix are the variances of the individual elements. The off-diagonal elements indicate the relationship between the elements of the vectors. Thus, if the variables are statistically independent, the covariance matrix will then be diagonal.

Equation (5-8) is the normal distribution for multiple variables. The vector $\vec{x}$ contains the input data variables, the vector $\vec{\mu}$ is the mean of the variables, Σ is the covariance matrix, and k is the number of dimensions of the data vectors. This equation is similar to Equation (5-5) in that the mean is subtracted from the data and squared. This is divided by the covariance matrix

$$g(\vec{x}) = \frac{1}{(2\pi)^{k/2}|\Sigma|^{1/2}} \exp\left\{-\frac{1}{2}(\vec{x} - \vec{\mu})^T (\Sigma)^{-1}(\vec{x} - \vec{\mu})\right\} \tag{5-8}$$

Consider an experiment that measures two variables. In this case the $\vec{x}$ and $\vec{\mu}$ will be vectors of length 2 and the covariance matrix will be 2×2. Code 5-12 depicts a case of independent variables. The first two lines generate 1,000 (x, y) data points. Since the data is independent, the covariance matrix can be assumed to be diagonal. These elements are the variances of the two variables. The variable *dt* is the determinant of the covariance matrix. For a diagonal matrix, it is simply the multiplication of the diagonal elements. The code shown here also works for nondiagonal matrices.

The value *A* is the amplitude term from Equation (5-5). In this case $k = 2$. The *covinv* is the inverse of the covariance matrix. For a diagonal matrix, the inverse is simply the same matrix with the diagonal terms inverted. The code shown here also works for nondiagonal matrices.

The array *g* is the output space. The range of the *x* and *y* values is 0 to 10, and the size of the *g* array is 100×100, thus there is a scaling factor of ten that will appear when creating the data vector, *xvec*. The *for* loop will consider every point in output space *g*. For each of these points, the value within the exponential term, *temp*, is computed. The last line of the *for* loop puts it all together.

```
>>> from numpy import linalg
>>> x = MakeNormData( 1000, 3.2, 2.1 )
>>> y = MakeNormData( 1000, 5., 3.)
>>> covmat = zeros( (2,2), float )
>>> covmat[0,0] = 2.1**2
>>> covmat[1,1] = 3.0**2
>>> dt = linalg.det( covmat )
>>> avg = array( [x.mean(), y.mean()] )
>>> A = 1./((2*pi)*sqrt(dt) )
>>> covinv = linalg.inv( covmat )
>>> g = zeros( (100,100), float )
>>> for i in range( 100 ):
        for j in range( 100 ):
            xvec = array( [i,j])/10.
            t = xvec-avg
            temp = -0.5* dot(dot( t,covinv),transpose(t))
            g[i,j] = A * exp( temp )
>>> akando.Plot2D( 'plot.txt', g ) # for the plotter
```

Code 5-12

The output is a two-dimensional surface that resembles a single mountain of which the peak is the center of the inner contour. The peak of this surface is at the average of the two variables (3.2, 5.0) as shown in Figure 5-9. The width of the mountain in the *x*

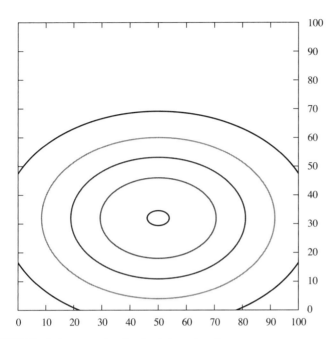

FIGURE 5-9 A contour plot showing independent data.

direction is controlled by the standard deviation of the *x* data. The case of dependent data is shown in Figure 5-10 and Code 5-13 shows how this data was generated. The vector *y2* is generated from a combination of the original random data and the *x*-data, thus linking the two variables. The matrix **M** stores the data in columns for plotting.

```
>>> x = MakeNormData( 1000, 3.2, 2.1 )
>>> y = MakeNormData( 1000, 5., 3.)
>>> y2 = 0.5*y + 0.5*x
>>> M = zeros( (1000,2), float )
>>> M[:,0] = x + 0
>>> M[:,1] = y2 + 0
>>> covmat = cov( M.transpose() )
>>> dt = linalg.det( covmat )
>>> A = 1./((2*pi)*sqrt(dt) )
>>> covinv = linalg.inv( covmat )
>>> g = zeros( (100,100), float )
>>> avg = M.mean(0)
>>> for i in range( 100 ):
        for j in range( 100 ):
            xvec = array( [i,j])/10.
            t = xvec-avg
            temp = -0.5* dot(dot( t,covinv),transpose(t))
            g[i,j] = A * exp( temp )
>>> akando.Plot2D( 'plot.txt', g ) # for the plotter
```

Code 5-13

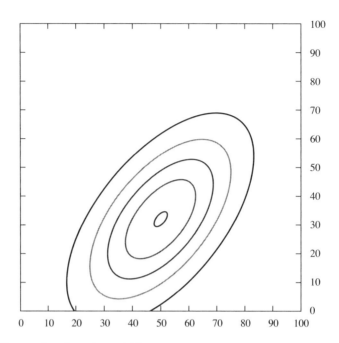

FIGURE 5-10 Contours from dependent variables.

If the deviations of the two variables were the same, then the contours would be circular. Differences in the standard deviations can make the contours ovoid in the horizontal direction or the vertical direction. However, it is not possible to make an oval with the long axis at an angle since the data is independent. If the contours are ovoid at an angle, then it would indicate that knowing the value of one variable would lead to information about the other variable. In this scenario, the variables would be linked, and the off-diagonal terms of the covariance matrix would no longer be 0.

5.5 Probabilities

Biology rarely offers exact measurements, which means that problems are usually cast in terms of probabilities. One example is to use simple probabilities to analyze DNA strings. There are four bases in a DNA string (A, C, G, T), and for a random string the probability of the first base being an "A" is 25%. This is expressed by

$$p(A) = 0.25. \tag{5-9}$$

The total probability of a system should always be exactly 1.0 except for unusual cases. This implies the trivial idea that there is a 100% chance of something happening. For example, for the first letter in a DNA string, there is a 25% chance of any of the four letters:

$$p(A) + p(C) + p(G) + p(T) = 1.00 \tag{5-10}$$

The probability of the first two letters being an "A" is computed first, and then once that condition is satisfied the probability of the second letter being an "A" is computed. This process can be described as a *conditional probability*, a situation where one probability relies on another. For the random string case, there is a 25% probability that the first letter is an "A." Once that is set, there is a 25% probability that the second letter will be an "A":

$$p(A|1)p(A|2) = .0625 \tag{5-11}$$

Let's consider a slightly different problem: to compute the probability that any letter in a random string of length N contains an "A" from a four-letter alphabet. This problem is slightly different than the previous. Equation (5-11) describes an AND case in which both letters are an "A." This new problem describes an OR case in which any of the letters is an "A." The OR problem is solved by converting it to an AND problem. The original problem states that the first character is an "A" or the second character is an "A" etc. The opposite problem is to state that the first character is not an "A" *and* the second character is not an "A." This opposite problem is a conditional probability.

The probability of the first character not being an "A" (equivalently the first character being "~A" is 0.75 (since there are only four letters in the DNA alphabet). Using Equation (5-11), the probability of the first two characters being "~A" is 0.75*0.75. Likewise, the probability of a 10-character string containing "~A" at all positions is 0.75^{10}. Recall that this is the opposite problem that is being solved. If the computation of the opposite problem produces an X percent chance, then the solution to the original problem is a $1 - X$ percent chance. Thus, for a four-letter alphabet, the chances of a randomly chosen string of length 10 having an "A" is

$$1 - \prod_{k=1}^{1} p(\sim A) = 0.9437 \tag{5-12}$$

Code 5-14 confirms this result. In this case, a random vector r of length 10 is generated. Since there is a 25% chance of an "A" being randomly chosen from an alphabet of length 4, any element in r that is less than or equal to 0.25 is considered an "A." There will be 10,000 vectors generated, and if any of the elements is below 0.25 then the variable *sm* is incremented. This is mathematically equivalent to choosing 10,000 random strings and searching for an "A." The variable *sm* counts the number of strings that would have contained an "A," and the final answer is the percentage of such strings. As can be seen, it is very close to the value calculated in Equation (5-12). A larger number of trials will produce a more accurate result.

```
>>> sm = 0
>>> for i in range( 10000 ):
        r = random.ranf( 10 )
        if len(nonzero( less(r,0.25))[0]) > 0:
            sm += 1
>>> sm/10000.
0.9464
```

Code 5-14

A slightly different problem is to consider a case of finite resources. In the previous case, there was no limit on the number of "A"s that could be used. Now, consider the case of selecting playing cards from a deck of 52 cards. The chances of the first card being an ace of spades is 1/52 = 0.0192. The chances of the second card being a king of spades is 1/51 since there are only 51 cards left in the deck. The chances of shuffling a deck and getting any specified order is 1/52, 1/51, 1/50, ..., 1/1 = 6.4e–67.

5.6 Odds

The probability of a random string being generated is dependent upon the length of a string, and for longer strings the probabilities can be rather small. For example, the probability of a string of length 4 having a specific arrangement of four letters is 0.0039 whereas the probability of a string of length 100 having a specified arrangement of four letters is 6e–61. Since genes vary widely in length, a measure that is sensitive to length is often undesirable.

The *odds* of an event occurring is defined in terms of the probability:

$$odds = \frac{p}{1 - p} \quad (5\text{-}13)$$

The range of the odds is still quite large, and so the *log-odds* are often employed in bioinformatics. This is simply the log of the odds.

A simple example is used in describing a *substitution matrix*, which describes the chances of one element in a DNA string (or protein) transforming into another element through the process of evolution. Basically, several sequences were considered and the probabilities of one base transforming into another were calculated. Since there are 20 amino acids, the substitution matrix for proteins is a matrix that is 20 × 20. Each cell of the matrix is the log-odds of the transition from one base to another.

5.7 Decisions from Distributions

Consider a case in which several biological samples are examined for the presence of phenotypes associated with cancer. It is the duty of the examiner to identify those samples that are characteristic of cancer from those that are normal.

This type of problem is a typical one in bioinformatics: the measurements of the samples do not precisely delineate the abnormal samples from the normal ones. Instead, measurements of the normal samples provide a distribution of values, and the measurements of the abnormal samples provide a different distribution. As in most realistic problems, the two distributions overlap. Figure 5-11 displays such a case in which the normal samples create a distribution centered at $x = 0$, and the abnormal samples create a distribution centered at 3.2. The initial thought is that samples that create a measurement near 3.2 are abnormal. However, as can be seen at $x = 3.2$ there can be a significant number of normal samples with a value greater than $x = 3.2$. In this example, more than 5% of the normal samples have a value of $x > 3.2$.

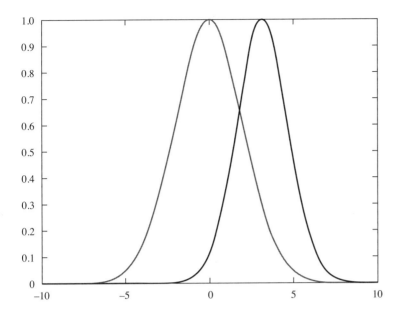

FIGURE 5-11 Overlapping distributions.

In reality the only samples that can classified with confidence are those with $x < -1$ or $x > 6$. The samples with x values between -1 and 6 can actually be normal or abnormal. Often the optimal decision is to find the location in which the distributions overlap. This sort of decision making is simplistic and works only if the numbers of samples for the two classes are somewhat the same. For cases in which the number of samples of one class far exceeds the number of samples in the other class, a new set of rules must be employed.

5.8 Summary

Statistics are an important component of bioinformatics. The NumPy and SciPy packages offer several statistical functions. This chapter reviewed basic statistics and their implementation in Python.

Problems

1. Create a vector of length N containing random numbers (using **numpy.random**). Compute the average. Repeat this experiment for different values of N. Plot the computed average versus N.

2. Write a single function that performs Problem 1. The vector should return a single vector that contains the values of computations.

3. The function **numpy.random.normal** creates a distribution of random numbers obeying a Gaussian distribution. It is called using *random.normal(m, s, N)*, where m is the mean, s is the standard deviation, and N is an optional

argument indicating the number of values to be computed. Without using a *for* or *while* loop, compute 1,000,000 numbers, and show that the mean and standard deviation of these numbers match m and s.

4. The **random.seed** function seeds the random number generator. Thus, *random.seed(0); print random.rand( )* will always return the same random number. For seeds ranging from 0 to 1,000,000, find all seeds that generate a random number less than 0.0001.

5. Compute 1,000,000 vectors of random numbers. Each vector should be three elements long. Compute the covariance matrix. Are the three elements of the vectors independent of each other?

6. Create a Python function that confirms Equation 5.11.

7. Create a function that creates a random string of DNA characters. The function should receive N the length of the string.

8. Using Problem 7, compute the probability of the string having the letter "A."

9. Using Problem 7, compute the probability of the string having the combination "AC."

10. Using Problem 7, compute the probability of the string having two "G"s and an "A" together in any combination.

6

Parsing DNA Data Files

Large databases of DNA information are being collected by several institutes. In the United States, a large repository is Genbank, which is under the sponsorship of the National Institutes of Health (http://www.ncbi.nlm.nih.gov/Genbank/index.html). The concern of this chapter is to develop programs capable of reading the files that are stored in three of the most popular formats: FASTA, Genbank, and ASN.1.

6.1 FASTA Files

The FASTA format is extremely simple, but it contains very little information aside from the sequence. A typical FASTA format is shown in Figure 6-1.

The first line contains a small header that may vary in content. In this case, the accession number and name of species and chromosome number are given. Some files may have comment lines after the first line that begin with a semicolon. The rest of the file is simply the DNA data.

Code 6-1 shows the commands that are needed to read in this file. The first version shown opens the file, reads the data, and closes the file. The second version performs all three operations in a single command. The **readlines** function will read all of the data and return a list. Each item in the list is a line of text ending in a newline character. In the FASTA file there is a newline character at the end of the header and one at the end of each line of DNA.

```
# version 1
>>> fp = file( 'data/nc_006046.fasta.txt')
>>> a = fp.readlines()
>>> fp.close()
# version 2
>>> a = file( 'data/nc_006046.fasta.txt').readlines()
```

Code 6-1

```
>gi|50428312|ref|NC_006046.1| Debaryomyces hansenii CBS767 chromosome
D, complete sequence
CCTCTCCTCTCGCGCCGCCAGTGTGCTGGTTAGTATTTCCCCAAACTTTCTTCGAATGATACAACAATCA
CACATGACGTCTACATAGGAGCCCCGGAAGCTGCATGCATTGGCGGCTGATGCGTCAGTGCCAGTGCTCA
AGC...
```

FIGURE 6-1 The first few lines of a FASTA file.

The first line *a[0]* is the header information. Excluding the comment lines, which are rarely used, the rest of the lines are the DNA. The extraction of the first part of the file *nc_006046.fasta* is shown in Code 6-2.

```
>>> a[0]
'>gi|50428312|ref|NC_006046.1| Debaryomyces hansenii CBS767 chromosome D,
complete sequence\n'
>>> a[1]
'CCTCTCCTCTCGCGCCGCCAGTGTGCTGGTTAGTATTTCCCCAAACTTTCTTCGAATGATACAACAATCA\n'
>>> a[2]
'CACATGACGTCTACATAGGAGCCCCGGAAGCTGCATGCATTGGCGGCTGATGCGTCAGTGCCAGTGCTCA\n'
```

Code 6-2

As can be seen, each line ends with the newline character "\n." So, the only tasks remaining are to combine all of the DNA lines into a long string and to remove the newline characters. Combining strings in a list is performed by the **join** function. The **join** function combines all but the first line of data, and the empty quotes indicate that there are no characters in between each line of DNA. Code 6-3 joins the strings and removes the newline characters.

```
>>> dna = "".join( a[1:] )
>>> dna = dna.replace( '\n',")
```

Code 6-3

In this case the DNA string is 1,602,771 characters long. Basically, it takes only three lines of Python code to read a FASTA file and extract the DNA. In actuality it could only take one line as shown in Code 6-4. However, because such code does not increase the speed of the program and is much more difficult to read, it should actually be avoided.

```
>>> dna = ("".join(file( 'data/nc_006046.fasta' ).readlines()[1:]) ).replace(
'\n',")
```

Code 6-4

6.2 Genbank Files

Genbank files are text-based and contain considerably more information than FASTA files. The typical files contain information about the source of the data, the researchers that created the file, the publication where it was presented, the DNA, the proteins, repeat regions, and more. However, some of these items are optional, and not all files contain every possible type of data.

Because Genbank files are text-based, they can be viewed with text editors, word processors, and even the IDLE editor. It is worth the time to load a file and examine its contents.

6.2.1 File Overview

For demonstration purposes, the file *data/NC_006046.gbk (NC_006046)* will be used. The first four lines of the file in Figure 6-2 display the locus identification, definition of the file, accession number, and version. As can be seen, the capitalized keywords are followed by the data and each entry ends with a newline.

As there are many items in this file, this chapter will not develop code to extract all of them. Instead code will be developed to extract the most important items, which will demonstrate how the rest of the items can be extracted. Although it is possible to develop code to completely automate the entire reading process, a different approach is adopted here. Because it is likely that a user wants only a small part of the file (just the DNA information, for example), functions will be built to extract the individual components. These functions can be called individually, or the user can easily build a driver program to call the desired functions.

The first function is simply to read the data. Code 6-5 displays the reading function **ReadGenbank**, which follows the simplicity of Code 6-1. The last line calls the function to read in this data.

```
# genbank.py
def ReadGenbank( filename ):
    fp = file( filename )
    data = fp.read()
    fp.close()
    return data
>>> data = ReadGenbank( 'data/nc_006046.gbk')
```

Code 6-5

6.2.2 Parsing the DNA

The DNA information is the last entry in the file, although it consumes more than half of the file. In this example the DNA information starts around line 15,394 in the file, which contains 42,110 lines of text. The first four lines at the beginning of the DNA section and the final four lines are shown in Figure 6-3.

The word "ORIGIN" begins the DNA section and each line contains six sections of 10 bases. The last line may be incomplete and the final line of the file is two slashes. In order to extract the DNA, three steps are necessary: (1) this information needs to be taken from the file, (2) the line numbers need to be removed, and (3) the groups of 10 bases need to be combined into a long string.

```
LOCUS       NC_006046               1602771 bp    DNA     linear   PLN 12-JAN-2007
DEFINITION  Debaryomyces hansenii CBS767 chromosome D, complete sequence.
ACCESSION   NC_006046
VERSION     NC_006046.1  GI:50428312
```

FIGURE 6-2 First few lines of a Genbank file.

74   CHAPTER 6   PARSING DNA DATA FILES

```
ORIGIN
        1 cctctcctct cgcgccgcca gtgtgctggt tagtatttcc ccaaactttc ttcgaatgat
       61 acaaeaatca cacatgacgt ctacatagga gccccggaag ctgcatgcat tggcggctga
      121 tgcgtcagtg ccagtgctca agccccagca cgctcacctc gcaggaggtt gggctgaagc
...
  1602601 gtcatttcta gctaataaat agttaacaga gaaagtgtgg aagtgctagg gaggttgaaa
  1602661 agatgagagt gatgttggta aggattatat tagtttcagg caggtaaagg ctttttccgg
  1602721 gaagggaaaa caagccaaca acttggtgcc aggcttttgg tcaaaaaacc c
//
```

FIGURE 6-3 The DNA portion of a Genbank file.

Code 6-6 shows the function **ParseDNA**, which performs these steps. Since the word "ORIGIN" exists only once in the file, it is a good trigger for finding the beginning of the DNA data. The next character is the "1", which always starts the next line. The variable *start* is the location of the "1" that follows "ORIGIN." The variable *end* is the location of the two slashes at the end of the file. This file was read in using the **read** function instead of **readlines**. The reason for this approach is to facilitate the large amount of searching that needs to be performed in the functions. Thus, the variable *a* is created, and it is the lines of DNA data from the file split according to the newline characters. In this example, the list *a* has 26,714 items and each one is a line from the file, as shown in Code 6-7.

```
# genbank.py
def ParseDNA( data ):
    # find ORIGIN
    orgn = data.find( 'ORIGIN' )
    # find the first row after that
    start = data.find( '1', orgn )
    # find the ending slashes
    end = data.find( '//', orgn )
    # split the substring into lines
    a = data[start:end].split('\n')
    dna = ''    # blank string
    for i in a:
        spl = i.split()
        dna += ''.join( spl[1:] )
    return dna

>>> dna = ParseDNA( data )
```

Code 6-6

```
a[0]
'1 cctctcctct cgcgccgcca gtgtgctggt tagtatttcc ccaaactttc ttcgaatgat'
a[1]
' 61 acaacaatca cacatgacgt ctacatagga gccccggaag ctgcatgcat tggcggctga'
```

Code 6-7

The string *dna* is initially empty but will soon be used to collect the DNA data. Inside of the *for* loop the variable *i* is a line from *a*. The call to the **split** function will create a new list that is the individual items in a single line of text. For example, the first item in list *a* is shown in Code 6-8.

```
>>> a[0].split()
['1', 'cctctcctct', 'cgcgccgcca', 'gtgtgctggt', 'tagtatttcc', 'ccaaactttc',
'ttcgaatgat']
```

Code 6-8

It is a list of seven items. The first is the line number and the next six are the DNA bases. For the extraction of DNA, the line numbers are not needed. Thus, the rest of the lines are joined and concatenated to the *dna* string. The call to this function is shown at the end of Code 6-6, and the returned string is a long string of only the DNA characters.

6.2.3 Gene and Protein Information

Consider the data in this file starting at line 60 shown in Figure 6-4. It indicates that there is a gene that begins at location 2657 and ends at location 3115. This particular gene is on the opposing strand of the double helix, and so the data in this file is the complement of the gene. This is actually an mRNA and other annotations are provided. This is not the complete list of information that is available. Some files will list genes and their protein translations, for example. This optional information will be explored in a later section in this chapter.

This section is concerned with the ability to identify the location of the gene information in the file. Obviously the information begins with the keyword 'gene' and so it should be identified. In this file the keyword "mRNA" is used, but in other files there are different keywords depending on the type of data. Some files indicate repeating regions, gaps, etc. Thus, it is necessary to find any type of keyword and then to extract the information following it.

Words used as keywords may also be used elsewhere in the file. For example, "gene" is commonly found in other locations. The keywords in the file are preceded by five space characters and then by several space characters depending on the length of the keyword. When the word "gene" is used elsewhere in the file, it is not preceded and followed by multiple spaces. The default keyword should be "CDS" or "gene".

```
gene            complement(2657..3115)
                /locus_tag="DEHA0D00132g"
                /db_xref="GeneID:2900904"
mRNA            complement(2657..3115)
                /locus_tag="DEHA0D00132g"
                /transcript_id="XM_458474.1"
                /db_xref="GI:50419882"
                /db_xref="GeneID:2900904"
CDs             complement(2657..3115)
                /locus_tag="DEHA0D00132g"
                /note="no similarity, possibly noncoding, hypothetical
                start"
                /codon_start=1
                /protein_id="XP_458474.1"
                /db_xref="GI:50419883"
                /db_xref="GeneID:2900904"
```

FIGURE 6-4 The tag portion of a Genbank file.

Unfortunately, the **find** function locates only the next occurrence of a substring instead of all of the occurrences. The number of times that the **find** function is called is the number of times that the substring is found, which is determined by the **count** function. Code 6-9 shows the function **FindKeywordLocs**, which is used to find all locations of a keyword in the *data*. It counts the number of occurrences and then creates a *for* loop that finds consecutive occurrences of the keyword. The variable *k* keeps track of where the searches should begin. Initially, it starts with $k = 0$ and the *k* becomes the location of the first occurrence of the keyword. The search for the next occurrence begins at $k + 1$. The call to this function is also shown along with the first ten results.

```
# genbank.py
def FindKeywordLocs( data, keyword = '     CDS      ' ):
    # number of occurences
    ngenes = data.count( keyword )
    # find the occurences
    keylocs = []
    k =0
    for i in range( ngenes ):
        t = data.find( keyword,k)
        keylocs.append( t )
        k = t + 10
    return keylocs

>>> keylocs = FindKeywordLocs( data )
>>> keylocs[:10]
[3411, 4412, 6523, 7426, 8612, 9688, 11536, 15772, 16992, 17884]
```

Code 6-9

The first location of the keyword in the string *data* is at 3411 in this example. The first 100 characters starting at this location are shown in Code 6-10.

```
>>> data[3411:3411+95]
'     CDS             complement(2657..3115)\n                     /locus_tag="DEHA0D00132g"\n          '
```

Code 6-10

It is the responsibility of the next functions to extract the proper information at each of the locations of the keywords and to decipher its meaning.

6.2.4 Gene Locations

A gene is a coding region in the DNA. It has at least one starting location and an ending location. The data may be in its complement form or the coding DNA may be composed of splices. The code developed in this section will extract the locations of the coding sequences and an indication if it is a complement.

There is a small difference between Python and Genbank indexing. The first DNA base in the Genbank file is at location 1, as shown in Code 6-7. Python, however, uses the index 0 for the first location. So, the locations of the coding splices will differ from the Python strings by a value of 1. This will be addressed in the function **GetCodingDNA**.

```
6293..9061
complement(2657..3115)
join(66124..66177,66220..67500)
complement(join(122306..123386,123440..124053))
join(193283..193290,193974..194064,194280..194906)
complement(333474..>333850)
```

FIGURE 6-5 Possibilities for gene tags.

The line that follows the keyword " CDS " has a few different forms, as shown in Figure 6-5. The first example is merely a start and end location of the coding sequence. The second example is a complemented string. The third demonstrates a splice in which the coding sequence is found in two sections. The fourth is a complemented splice. The fifth shows multiple splice locations. The final example shows a ">" symbol, which indicates that the exact location is not known. There is no limit on the number of splices that a coding sequence may have. Thus, a function that is to extract the locations of the coding region(s) for a single gene needs to be able to handle all of these situations.

6.2.5 Normal and Complement

For the purpose of extracting gene locations, the complement flag needs to be noted as the actual complement operation is performed later. The beginning and ending of a splice location are two numbers separated by two periods. When there is a "complement" or a "join," the first and last splice will be enclosed in parentheses.

For normal or complement-only cases, there will be two numbers separated by two periods. These two cases are considered in the function **EasyStartEnd** (Code 6-11), which receives the data string and the location of the keyword. It also receives a variable *cflag*, which is 1 if it is a complement. The program starts by finding the location of the two periods. The variable *b1* finds the beginning of the first number and *b2* finds the location of the second number. The first number precedes the location of the *dots* and is preceded by one of two characters. It is preceded by a blank space if there is no complement and a "(" if there is a complement. The *if* statements consider both options. The first number lies between *b1+1* and *dots* whereas the second number lies between *dots+2* and *b2*. Recall that *dots* is the location of the first of two periods and that the second number starts two locations after that. The rest of the code is used to remove any "<" or ">" symbols. The final function **int** converts a string of numerical characters into an integer. If there are any nonnumerical characters in the string, then an error will occur. It is thus important to remove the "<" and ">" symbols. This function returns two integers that are the beginning and ending locations of the gene.

```
# genbank.py
def EasyStartEnd( data, loc, cflag=0 ):
    # find the ..
    dots = data.find( '..', loc )
    # targets
    if cflag==0:
        t1, t2 = ' ','\n'
    if cflag==1:
        t1, t2 = '(',')'
```

```
        # find the first preceding blank
        b1 = data.rfind( t1,loc, dots )
        # find the first following newline
        b2 = data.find( t2, dots )
        # extract the numbers
        temp = data[b1+1:dots]
        temp = temp.replace( '>', " )
        temp = temp.replace( '<', " )
        start = int( temp )
        temp = data[dots+2:b2]
        temp = temp.replace( '>', " )
        temp = temp.replace( '<', " )
        end = int( temp )
        return start, end
```

Code 6-11

Two examples are used in Code 6-12 to demonstrate this program. The first one extracts the 15th gene, which does not have a complement or a join. The second uses the first gene that has a complement. In both cases the pertinent segment of *data* is also printed.

```
# first example
>>> keylocs[14]
24899
>>> data[24899:24899+33]
'     CDS             33212..33418\n '
>>> start, end = EasyStartEnd( data, 24899 )
>>> start, end
(33212, 33418)

# second example
>>> keylocs[0]
3411
>>> data[3411:3454]
'     CDS             complement(2657..3115)\n'
>>> start, end = EasyStartEnd( data, 3411, 1 )
>>> start, end
(2657, 3115)
```

Code 6-12

6.2.6 Splices

Splices will always have the keyword "join," and therefore the first splice will always begin with a "(" and the last will end with a ")". If there are intermediate splices, they will be of the form, *nnn.mmm*, where they begin and end with commas and have two periods in the middle.

Code 6-13 is a rather lengthy function that finds the locations of the parentheses, $p1$ and $p2$, in the data string. The node is the number of double periods that exist between $p1$ and $p2$, which is also the number of splices. It then isolates the first splice by finding the first comma after the "(". It converts the numerical strings to integers and stores them as a tuple in the list *numbs*. The *for* loop considers all intermediate splices and after the loop the final splice is extracted. The result of the function is a list, and each item in the list is a tuple. Each tuple contains two integers that are the beginning and end of each splice. The number of splices is *len(numbs)*.

```
# genbank.py
def Splices( data, loc ):
    # loc is the location of '
    join = data.find( 'join', loc )
    # find the parentheses
    p1 = data.find( '(', join )
    p2 = data.find( ')', join )
    # count the number of dots
    ndots = data.count( '..', p1, p2 )
    # extract the numbers
    numbs = []
    # the first has a ( .. ,
    dots = data.find( '..', p1 )
    t = data[p1+1:dots]
    t = t.replace( '<',"")
    st = int( t )
    comma = data.find( ',',p1 )
    en = int( data[dots+2:comma] )
    numbs.append( (st,en) )
    # consider the rest except first and last
    # code is , .. ,
    for i in range( ndots - 2):
        dots = data.find( '..', comma )
        comma2 = data.find( ',', dots )
        st = int( data[comma+1:dots] )
        en = int( data[dots+2:comma2] )
        numbs.append( (st,en) )
        comma = comma2
    # last one has code , .. )
    dots = data.find( '..', comma )
    st = int( data[comma+1:dots] )
    en = int( data[dots+2: p2] )
    numbs.append( (st,en) )
    return numbs
```

Code 6-13

In this example the 41st gene is a spliced gene, and Code 6-14 calls **Splices** to extract the splice information. The display from the original data file is shown first and then results are extracted.

```
>>> k = keylocs[40]
>>> data[k:k+100]
'     CDS             join(90402..90907,91036..92020)\n
/locus_tag="DEHA0D01705g"\n   '
>>> spl = Splices( data, keylocs[40] )
>>> spl
[(90402, 90907), (91036, 92020)]
```

Code 6-14

6.2.7 Extracting All Gene Locations

The previous two sections provide the tools necessary to extract all the locations of the genes; this section will create the driver function. For each CDS, it is necessary to determine if it is a complement or a splice. The appropriate function is then called and the data returned. The normal or complement cases return a single tuple of the start and end locations. The splice case returns a list of tuples. Because these are slightly different data types, the tuple returned from **EasyStartEnd** is placed within a list.

The final output is a list that contains lists of the locations and a complement flag for each gene. Code 6-15 shows the function **GeneLocs**, which receives the original *data* and the *keylocs*. The *for* loop iterates over each entry in *keylocs*. Within this loop, the function sets two flags—*joinf* and *compf*—to True if this CDS has the "join" or "complement" keywords. It then selects which function to call to extract the starting and ending locations of the coding regions. The results are stored in a list named *genes* and returned.

```
# genbank.py
def GeneLocs( data, keylocs ):
    genes = []
    for i in range( len( keylocs )):
        # get this line and look for join or complement
        end = data.find( '\n', keylocs [i] )
        temp = data[keylocs [i]: end ]
        joinf = 'join' in temp
        compf = 'complement' in temp
        # get the extracts
        c = 0
        if compf == True: c = 1
        if joinf:
            numbs = Splices( data, keylocs [i] )
            genes.append( (numbs, compf ))
        else:
            sten = EasyStartEnd( data, keylocs [i], c )
            genes.append( ([sten],compf) )
    return genes
```

Code 6-15

Code 6-16 shows a call to this function and the results for this data file. This file contains 858 genes and two of them are shown.

```
>>> genelocs = GeneLocs( data, keylocs )
>>> len(genelocs)
858
>>> genelocs[0]
([(2657, 3115)], True)
>>> genelocs[40]
([(90402, 90907), (91036, 92020)], False)
```

Code 6-16

6.2.8 Coding DNA

All of the information is available to extract the coding splices from the *dna* string. For each case, the appropriate substrings are extracted and concatenated. If the complement flag is set, then the string is sent to the **Complement** function shown in Code 6-17.

```
# genbank.py
# create the complement
def Complement( st ):
    st = st.replace( 'a', 'T' )
    st = st.replace( 't', 'A' )
    st = st.replace( 'c', 'G' )
    st = st.replace( 'g', 'C' )
    st = st.lower()[::-1]
    return st
```

Code 6-17

A more efficient method of replacing letters in a string is to use a translation table. The table is constructed using the **maketrans** function, and the actual replacement is called by the **translate** function. These commands are shown in Code 6-18, which creates the complement string. Since there are only four characters in a DNA string, either function can be used. However, if a situation were to use a much larger alphabet, this latter version might be simpler to write.

```
# genbank.py
import string
def Complement( st ):
    table = string.maketrans( 'acgt', 'tgca' )
    st = st.translate( table )
    st = st[::-1]
    return st
```

Code 6-18

Code 6-19 contains the function **GetCodingDNA**, which extracts the coding DNA for a single gene. Thus, it receives the original *dna* string and just one member of the list *genelocs*. The *for* loop considers each splice, and the start and end locations for each are placed into variables *st* and *en*. Recall that the Genbank file starts with element 1 while a Python string starts with element 0. Thus, the use of *st–1* is used to get the correct location in the string. Also recall that in a Python string slice the second index is excluded. In this case the *dna[st–1:en]* includes the character at location *en–1* but not *en*. That is why the code uses *st–1* but not *en–1*. A sample call is also shown in Code 6-19. To get all of the coding DNA strings for all of the CDS entries, simply call **GetCodingDNA** for each item in the *genelocs* list.

```
# genbank.py
def GetCodingDNA( dna, genesi ):
    # dna from ParseDNA
    # genesi is genelocs[i] from GeneLocs
    codedna = ''
    N = len( genesi[0] ) # number of splices
    for i in range( N ):
        st, en = genesi[0][i]
        codedna += dna[st-1:en]
```

```
        # complment flag
        if genesi[1]:
            codedna = Complement( codedna )
    return codedna

>>> cdna = GetCodingDNA( dna, genelocs[0] )
>>> cdna[:10]
'atgtcgtatc'
```

Code 6-19

6.2.9 Proteins

Converting the DNA string into an amino acid string is simple. A lookup table for this conversion is created, and each codon in the string is used to create a single amino acid. The lookup table is shown in Code 6-20. The function **Codons** creates a tuple with 64 entries. Each item is a tuple of a codon and an amino acid. The *for* loop forces the codon to lowercase and creates a dictionary of the 64 items. The return is a dictionary that can look up any codon and return its associated amino acid as demonstrated.

```
# genbank.py
# codon dictionary
# stop = 'p'
# start = 'M'
def Codons( ):
    codons = {}
    CODONS = ( ('TTT','F'), ('TTC','F'), ('TTA','L'),('TTG','L'),\
               ('TCT','s'), ('TCC','s'), ('TCA','s'),('TCG','s'),\
               ('TAT','Y'), ('TAC','Y'), ('TAA','p'),('TAG','p'),\
               ('TGT','C'), ('TGC','C'), ('TGA','p'),('TGG','W'),\
               ('CTT','L'), ('CTC','L'), ('CTA','L'),('CTG','L'),\
               ('CCT','P'), ('CCC','P'), ('CCA','P'),('CCG','P'),\
               ('CAT','H'), ('CAC','H'), ('CAA','Q'),('CAG','Q'),\
               ('CGT','R'), ('CGC','R'), ('CGA','R'),('CGG','R'),\
               ('ATT','I'), ('ATC','I'), ('ATA','I'),('ATG','M'),\
               ('ACT','T'), ('ACC','T'), ('ACA','T'),('ACG','T'),\
               ('AAT','N'), ('AAC','N'), ('AAA','K'),('AAG','K'),\
               ('AGT','s'), ('AGC','s'), ('AGA','R'),('AGG','R'),\
               ('GTT','V'), ('GTC','V'), ('GTA','V'),('GTG','V'),\
               ('GCT','A'), ('GCC','A'), ('GCA','A'),('GCG','A'),\
               ('GAT','D'), ('GAC','D'), ('GAA','E'),('GAG','E'),\
               ('GGT','G'), ('GGC','G'), ('GGA','G'),('GGG','G'))
    for i in CODONS:
        codons[i[0].lower()] = i[1]
    return codons

>>> codons = Codons()
>>> codons['tgt']
'C'
```

Code 6-20

To convert a DNA string into a protein, the DNA bases are considered in groups of three, and each codon is converted into an amino acid. Code 6-21 builds a protein string from a coding DNA string.

```
# genbank.py
def Codons2Protein( codedna, codons ):
    protein = ''
    for i in range( 0, len(codedna), 3 ):
```

```
        codon = codedna[i:i+3]
        protein += codons[ codon ]
    return protein
```

Code 6-21

Code 6-22 shows a reading of Genbank file *bc123620*, which contains protein information. The result of this program is that the DNA was read from a Genbank file, the location of the first gene was extracted, and the DNA in this region was collected and converted into an amino acid sequence. The results from Code 6-22 are compared directly to the protein as printed in the original data file. They are the same, which validates the functions.

```
>>> data = ReadGenbank( 'data/bc123620.txt')
>>> dna = ParseDNA( data )
>>> keylocs = FindKeywordLocs( data )
>>> glocs = GeneLocs( data, keylocs )
>>> cdna = GetCodingDNA( dna, glocs[0] )
>>> protein = Codons2Protein( cdna, codons )
>>> protein
'MSYQNLRLYFNVVSKCIDYCNVILALPSHSSGGRSLTTFILQSNFGLLGGYVNMVAAQEATYYIIGHSKK
RCKGAINWEMDLSASGECENTGSLGGDTAKIWNNGGAQFLQLNTWQRHDFRKDSVCNYAIPNNGRGVCSED
GVLSYEDIGTKp'
```

Code 6-22

6.2.10 Extracting Translations

The *translation* in this file is the amino acid sequence, and it is located after the splice locations. Code 6-23 reads this file following the same philosophy of searching for a keyword and extracting the information after the keyword. In the case of a translation, the data follows and is contained between two double quotes. It is also necessary to remove newline characters and spaces.

```
# genbank.py
def Amino( data, loc ):
    # get the amino acids
    trans = data.find( '/translation', loc )
    # find the second "
    quot = data.find( '"', trans + 15 )
    # extract
    prot = data[trans+14:quot]
    # remove newlines and blanks
    prot = prot.replace( '\n'," " )
    prot = prot.replace( ' ',"")
    return prot

>>> prot = Amino( data, keylocs[0] )
```

Code 6-23

The result of Code 6-22 is the conversion of the coding DNA to an amino acid chain. The result of Code 6-23 is the protein as listed in the Genbank file. The two results should be the same. The only difference is that Code 6-22 includes the stop codon at the end of the string. Code 6-24 thus compares these two strings, excluding the stop codon, and shows that they are the same.

```
>>> protein[:-1] == prot
True
```

Code 6-24

The Genbank file contains many more types of data, and programs can easily be written to extract other types of data by following the same recipe as in Code 6-23. Alterations to the program are necessary as parsing the data from the file requires different procedures.

6.3 ASN.1 File Format

The ASN.1 format is another type of file that contains several different classes of information about sequence data, which is encapsulated within curly braces. The first part of a file is shown in Figure 6-6. The sequence data starts with a "{" and ends with a "}" (which is not shown in the code). Within this set of braces are other items. The code also shows the *id* data, which encloses *other* and *general*. There are several different types of entries in this file, but only the data will be examined here.

The start of the DNA data for this particular file is shown in Figure 6-7. The actual DNA string starts just after *ncbi2na*, but the data is compressed to reduce file size.

```
Seq-entry ::= seq {
  id {
    other {
      accession "NC_006046" ,
      version 1 } ,
    general {
      db "NCBI_GENOMES" ,
      tag
        id 435 } ,
    gi 50428312 } ,
```

FIGURE 6-6 A portion of an ASN.1 file.

```
inst {
    repr delta ,
    mol dna ,
    length 1602771 ,
    ext
      delta {
        literal {
          length 131072 ,
          seq-data
            ncbi2na '5DD7766594BB9EBCB3F5501FDF60E31043444E1B713289568279393E9
A78E6D2E52E74255246745D928AFA9E09E76DFE165241D4D1F46818493FD8463A5EA6CFDD02049
2E830ECBFF89AE76BB114698B58ED72E4E543589BBC782EB2D9D336CB63A9FA514545AE76C1E07
5A96C2FF76CF7D2924C98FFB9922100CB035FA58A3FD0103330A27CD80B2E0111010C31608F089
7E00C7F54112E8FC0CDF55F5CC30D0C3313888AE3F02E0E29CB0CB31778CECE2DCC1924DCBDF03
```

FIGURE 6-7 The compressed DNA encoding of an ASN.1 file.

6.3 ASN.1 File Format

Table 6-1 The Encoding Used in ASN.1 Files.

DNA	Code
A	00
C	01
G	10
T	11

Since there are only four different letters that are used for DNA (A, C, G, T), it is inefficient to store each letter as a single byte of information. Since there are only four items, it takes only two bits to encode them. Table 6-1 shows the encoding used in ASN.1 files. Thus, a string such as ATTG would be encoded as 00111110. This binary string is converted into a standard hexadecimal string, as shown in Table 6-2. The binary string 00111110 would then be converted to 3E.

Decoding the ASN.1 format is simply the reverse of this process. The lookup table is created as a Python dictionary, as shown in Code 6-25. The codes in this section are stored in the file *asn1.py*.

Table 6-2 The Conversion of a Binary String into a Standard Hexadecimal String.

Binary	Hex
0000	0
0001	1
0010	2
0011	3
0100	4
0101	5
0110	6
0111	7
1000	8
1001	9
1010	A
1011	B
1100	C
1101	D
1110	E
1111	F

```
# create a decoding dictionary
def DecoderDict( ):
    ddct = {}
    ddct['0'] = 'AA'; ddct['1'] = 'AC'; ddct['2'] = 'AG'
    ddct['3'] = 'AT'; ddct['4'] = 'CA'; ddct['5'] = 'CC'
    ddct['6'] = 'CG'; ddct['7'] = 'CT'; ddct['8'] = 'GA'
    ddct['9'] = 'GC'; ddct['A'] = 'GG'; ddct['B'] = 'GT'
    ddct['C'] = 'TA'; ddct['D'] = 'TC'; ddct['E'] = 'TG'
    ddct['F'] = 'TT'
    return ddct
```

Code 6-25

Code 6-26 illustrates function **DNAFromASN1** and the call to it. The function finds "ncbi2na" and then the single quotes that follow it. These quotes surround the compressed DNA string. The string is extracted and the newlines are removed. The *for* loop considers each letter in the string and uses the dictionary to look up the conversion.

```
def DNAFromASN1( filename, ddct ):
    # read in data
    fp = file( filename )
    a = fp.read()
    fp.close()
    # extract DNA
    loc = a.find( 'ncbi2na' )
    start = a.find( "'",loc )+1
    end = a.find( "'", start+2)
    cpdna = a[start:end] # compressed dna
    cpdna = cpdna.replace( '\n','')
    # decode
    dna = ''
    for i in range( len( cpdna )):
        dna += ddct[ cpdna[i] ]
    return dna
>>> dna = DNAFromASN1( 'c20/nc_006046.asn1', ddct)
>>> dna[:100]
'CCTCTCCTCTCGCGCCGCCAGTGTGCTGGTTAGTATTTCCCCAAACTTTCTTCGAATGATA
CAACAATCACACATGACGTCTACATAGGAGCCCCGGAAG'
```

Code 6-26

The ASN.1 format also contains the locations of coding regions. One example is shown in Figure 6-8. This is also very easy to extract. By simply finding keywords such as "location," "from," and "to," the beginning and end of a coding region can be extracted.

```
{
  data
    gene {
      locus-tag "DEHA0D09196g" } ,
  location
    int{
      from 669337 ,
      to 670620 ,
      strand minus ,
      id
        gi 50428312 } ,
  dbxref {
    {
      db "GeneID" ,
      tag
        id 2901637 } } } ,
```

FIGURE 6-8 Tag information in an ASN.1 file.

6.4 Summary

DNA information is stored in several formats. Two of the most popular are FASTA and Genbank. The FASTA files are very easy to read and require only a few lines of code. The Genbank files are considerably more involved and store significantly more information beyond the DNA sequence. They can store identifying information, publication and author information, proteins, identified repeats, and much more. Thus, reading these files requires a bit more programming. These programs, however, are not complicated.

Bibliography

Biopython. (2007). Retrieved from http://biopython.org.
NC_006046. (2007). Retrieved from http://www.ncbi.nlm.nih.gov/entrez/-viewer.fcgi?db=nuccore&id=50428312.

Problems

1. Extract the start and stop locations of the DNA for each gene using "CDS" as the keyword.

2. Repeat Problem 1 using "gene" as a keyword. Do the results differ from Problem 1?

3. Extract the author from the Genbank file.

4. Extract the accession number from the Genbank file.

5. Each gene may have a "product". Extract the product for each gene.

7 Sequence Alignment

DNA sequences are complicated structures that have been difficult to decode. A strand of DNA contains coding regions that produce genes and contains noncoding regions that may or may not have functionality. As systems evolve, genes are sometimes passed on with small alterations or relocations. Since the noncoding regions are less important in many respects, they are often passed on with more alterations. These similarities allow us to infer the functionality of a gene by relating it to other genes with known functions.

The main computational engine for accomplishing this comparison is to align sequences. The purpose of alignment is to demonstrate the similarity of two (or more) sequences. At first this sounds like an easy job. Each sequence has only four bases and it should not be too hard to determine if the sequences are similar.

As in most real-world problems, the process is not that easy. Two sequences may differ because of base differences, or because they have extra or missing bases. Computationally, this becomes a more difficult problem to solve since smaller chunks of the sequences will need to align differently. Another problem is that in a DNA strand we may not know which parts of the strand are coding regions. Thus, between two strands the important parts may be similar and the unimportant parts could be dissimilar. A program to align these sequences may need to discount the unimportant regions. Another problem is that segments of the coding regions may be located in different regions of a strand. For example, a gene may be constructed from two different subsections of the strand. There is no guarantee that these two sections will be located in the same regions of the two strands. Still, a computer program will need to find similarities among the strings.

This chapter will consider simple alignment algorithms and review the highly used dynamic programming approach. More complicated approaches will be discussed but not replicated.

7.1 Alphabets

The algorithms presented here may be applied to strings from any source. Before they are presented, it is necessary to provide a few definitions. A string is an array of characters from an alphabet. For DNA cases, the alphabet has only four characters (ACGT). Protein sequences are made from an alphabet of 20 characters (ARNDCQEGHILKMFPSTWYV). Certainly, we can consider strings from English (a 26-letter alphabet) or any other language. Usually, the alphabet is represented with Σ.

$$\Sigma = \{A, C, G, T\} \tag{7-1}$$

7.2 Matching Sequences

This chapter will primarily be concerned with aligning two sequences. This requires that a metric of performance be created in order to measure the quality of alignment. There are also several types of deviations from a perfect alignment that also need to be considered when creating alignment programs.

7.2.1 Perfect Matches

A perfect match occurs when aligned letters from two strings are the same. Even in situations having this simplicity, questions of measuring the quality of the match need to be considered. In the following case, two simple sequences are perfectly aligned. Should the measure of alignment treat all of the letters equally? Is it more important for a sequence AATT to align with itself or with ACGT?

These questions can be further complicated by considering the function of the DNA. As an example consider the alignment of two codons: CGA and CGG. While the bases do not align perfectly, they both code for the same amino acid. Should this mismatch count the same as other mismatches that create different amino acids?

7.2.2 Insertions and Deletions

Insertions and deletions (*indel*s) are cases in which a base is added or removed from a sequence. When identified these are denoted by a dash as in the following case:

```
ACGT
AC-T
```

The indels can arise from biological causes in which an offspring actually removes or inserts a base. Other times indels can be caused by difficulties arising from the sequencing process. It is possible that a base was not called correctly or that the signal was too weak/noisy/imperfect to call the base. In any case the alignment process needs to consider the possibility of indels. In the previous case, the computer program would receive two strings (ACGT and ACT) and would have to figure out that the deletion of a *G* has occurred.

This is a serious matter. If the sequences are very long (perhaps thousands of bases), then there are thousands of locations where the indel may occur. Furthermore, the sequences may have several indels, and one location may have multiple indels. In other words, there may be consecutive dashes. For sequences of significant length, it is not feasible to consider all possible indels in a brute force computing fashion.

7.2.3 Rearrangements

Genes are encoded within DNA strands, but a single gene may be coded in more than one region of the strand. Thus, coding regions may contain noncoding regions within their boundaries. Consider an example that has a strand consisting of xNxMx where x is a noncoding region and the N and M are coding regions. It is possible that the distance between N and M may change in another sequence. It is also possible that the new sequence may be of the form xMxNx. Thus, it may be necessary to identify noncoding regions during the alignment and to lower their importance.

7.2.4 Global Versus Local Alignments

It is also necessary to note the difference between global and local alignments. In the global alignment case, the task is to align an entire sequence to another entire sequence. In the local alignment case the task is to find segments (such as N and M) that align with other segments in another sequence.

Given two sequences and the task of global alignment, it is still necessary to be concerned with the beginning and ending of the sequences. The sequencing technology tends to have problems calling the very beginning and very end of sequences. Thus, the actual sequence may be longer than necessary, which means sequences would have the form xNx. The leading and trailing x part of the sequence may be any length, and thus it may still be necessary to exclude leading and trailing portions of a sequence during the global alignment.

7.2.5 Sequence Length

Another complication involves sequence length. Often the alignment algorithms are based on the number of matches. Consider a case in which the sequences are 100 elements long and 90 of them align. Consider a second case in which the sequences are 1,000 elements long and 800 of them align. In the second case, the score can be higher since many more elements aligned, but the percentage of alignment is greater in the first case. For this reason, some algorithms consider the sequence length when producing an alignment score.

7.3 Simple Alignments

Aligning two strings sounds like a simple process, but it has long ago been mostly abandoned in a majority of bioinformatics applications. This section will explore simple alignment algorithms and the reasons for more complicated engines.

7.3.1 Direct Alignment

Direct alignment is an extremely simple concept. Given two sequences, a score is computed by adding a positive number for each match and a negative number for a mismatch. For the sake of argument, let's score a +1 for each match and a -1 for each mismatch. A simple example with a total score of 4 is

```
RNDKPKFSTARN
RNQKPKWWTATN
++-+++-++-+
```

An alignment between two different letters in this case counts as -1. An alignment of a letter with a gap is also a mismatch but perhaps should be counted as a bigger penalty—for example, -2. In this fashion, alignment with gaps is more discouraged than just mismatched letters. Code 7-1 displays the function **SimpleScore**, which performs this comparison. In previous versions of the numerical Python package, it was possible to compare two strings directly using *equal(s1, s2)*. The returned result was a vector of 1's and 0's indicating locations in which the two strings had the same character or different characters. In current versions of *numpy*, this option is not implemented. Thus, a slightly more complicated route is chosen.

The **ord** function returns the ASCII value for a given character. For example, the command *ord("R")* returns 82. The **map** function will send all elements of a list to a particular function and return the results. The **list** function will convert a string to a list in which each character of a string is an element in the list. Thus, *a1* is a list and each element of that list is the ASCII value for the elements of *s1*. The *score* variable first counts the number of matches and then subtracts the number of mismatches. This includes mismatches with gaps. So, at this point gaps are counted as -1. To count the gap mismatches as a -2 they are counted and the number of gaps is subtracted from the score.

This function assumes that the two sequences do not align two gaps. When considering only two strings, this situation should not occur. However, in more complicated applications such as multisequence alignment it may be possible to generate sequences that align gaps.

```
from numpy import equal, not_equal

# easyalign.py
def SimpleScore( s1, s2 ):
    a1 = map( ord, list(s1) )
    a2 = map( ord, list(s2) )
    # count matches
    score = ( equal( a1, a2 )).astype(int).sum()
    # count mismatches
    score = score -(not_equal( a1,a2 )).astype(int).sum()
    # gaps
    ngaps = s1.count( '-' ) + s2.count('-')
    score = score - ngaps
    return score

>>> SimpleScore( 'AGTCGATCGATT', 'AGTCGATCGATT')
12
>>> SimpleScore( 'AGTCGATCGATT', 'AGTCGATCGAAT')
10
>>> SimpleScore( 'AGTCGATCGATT', 'AGTCGATCGA-T')
9
```

Code 7-1

7.3.2 Statistical Alignment

The previous style of alignment was very simple, but it was mostly useless. Such alignments would require sequences to be already aligned and any gaps identified, and that is rarely the case. Even if we had that information, it still does not take into consideration that some matches are more important than others.

More formally, consider a sequence that contains letters from an alphabet Σ. An element (or *residue*) from this sequence is denoted by a. The probability of observing a residue is q_a. Of course, the sum of the probabilities for all residues is 1.0.

Following an example described in Durbin et al. (1998), let $p(a, b \mid M)$ be the probability that a will align with residue b given a model M. The model M can be considered as a finite set of sequence alignments from which the statistics are computed. This probability is the number of times that a aligns with b divided by the total number of aligned residues.

A random case, signified by the model R, is one in which the residues have an equal probability of being at any location. The probability of a particular sequence, x, occurring is

$$P(x|R) = \prod_i q_{x_i} \tag{7-2}$$

Thus, the probability of two sequences is the multiplication of the probability of each sequence, as in

$$P(x, y|R) = \prod_i q_{x_i} \prod_j q_{y_j} \tag{7-3}$$

DNA sequences are not random, and through some segments the probabilities are not the same for all residues. The model M indicates which type of segment is being considered, and the probability of a particular alignment is

$$P(x, y|M) = \prod_i p_{x_i y_j} \tag{7-4}$$

Now, for an aligned residue the ratio of its statistical alignment with the random case is

$$\frac{P(x, y|M)}{P(x, y|R)} = \prod_i \frac{p_{x_i y_i}}{q_{x_i} q_{y_i}} \tag{7-5}$$

The values of these probabilities may have quite a range, and thus it is easier to manipulate the *log-odds*:

$$S(x, y) = \sum_i \log\left(\frac{p_{x_i y_i}}{q_{x_i} q_{y_i}}\right) \tag{7-6}$$

The computation of the probabilities is tedious in that it takes several aligned sequences through an evolutionary chain. One complication is that it is possible to use different evolutionary time steps. Thus, there are several matrices that could be used for alignment. One such example is the BLOSUM50 matrix shown in Table 7-1. The entries are the log-odds of substitutions. In other words, the log-odds of an *I* being

Table 7-1 The BLOSUM50 Matrix.

	A	R	N	D	C	Q	E	G	H	I	L	K	M	F	P	S	T	W	Y	V
A	5	-2	-1	-2	-1	-1	-1	0	-2	-1	-2	-1	-1	-3	-1	1	0	-3	-2	0
R	-2	7	-1	-2	-1	1	0	-3	0	-4	-3	3	-2	-3	-3	-1	-1	-3	-1	-3
N	-1	-1	7	2	-2	0	0	0	1	-3	-4	0	-2	-4	-2	-1	0	-4	-2	-3
D	-2	-2	2	8	-4	0	2	-1	-1	-4	-4	-1	-4	-5	-1	0	-1	-5	-3	-4
C	-1	-4	-2	-4	13	-3	-3	-3	-3	-2	-2	-3	-2	-2	-4	-1	-1	-5	-3	-1
Q	-1	1	0	0	-3	7	2	-2	1	-3	-2	2	0	-4	-1	0	-1	-1	-1	-3
E	-1	0	0	2	-3	2	6	-3	0	-4	-3	1	-2	-3	-1	-1	-1	-3	-2	-3
G	0	-3	0	-1	-3	-2	-3	8	-2	-4	-4	-2	-3	-4	-2	0	-2	-3	-3	-4
H	-2	0	1	-1	-3	1	0	-2	10	-4	-3	0	-1	-1	-2	-1	-2	-3	-1	4
I	-1	-4	-3	-4	-2	-3	-4	-4	-4	5	2	-3	2	0	-3	-3	-1	-3	-1	4
L	-2	-3	-4	-4	-2	-2	-3	-4	-3	2	5	-3	3	1	-4	-3	-1	-2	-1	1
K	-1	3	0	-1	-3	2	1	-2	0	-3	-3	6	-2	-4	-1	0	-1	-3	-2	-3
M	-1	-2	-2	-4	-2	0	-2	-3	-1	2	3	-2	7	0	-3	-2	-1	-1	0	1
F	-3	-3	-4	-5	-2	-4	-3	-4	-1	0	1	-4	0	8	-4	-3	-2	1	4	-1
P	-1	-3	-2	-1	-4	-1	-1	-2	-2	-3	-4	-1	-3	-4	10	-1	-1	-4	-3	-3
S	1	-1	1	0	-1	0	-1	0	-1	-3	-3	0	-2	-3	-1	5	2	-4	-2	-2
T	0	-1	0	-1	-1	-1	-1	-2	-2	-1	-1	-1	-1	-2	-1	2	5	-3	-2	0
W	-3	-3	-4	-5	-5	-1	-3	-3	-3	-3	-2	-3	-1	1	-4	-4	-3	15	2	-3
Y	-2	-1	-2	-3	-3	-1	-2	-3	2	-1	-1	-2	0	4	-3	-2	-2	2	8	-1
V	0	-3	-3	-4	-1	-3	-3	-4	-4	4	1	-3	1	-1	-3	-2	0	-3	-1	5

replaced by a C is –2. The negative sign indicates that this occurs less frequently than the random case. The log-odds provide computational advantages in that all of the values are integers and within a small range (from −5 to +15). These are much easier to use than the wide range of probabilities.

Using the BLOSUM50 matrix as the reference, the score for the previous case is 49:

```
R N D K   P K F   S T A   R N
R N Q K   P K W   W T A   T N
7 7 0 6  10 6 1  -4 5 5  -1 7 0
```

There are several BLOSUM matrices that have been computed using different evolutionary time steps. Another popular set of matrices is named PAM, and it uses a different method of calculation. However, for the purposes of alignment the matrices are used in the same fashion.

Code 7-2 converts the *BLOSUM50* matrix into a Python matrix, but the entire command is too tedious to print in its entirety but B50 can be loaded easily as shown in Code 8-8. The *BLOSUM50* matrix does not actually contain information about which amino acid is associated with which row or column. As it is quite possible to find BLOSUM50 matrices on the Web with a different sorted order of amino acids, it

7.3 Simple Alignments

is necessary to have the string *PBET*, which contains the amino acids in the order that corresponds to *B50*.

```
B50 =array( {[ 5,-2,-1,-2,-1,...,-3,-2, 0],
             [-2, 7,-1,-2,-1,...,-3,-1,-3],
             ...
             [ 0,-3,-3,-4,-1,...,-3,-1, 5]})
PBET ='ARNDCQEGHILKMFPSTWYV'
```

Code 7-2

In the previous example, the first alignment is an *R* with an *R*. The task is then to extract the element in *B50* that corresponds with this substitution. The first step is to convert the amino acid letters to row and column indices. Code 7-3 shows that by using the **index** function the amino acid is converted to a location in the string *PBET*, which is equivalent to the row or column desired in the matrix. In this example, the amino acid *R* is the second letter in the string, and it is thus converted to the integer 1. Since both residues in this alignment are *R*, both the column and the row of the matrix are 1.

```
>>> PBET.index( 'R' )
1
>>> B50[1,1]
7
```

Code 7-3

Code 7-4 shows the alignment of *D* with *Q*. Since the *B50* matrix is symmetric, it does not matter which letter is associated with the row and which is associated with the column.

```
>>> PBET.index( 'D' )
3
>>> PBET.index( 'Q' )
5
>>> B50[3,5]
0
```

Code 7-4

The function **BlosumScore** in Code 7-5 computes the alignment score for two strings. It receives the scoring matrix, its associated alphabet, the two strings, and an optional gap penalty. There are two additional considerations in this function. The first is that it is possible that the two strings are not the same length. The alignment score therefore applies only over those parts of the sequence that align, and so the variable n is the length of the shortest sequence. The second consideration is that either sequence may have a gap. There will thus be a gap penalty of -8, which is more negative than any number in the BLOSUM50 matrix. The *PBET* string does not have a gap, and the **index** function will return an error message if its argument is not in the string. The *if* statement considers the case of either string having a gap. If neither has a gap, then the process of Code 7-4 is employed. The example call demonstrates the previous example. The *elif* statement is explained in the next section.

```
# easyalign.py
def BlosumScore( mat, abet, s1, s2, gap=-8 ):
    sc = 0
    n = min( [len(s1), len(s2)] )
    for i in range( n ):
        if s1[i] == '-' or s2[i] == '-' and s1[i] != s2[i]:
            sc += gap
        elif s1[i] == '.' or s2[i] == '.':
            pass
        else:
            n1 = abet.index( s1[i] )
            n2 = abet.index( s2[i] )
            sc += mat[n1,n2]
    return sc

>>> a = 'RNDKPKFSTARN'
>>> b = 'RNQKPKWWTATN'
>>> BlosumScore( B50, PBET, a, b )
49
```

Code 7-5

7.3.3 Brute Force Alignment

Because the two input sequences are not usually aligned, the alignment needs to be determined. The most simplistic and costliest method is to consider all possible alignments. A brief example would be

```
Seq1 = 'abc'
Seq2 = 'bcd'
All possible shifts
abc..        abc.        abc        .abc        ..abc
..bcd        .bcd        bcd        bcd.        bcd..
```

There are actually three types of shifts shown. The first two examples shift the second sequence toward its right, the third example has neither shifted, and the last two examples have the first sequence shifted to the right. It is cumbersome to create a program that considers these three different types of shift. An easier approach is to create two new strings that have the original data and empty elements represented by dots. In this case the new string *t1* contains the old string *Seq1* and *N2* empty elements (where *N2* is the length of *Seq2*). String *t2* is created from *N1* empty elements and the string *Seq2*.

```
t1 = 'abc...'
t2 = '...bcd'
```

Finding all possible shifts for *t1* and *t2* is quite easy. By sequentially removing the first character of *t2* and the last character of *t1*, all possible shifts are considered. Table 7-2 shows the iterations and the strings used for all possible shifts. In this case, iteration 2 would provide the best alignment.

The number of iterations is $N1+N2-1$, and the result of the computation is now a vector with the alignment scores for all possible alignments. Code 7-6 shows the function **BruteForceSlide**, which creates the new strings and considers all possible shifts. The scoring of each shift is performed by **BlosumScore**, but certainly other scoring functions can be used instead. Since **BlosumScore** is capable of handling strings of different lengths, it is not necessary to actually create *t2* in this function.

```
# easyalign.py
from numpy import zeros
def BruteForceSlide( mat, abet, seq1, seq2 ):
    # length of strings
    l1, l2 = len( seq1 ), len( seq2 )
    # make new string with leader
    t1 = len( seq2) * '.' + seq1
    lt = len( t1 )
    answ = zeros( lt, int )
    for i in range( lt ):
        answ[i] = BlosumScore( mat, abet, t1[i:], seq2 )
    return answ

>>> v = BruteForceSlide( B50, PBET, a,b )
```

Code 7-6

Table 7-2 The Alignments of Two Strings for Each Iteration.

Iteration	Strings
0	abc...
	...bcd
1	abc..
	..bcd
2	abc.
	.bcd
3	abc
	bcd
4	ab
	cd
5	a
	d

The example call to this function returns vector v, which is 24 elements in length. The first element is the score for the first alignment and so on. The first alignment corresponds to the first iteration (see again Table 7-2). Consider the example in Code 7-7, which uses two identical sequences. The first element of the answer is the score for the alignment of "... TCN" with "TCN...". The alignment of "TCN" with "TCN" is actually the fourth element in the vector. The alignment of the first character of the first string is located in the $N2+1$ location of the answer.

```
>>> BruteForceSlide( B50, PBET, 'TCN','TCN' )
array([ 0,  0, -3, 25, -3,  0])
```

Code 7-7

The vector returned by **BruteForceSlide** indicates the score of the best alignment and the relative shift between the two sequences that create the best alignment. Code 7-8 shows a different example that requires a relative shift. In this case, the score of the best alignment is 37, and the alignment of -2 indicates that the first sequence needs to be shifted two locations to the right to align with the second sequence. A positive alignment would require that the second sequence be shifted to the right in order to align with the first.

```
>>> v = BruteForceSlide( B50, PBET, 'TCAQN','TTTCAQN' )
>>> v
array([ 0,  0, -3, -4, -3, 37,  2,  0, -4, -1, -1,  0])
>>> v.max()
37
>>> v.argmax() - len( 'TTTCAQN' )
-2
```

Code 7-8

7.4 Summary

DNA sequences are strings from a restricted alphabet. An obvious method of aligning two strings is to find segments in which the strings have common sequences. However, realistic applications are more involved. Mismatches between characters are weighted differently. Thus, it is necessary to use a scoring matrix that contains the mismatched scores.

Bibliography

Durbin, R., S. Eddy, A. Krogh, and G. Mitchison. (1998). *Biological Sequence Analysis*. Cambridge, England: Cambridge University Press.

Problems

1. Write a program that converts a DNA string into an amino acid string.

2. Create a scoring matrix that is

$$M[i,j] = \begin{cases} 5 & i = j \\ -1 & i \neq j \end{cases}$$

   Align two amino acid sequences (of at least 100 characters) using the BLOSUM50 and the above M matrices. Are the alignments significantly different?

3. Modify the **BlosumScore** algorithm to align DNA strings so that the third element in each codon is weighted half as much as the other two.

8 Dynamic Programming

The functions discussed in the previous chapter required users to insert gaps manually into sequences. Information about where and how many gaps are needed is not generally available. A commonly executed task is to align two sequences and to determine the locations of the gaps that provide the optimal alignment. Because brute force alignment—a technique that considers all of the possibilities for the location of gaps—is no longer a viable option, the field has adopted dynamic programming as a solution.

8.1 The Problem with the Brute Force Approach

For the purposes of alignment, it is not possible to distinguish between an insertion and a deletion. Consider the following two sequences:

```
A = 'TGCGTAG'
B = 'TG-GTAG'
```

Has a 'C' been inserted into string A or has a letter been removed from string B? Since they are indistinguishable, the event is termed a *gap* or an *indel*. Before two strings are aligned, the number of gaps is not known, and even worse the number of consecutive gaps is also not known. For a sequence with n elements, the number of possible combinations of alignments between two sequences is

$$\binom{2n}{n} = \frac{(2n)!}{(n!)^2} \approx \frac{2^{2n}}{\sqrt{\pi n}}. \tag{8-1}$$

Figure 8-1 shows the astronomical behavior of Equation (8-1). Since many sequences can be hundreds of elements long, it is quite obvious that the number of possible combinations of inserting gaps into a sequence is prohibitive.

8.2 The Dynamic Programming Algorithm

The *dynamic programming* approach attempts to find an optimal alignment by considering three options for each location in the sequence and selecting the best option before considering the next location. Each iteration considers the alignment of two bases (one from each string) or the insertion of a gap in either string. The best of the three is chosen and the system then moves on to the next elements in the sequence.

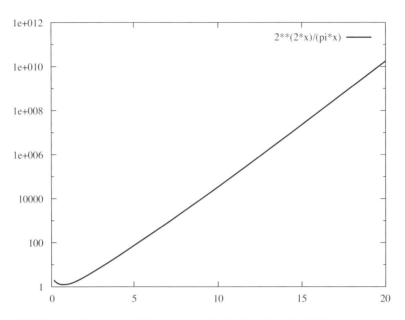

FIGURE 8-1 An illustration of the astronomical behavior of Equation (8-1).

To handle this efficiently, the computer program maintains a *scoring matrix* and an *arrow matrix*. This program will also use a substitution matrix such as BLOSUM or PAM.

8.2.1 The Scoring Matrix

The scoring matrix, **S**, maintains the current alignment score for the particular alignment. Since it is possible to insert a gap at the beginning of a sequence, the size of the scoring matrix is $(N1 + 1) \times (N2 + 1)$ where $N1$ is the length of the first sequence. Consider two sequences that contain some similarities. The lengths of the sequences are $N1 = 15$ and $N2 = 14$, and thus the scoring matrix is 16×15:

```
S1 = 'IQIFSFIFRQEWNDA'
S2 = 'QIFFFFRMSVEWND'
```

Alignment with a gap is usually penalized more than any mismatch of amino acids. For this example, $gap = -8$, but certainly other values can be used to adjust the performance of the system. The alignment of the first character with a gap is a penalty of -8, and the alignment with the first two characters with two gaps is -16, and so on. The scoring matrix is configured so that the first row and first column consider runs of gaps aligning with the beginning of one of the sequences. Thus, the first step in the construction of the scoring matrix is to establish the first row and first column as shown in Table 8-1.

The next step is to fill in each cell in a raster scan. The first undefined cell considers the alignment of I with Q or either one with a gap. There are three choices, and the selection is made by choosing the option that provides the maximum value,

$$S_{m,n} = \begin{cases} S_{m-1,n} + gap \\ S_{m,n-1} + gap \\ S_{m-1,n-1} + B(a,b) \end{cases} \qquad (8\text{-}2)$$

8.2 The Dynamic Programming Algorithm

Table 8-1 The Initial Row and Column Are Filled with Incremental Gap Penalties.

	-	Q	I	F	F	F	F	R	M	S	V	E	W	N	D
-	0	-8	-16	-24	-32	-40	-48	-56	-64	-72	-80	-88	-96	-104	-112
I	-8														
Q	-16														
I	-32														
F	-24														
S	-32														
F	-40														
I	-48														
F	-56														
R	-64														
Q	-72														
E	-80														
W	-88														
N	-96														
D	-104														
A	-112														

The $B(a, b)$ indicates the entry from the scoring matrix for residues a and b.

Normally, the first entry in the matrix is denoted as $S_{1,1}$. In order to be congruent with Python, the first cell in the matrix is $S_{0,0}$ and the first cell that needs to be computed is $S_{1,1}$. In order to be clear, it should be noted that this cell aligns the first characters in the two strings $S1[0]$ and $S2[0]$. Thus the indexing of the strings is slightly different than the matrix locations. In the example, the cell $S_{1,1}$ considers the alignment of the first two letters in each sequence. With $m = 1, n = 1, a =$ "I," and $b =$ "Q," the first cell has the following computation:

$$S_{1,1} = \begin{cases} -8 - 8 \\ -8 - 8, \\ 0 - 3 \end{cases} \quad (8\text{-}3)$$

The obvious choice is the third option. The results are shown in Table 8-2.

8.2.2 The Arrow Matrix

It is necessary to keep track of the choices made for each cell. Once the entire scoring matrix is filled out, it will be necessary to use it to extract the optimal alignment. This process uses the arrow matrix to find out which cell was influential in determining the value of the subsequent cell. In the previous example, the third choice was selected, which indicates that $S_{0,0}$ was the cell that influenced the value of $S_{1,1}$. The arrow matrix will place one of three integers (0, 1, 2) in the respective cells, and so $R_{1,1}$ would contain a 2. The scoring matrix and the arrow matrix are created concurrently. Code 8-1 displays the routine to compute these matrices.

The function **ScoringMatrix** receives the substitution matrix, its associated alphabet, the two sequences, and the gap penalty. The example call uses the BLOSUM50 matrix and the two example sequences. In this program the vector f contains

Table 8-2 The First Alignment Cell Is Filled with the Maximum of Three Possible Values.

	–	Q	I	F	F	F	F	R	M	S	V	E	W	N	D
–	0	-8	-16	-24	-32	-40	-48	-56	-64	-72	-80	-88	-96	-104	-112
I	-8	-3													
Q	-16														
I	-32														
F	-24														
S	-32														
F	-40														
I	-48														
F	-56														
R	-64														
Q	-72														
E	-80														
W	-88														
N	-96														
D	-104														
A	-112														

the three choices of Equation (8-2). The *scormat* is the scoring matrix that is being populated and the *arrowmat* is the partner arrow matrix.

```
from numpy import *
# dynprog.py
def ScoringMatrix( mat, abet, seq1, seq2, gap=-8 ):
    l1, l2 = len( seq1), len(seq2)
    scormat = zeros( (l1+1,l2+1), int )
    arrow = zeros( (l1+1,l2+1), int )
    # create first row and first column
    scormat[0] = arange(l2+1)* gap
    scormat[:,0] = arange( l1+1)* gap
    arrow[0] = ones(l2+1)
    for i in range( 1, l1+1 ):
        for j in range( 1, l2+1 ):
            f = zeros( 3 )
            f[0] = scormat[i-1,j] + gap
            f[1] = scormat[i,j-1] + gap
            n1 = abet.index( seq1[i-1] )
            n2 = abet.index( seq2[j-1] )
            f[2] = scormat[i-1,j-1] + mat[n1,n2]
            scormat[i,j] = f.max()
            arrow[i,j] = f.argmax()
    return scormat, arrow

>>> scmat, arrow = ScoringMatrix( B50, PBET, S1, S2 )
```

Code 8-1

8.2.3 Extracting the Aligned Sequences

The final step is to extract the aligned sequences from the arrow matrix. The process starts at the lower-right corner of the arrow matrix and works toward the upper-left corner. Basically, the aligned sequences are created from back to front. Code 8-2 displays the arrow matrix for the current example. In the lower-right corner, the entry is a 0, which indicates that this cell was created from the first choice of Equation (8-2). It aligned the last character of the first string with a gap. The current alignment is thus,

```
Q1 = 'A'
Q2 = '-'
```

The value of 0 also indicates that the next cell to be considered is the one above the current position since a letter from *S2* was not used. This next location in the *arrowmat* contains a 2, which indicates that two letters are aligned:

```
Q1 = 'DA'
Q2 = 'D-'
```

A value of 2 indicates that the backtrace moves up and to the left. Code 8-2 shows the arrow matrix, and the locations used in the backtrace are indicated in boldface. Each time a 0 is encountered a letter from *S1* is aligned with a gap, and the backtrace moves up one location. Each time a 1 is encountered the letter from *S2* is aligned with a gap and the backtrace moves to the left. Each time a 2 is encountered a letter from both sequences is used and the backtrace moves up and to the left.

```
>>> arrow
array([[1, 1, 1, 1, 1, 1, 1, 1, 1, 1, 1, 1, 1, 1, 1],
       [0, 2, 2, 1, 1, 1, 1, 1, 1, 1, 1, 1, 1, 1, 1],
       [0, 2, 2, 2, 1, 1, 1, 2, 2, 1, 1, 2, 1, 1, 1],
       [0, 0, 2, 1, 2, 1, 1, 1, 2, 1, 2, 1, 1, 1, 1],
       [0, 0, 0, 2, 1, 2, 1, 1, 1, 1, 1, 1, 2, 1, 1],
       [0, 0, 0, 0, 2, 1, 2, 2, 1, 2, 1, 1, 1, 1, 1],
       [0, 0, 0, 0, 2, 2, 1, 1, 1, 2, 1, 2, 1, 1, 1],
       [0, 0, 0, 0, 0, 2, 2, 1, 2, 1, 2, 1, 1, 1, 1],
       [0, 0, 0, 0, 0, 2, 2, 2, 2, 1, 2, 1, 2, 1, 1],
       [0, 0, 0, 0, 0, 0, 0, 2, 1, 1, 1, 1, 1, 2, 1],
       [0, 0, 0, 0, 0, 0, 0, 0, 2, 1, 1, 2, 1, 1, 1],
       [0, 0, 0, 0, 0, 0, 0, 0, 0, 2, 1, 2, 1, 1, 1],
       [0, 0, 0, 0, 0, 0, 2, 0, 0, 0, 2, 1, 2, 1, 1],
       [0, 0, 0, 0, 0, 0, 0, 0, 0, 0, 0, 0, 2, 0, 2, 1],
       [0, 0, 0, 0, 0, 0, 0, 0, 2, 0, 2, 2, 0, 2],
       [0, 0, 0, 0, 0, 0, 0, 0, 2, 2, 0, 2, 0, 0]])
```

Code 8-2

Code 8-3 shows the **Backtrace** function. The two strings that are being aligned are *st1* and *st2*. The backtrace starts at the lower-right corner and works its way up to the upper-left in the *while* loop. For each cell, it appends a letter or gap to each sequence and moves the backtrace to a new location. The strings are constructed in reverse order, which means that the last two lines of code are used to reverse the strings into the correct order. The example call shows the alignment of the two test strings.

```
# dynprog.py
def Backtrace( arrow, seq1, seq2 ):
    st1, st2 = ","
    v,h = arrow.shape
    ok = 1
    v -=1
    h -=1
    while ok:
        if arrow[v,h] == 0:
            st1 += seq1[v-1]
            st2 += '-'
            v -= 1
        elif arrow[v,h] == 1:
            st1 += '-'
            st2 += seq2[h-1]
            h -= 1
        elif arrow[v,h] == 2:
            st1 += seq1[v-1]
            st2 += seq2[h-1]
            v -= 1
            h -= 1
        if v==0 and h==0:
            ok = 0
    # reverse the strings
    st1 = st1[::-1]
    st2 = st2[::-1]
    return st1, st2

>>> t1, t2 = Backtrace( arrow, S1, S2 )
>>> t1
'IQIFSFIFRQ-EWNDA'
>>> t2
'-QIF-FFFRMSVEWND-'
```

Code 8-3

The quality of the alignment can be scored using the **BlosumScore** function from *easyalign.py*. Code 8-4 shows the results. In this case, the score of this alignment was 39. Without gaps, the original sequences produced an alignment score of 35. We can thus conclude that the new gapped sequences provided a better alignment.

```
>>> easyalign.BlosumScore( B50, PBET, t1, t2 )
39
>>> easyalign.BlosumScore( B50, PBET, S1,S2 )
35
```

Code 8-4

8.3 Efficient Programming

Code 8-1 does create the scoring and arrow matrices according to the prescription. However, it is not the most efficient manner in which to create these matrices. Recall that Python is an interpreted language and that nested loops are slow. The problem with this program is clear: in order to compute the value of $S_{m,n}$, it is necessary to know the value of the cell above and to the left of it. Thus, it is not possible to compute a row or column in a single Python command.

8.3.1 Flowing Along the Diagonals

The creation of the matrices is indeed a nested loop operation, but efficiency is gained by performing only one loop in Python and the other with commands from Numpy. While it is not possible to compute a single row or column concurrently, it is possible to compute a diagonal as shown in Figure 8-2. This image shows the first 10 × 10 elements of the scoring matrix. The elements connected by a line can be computed concurrently. Only four are shown, but the ideal holds true for the entire matrix. The efficiency is gained by having the Python *for* loop iterate from diagonal to diagonal and the Numpy commands compute the values of all of the cells along a single line. The speed savings is more than an order of magnitude.

Because the current example is too large to display, a smaller example is used to demonstrate this efficient algorithm. The new sequences are

```
>>> sq1 = 'ADC'
>>> sq2 = 'ARDC'
```

```
>>> scmat[:10,:10]
array([[  0,  -8, -16, -24, -32, -40, -48, -56, -64, -72],
       [ -8,  -3,   3, -11, -19, -27, -35, -43, -51, -59],
       [-16,  -1,  -6,  -7, -15, -23, -31, -36, -43, -51],
       [-24,  -9,   4,  -4,  -7, -15, -23, -31, -34, -42],
       [-32, -17,  -4,  12,   4,   1,  -7, -15, -23, -31],
       [-40, -25, -12,   4,   9,   1,  -2,  -8, -16, -18],
       [-48, -33, -20,  -4,  12,  17,   9,   1,  -7, -15],
       [-56, -41, -28, -12,   4,  12,  17,   9,   3,  -5],
       [-64, -49, -36, -20,  -4,  12,  20,  14,   9,   1],
       [-72, -57, -44, -28, -12,   4,  12,  27,  19,  11]])
```

FIGURE 8-2 Lines indicating cells to be computed concurrently.

8.3.2 Slicing Matrices

NumPy provides efficient methods of extracting elements from arrays, including random access in a matrix. Code 8-5 shows an advanced slicing technique. Matrix A, which is composed of random values, is created in order to extract values from four locations: (3, 0), (4, 1), (4, 2), and (1, 4). Lists x and y contain the vertical and horizontal indices of the desired locations in order. The command *a[x, y]* extracts these values. The final usage shown is that the values of the matrix can also be set using this method of slicing.

```
>>> from numpy import set_printoptions, random
>>> set_printoptions( precision = 3)
>>> a = random.ranf( (5,6) )
>>> a
array([[ 0.17 , 0.548, 0.596, 0.74 , 0.572, 0.088],
       [ 0.626, 0.676, 0.57 , 0.014, 0.809, 0.996],
       [ 0.448, 0.751, 0.523, 0.987, 0.273, 0.709],
       [ 0.277, 0.211, 0.854, 0.01 , 0.751, 0.273],
       [ 0.908, 0.618, 0.165, 0.083, 0.855, 0.484]])
>>> x = [3,4,4,1]
>>> y = [0,1,2,4]
>>> a[x,y]
array([ 0.277, 0.618, 0.165, 0.809])

>>> a[x,y] = -1, -2, -3, -4
>>> a
array([[ 0.17 , 0.548, 0.596, 0.74 , 0.572, 0.088],
       [ 0.626, 0.676, 0.57 , 0.014, -4.  , 0.996],
       [ 0.448, 0.751, 0.523, 0.987, 0.273, 0.709],
       [-1.  , 0.211, 0.854, 0.01 , 0.751, 0.273],
       [ 0.908, -2.  , -3.  , 0.083, 0.855, 0.484]])
```

Code 8-5

8.3.3 Extracting Diagonal Element Locations

Given the correct set of indices, it is possible to access all of the elements along a diagonal. The function **CreateIlist** in Code 8-6 determines each diagonal within a *for* loop. Each iteration determines the leftmost point as (*st1*, *st2*) and the rightmost point as (*sp1*, *sp2*). These are the locations within the matrix where the diagonal starts and ends. All of the matrix cells in between these points are determined by the two **arange** functions, and at the end of the iteration there are two lists (*x*, *y*) that are set in a tuple and appended to *ilist*. Code 8-7 creates a list for a case in which the two sequences have lengths 5 and 6. Each element in *ilist* is a tuple that contains the locations of the points for a single diagonal.

```
# dynprog.py
def CreateIlist( l1, l2 ):
    ilist = []
    for i in range( l1 + l2 -1 ):
        st1 = min( i+1, l1 )
        sp1 = max( 1, i-l2+2 )
        st2 = max( 1, i-l1+2 )
        sp2 = min( i+1, l2 )
        #print st1, sp1, st2, sp2
        ilist.append( (arange(st1,sp1-1,-1),arange(st2,sp2+1)))
    return ilist
```

Code 8-6

```
>>> ilist = CreateIlist( 5,6)
>>> for i in ilist:
print i

(array([1]), array([1]))
(array([2, 1]), array([1, 2]))
(array([3, 2, 1]), array([1, 2, 3]))
(array([4, 3, 2, 1]), array([1, 2, 3, 4]))
(array([5, 4, 3, 2, 1]), array([1, 2, 3, 4, 5]))
(array([5, 4, 3, 2, 1]), array([2, 3, 4, 5, 6]))
(array([5, 4, 3, 2]), array([3, 4, 5, 6]))
(array([5, 4, 3]), array([4, 5, 6]))
(array([5, 4]), array([5, 6]))
(array([5]), array([6]))
```

Code 8-7

8.3.4 Extracting Values from the Substitution Matrix

In order to make a faster program, it is necessary to remove the nested *for* loops from all parts of the Python script. This includes the extraction of values from the substitution matrix (such as a BLOSUM matrix). In the function **FastSubValues** in Code 8-8, the first two *for* loops convert each string to a set of integers that represent the locations of the sequence letters in the alphabet. For example, the sequence "ADC" is converted to [0 3 4] since those are locations of the three letters in *PBET*. The final *for* loops fill in one row of the substitution matrix, at a time. To match the scoring matrix, an extra row and column are placed at the front of the matrix. The indexing thus requires +1's and −1's to align the computed matrix with the scoring matrix that will be generated in the next section.

The output *subvals* is a matrix that contains the scores from BLOSUM50 for the two sequences *sq1* and *sq2* aligned with the scoring matrix. As an example, consider the computation of *submat[1,1]* which requires as one of the choices *submat[0,0]+SV[sq1[0], sq2[0]* where *SV* is the substitution value for the two letters *sq1[0]* and *sq2[0]*. In this case the two letters are both "A" and the substitution value from BLOSUM50 is 5. This value is now stored as *subvals[1,1]*, and so the computation for the third choice of Equation (8-2) is *submat[0, 0]+subvals[1, 1]*.

```
# dynprog.py
def FastSubValues( mat, abet, seq1, seq2 ):
    l1, l2 = len( seq1 ), len(seq2)
    subvals = zeros( (l1+1,l2+1), int )
    # convert the sequences to numbers
    si1 = zeros( l1, int )
    si2 = zeros( l2, int )
    for i in range( l1 ):
        si1[i] = abet.index( seq1[i] )
    for i in range( l2 ):
        si2[i] = abet.index( seq2[i] )
    for i in range( 1, l1+1 ):
        subvals[i,1:] = mat[ [si1[i-1]]*l2,si2]
    return subvals

>>> from blosum import PBET
>>> from blosum import BLOSUM50 as B50
>>> subvals = FastSubValues( B50, PBET, sq1, sq2 )
>>> subvals
array([[ 0,  0,  0,  0,  0],
       [ 0,  5, -2, -2, -1],
       [ 0, -2, -2,  8, -4],
       [ 0, -1, -4, -4, 13]])
```

Code 8-8

8.3.5 Computing the Scoring Matrix Values for a Single Diagonal

In Code 8-1 the vector *f* was three values for the three choices for a single element in the scoring matrix. This computation must now be removed from inside a nested *for* loop. Thus, it is necessary to compute the three values of *f* for all elements of the diagonal in a single loop. The first change is that *f* must now be a matrix of size 3, L, where L is the length of the diagonal. Given a set of points in a diagonal x, y, the first choice of Equation (8-2) is then *scormat[x-1, y] + gap*. Since both x and y are vectors, the *scormat[x, y]* is a set of values along the diagonal and therefore *scormat[x-1, y]* is the set of values directly above *scormat[x, y]*. Likewise the second choice of Equation (8-2) uses *scormat[x, y-1]*, which represents all of the values directly to the left of the diagonal. The third choice uses *scormat[x-1, y-1]+subvals[x, y]*, which adds the values above and to the left of the diagonal to the correct substitution values computed from **FastSubValues**.

8.3.6 An Efficient Computation of the Scoring Matrix

The function **FastNW** in Code 8-9 computes the scoring matrix and the arrow matrix with only a single loop in the Python code. Similar to its predecessor **ScoringMatrix**, this function will create the *scormat* and *arrow* and populate the first row and column. At this point the two functions differ. First, the input to the **FastNW** function includes *subvals* that are the substitution values from **FastSubValues**. Each iteration inside the *for* loop considers a single diagonal, which is represented by the index i, a tuple that is separated into two lists x and y. The matrix *f* will contain the three values for all of the elements along the diagonal.

The matrix *f* stores the three values as columns, and the command *f.max(0)* computes the maximums of each column. The result is that *mx* is a vector that contains the max value

8.3 Efficient Programming

for each element in the diagonal. The command *scoremat[i] = mx+0* copies these multiple values into all of the elements of the diagonal. The result of this function is that it computes the same scoring matrix and arrow matrix as **ScoringMatrix**, but it is much faster.

```
# dynprog.py
def FastNW(subvals, seq1, seq2, gap=-8 ) :
    l1, l2 = len( seq1), len(seq2)
    scormat = zeros( (l1+1,l2+1), int )
    arrow = zeros( (l1+1,l2+1), int )
    # create first row and first column
    scormat[0] = arange(l2+1)* gap
    scormat[:,0] = arange( l1+1)* gap
    arrow[0] = ones( l2+1 )
    # compute the ilist
    ilist = CreateIlist( l1, l2 )
    # fill in the matrix
    for i in ilist:
        LI = len( i[0] )
        f = zeros( (3,LI), float )
        x,y = i[0]-1, i[1]+0
        f[0] = scormat[x,y] + gap
        x,y = i[0]+0,i[1]-1
        f[1] = scormat[x,y] + gap
        x,y = i[0]-1,i[1]-1
        f[2] = scormat[x,y] + subvals [i]
        f += 0.1 * sign(f) * random.ranf( f.shape ) # to randomly select
from a tie
        mx = (f.max(0)).astype(int)    # best values
        maxpos = f.argmax( 0 )
        scormat[i] = mx + 0
        arrow[i] = maxpos + 0
    return scormat, arrow
```

Code 8-9

Code 8-10 shows the example with the resultant substitution matrix and arrow matrix. The function **Backtrace** did not have a nested *for* loop and therefore does not require an alteration.

```
>>> scmat, arrow = FastNW( subvals, sq1, sq2 )
>>> scmat
array([[  0,  -8, -16, -24, -32],
       [ -8,   5,  -3, -11, -19],
       [-16,  -3,   3,   5,  -3],
       [-24, -11,  -5,  -1,  18]])
>>> arrow
array([[1, 1, 1, 1, 1],
       [0, 2, 1, 1, 1],
       [0, 0, 2, 2, 1],
       [0, 0, 0, 2, 2]])
>>> t1, t2 = Backtrace( arrow, sq1, sq2 )
>>> t1
'A-DC'
>>> t2
'ARDC'
```

Code 8-10

8.4 Global Versus Local Alignments

The previous example illustrates a *global alignment algorithm*—also known as the Needleman-Wunsch algorithm—which attempts to align two strings from tip to tail. The backtrace begins with attempts to align the last two characters in the strings and ends with attempts to align the first two characters.

A *local alignment algorithm*—also known as the Smith-Waterman algorithm—attempts to find the best substrings within the two strings that align. It accomplishes this through only a couple of modifications. The first is to adjust the selection equation such that no negative numbers are accepted:

$$S_{m,n} = \begin{cases} S_{m-1,n} + gap \\ S_{m,n-1} + gap \\ S_{m-1,n-1} + B(a,b) \\ 0 \end{cases} \tag{8-4}$$

The other modification is the backtrace algorithm. Instead of starting in the lower-right corner of the arrow matrix, the trace starts at the location with the largest value in S. The trace continues until the fourth choice in Equation (8-4) is reached. Code 8-11 shows **FastSW**, which creates the scoring matrix and arrow matrix for the Smith-Waterman approach. The differences are that the lines that set up the first row and column of *scormat* are removed, and *scormat* is initialized similarly to *arrow*. The matrix f now has four rows to account for the four choices in Equation (8-4).

```
# dynprog.py
def FastSW( subvals, seq1, seq2, gap=-8 ) :
    l1, l2 = len( seq1), len(seq2)
    scormat = zeros( (l1+1,l2+1), int )
    arrow = zeros( (l1+1,l2+1), int )
    # create first row and first column
    arrow[0] = ones( l2+1 )
    # compute the ilist
    ilist = CreateIlist( l1, l2 )
    # fill in the matrix
    for i in ilist:
        LI = len( i[0] )
        f = zeros( (4,LI), float )
        x,y = i[0]-1, i[1]+0
        f[0] = scormat[x,y] + gap
        x,y = i[0]+0,i[1]-1
        f[1] = scormat[x,y] + gap
        x,y = i[0]-1,i[1]-1
        f[2] = scormat[x,y] + subvals[i]
        f += 0.1 * sign(f) * random.ranf( f.shape )
        mx = (f.max(0)).astype(int)   # best values
        maxpos = f.argmax( 0 )
        scormat[i] = mx + 0
        arrow[i] = maxpos + 0
    return scormat, arrow
```

Code 8-11

The scoring matrix is shown in Code 8-12 for the example. The maximum value is 26, and it is located at coordinates (6, 6). The backtrace starts at this location and follows values from the *arrow* matrix, in a similar manner to the Needleman-Wunsch approach until a value of 3 is reached in the *arrow* matrix. The bold characters show the simple path for this case.

```
>>> sq1 = 'KMTIFFMILK'
>>> sq2 = 'NQTIFF'
>>> subvals = FastSubValues( B50, PBET, sq1, sq2 )
>>> scmat, arrow = FastSW( subvals, sq1, sq2 )
>>> scmat
array([[ 0,  0,  0,  0,  0,  0,  0],
       [ 0,  0,  2,  0,  0,  0,  0],
       [ 0,  0,  0,  1,  2,  0,  0],
       [ 0,  0,  0,  5,  0,  0,  0],
       [ 0,  0,  0,  0, 10,  2,  0],
       [ 0,  0,  0,  0,  2, 18, 10],
       [ 0,  0,  0,  0,  0, 10, 26],
       [ 0,  0,  0,  0,  2,  2, 18],
       [ 0,  0,  0,  0,  5,  2, 10],
       [ 0,  0,  0,  0,  2,  6,  3],
       [ 0,  0,  2,  0,  0,  0,  2]])
>>> scmat.max()
26
>>> divmod( scmat.argmax(), 7 )
(6, 6)
>>> arrow
array([[0, 0, 0, 0, 0, 0, 0],
       [0, 3, 2, 3, 3, 3, 3],
       [0, 3, 3, 2, 2, 2, 2],
       [0, 2, 3, 2, 2, 2, 3],
       [0, 3, 3, 3, 2, 1, 3],
       [0, 3, 3, 3, 0, 2, 1],
       [0, 3, 3, 3, 2, 2, 2],
       [0, 3, 2, 3, 2, 0, 0],
       [0, 3, 3, 3, 2, 2, 0],
       [0, 3, 3, 3, 2, 2, 2],
       [0, 2, 2, 3, 3, 3, 2]])
```

Code 8-12

Code 8-13 displays the backtrace algorithm **SWBacktrace**, which is slightly different than its cousin. This program must receive the scoring matrix because it needs to know where the maximum value occurs. The location of the maximum is determined in the **divmod** statement. For a two-dimensional array, the **argmax** function will return the position as a single number, which is $v*H+h$, where v and h are the locations of the maximum, and H is the horizontal dimension of the array. The horizontal dimension of this array is 1+ the length of the *sq2*. The **divmod** function returns the division and remainder values—for example, divmod(13, 4) returns (3, 1)—and thus is the vertical and horizontal location of the maximum in the matrix.

```
# dynprog.py
def SWBacktrace( scormat, arrow, seq1, seq2 ):
    st1, st2 = ","
    v,h = arrow.shape
    ok = 1
    v,h = divmod( scormat.argmax(), len(seq2)+1 )
    while ok:
        if arrow[v,h] == 0:
            st1 += seq1[v-1]
```

```
                st2 += '-'
                v -= 1
            elif arrow[v,h] == 1:
                st1 += '-'
                st2 += seq2[h-1]
                h -= 1
            elif arrow[v,h] == 2:
                st1 += seq1[v-1]
                st2 += seq2[h-1]
                v -= 1
                h -= 1
            elif arrow[v,h] == 3:
                ok = 0
            if v==0 and h==0 or scormat[v,h]==0:
                ok = 0
        # reverse the strings
        st1 = st1[::-1]
        st2 = st2[::-1]
        return st1, st2
```

Code 8-13

Code 8-14 shows an example run of the codes for the Smith-Waterman approach. Only the aligning characters are extracted during the backtrace function. Gaps will be inserted in other examples as necessary.

```
>>> sq1 = 'KMTIFFMILK'
>>> sq2 = 'NQTIFF'
>>> subvals = FastSubValues( B50, PBET, sq1, sq2 )
>>> scmat, arrow = FastSW( subvals, sq1, sq2 )
>>> t1, t2 = SWBacktrace( scmat, arrow, sq1, sq2 )
>>> t1
'TIFF'
>>> t2
'TIFF'
```

Code 8-14

8.5 Gap Penalties

The programs as presented use a standard gap penalty. The cost of each gap is the same independent of its location and independent of a consecutive run of gaps. In some views a consecutive run of gaps should be more costly than isolated gaps. These approaches use an *affine gap*, which adds extra penalties for consecutive gaps. This does complicate the program somewhat because it is now necessary to keep track of the number of gaps when computing Equation (8-4). This option will not be explored in this book.

8.6 Does Dynamic Programming Find the Best Alignments?

Dynamic programming can provide a good alignment, but is it the very best? Consider a case in which two random sequences are generated, each of them 100 elements in length. A substring of 10 elements is copied from the first string and replaces 10 elements in the second string. This results in two random strings, except for 10 elements

8.6 Does Dynamic Programming Find the Best Alignments?

that are exactly the same and the Smith-Waterman algorithm should align these 10 elements only.

Code 8-15 shows an example that returns sequences that are much longer than the expected length of 10. The last elements match but the random letters in front of the matching sequence do not.

```
>>> random.seed( 5290 )
>>> r = (random.rand( 100 )*20).astype(int)
>>> s1 = take( list(PBET), r ).tostring()
>>> r = (random.rand( 100 )*20).astype(int)
>>> s2 = take( list(PBET), r ).tostring()
>>> s2 = s2[:30] + s1[10:20] + s2[40:]
>>> subvals = FastSubValues( B50, PBET, s1, s2 )
>>> scmat, arrow = FastSW( subvals, s1, s2 )
>>> t1, t2 = SWBacktrace( scmat, arrow, s1, s2 )
>>> t1
'LGYTWFVTIQRMVQVDPLGPI'
>>> t2
'MAQLWNCSDMRMVQVDPLGPL'
```
Code 8-15

Code 8-16 shows a worse example, with two sequences generated randomly as in Code 8-15. Code 8-16 shows what was generated and then the process of aligning the sequences through Smith-Waterman. The returned sequences are much longer than 10 elements this time. The first 10 elements match but the rest do not. This has two implications: First, the largest value in the scoring matrix was not at the end of the 10-element alignment but in some other place—47 elements away from the end of the aligning strings. Second, there were no 0 values in the scoring matrix from this peak to the beginning of the aligning elements.

```
>>> s1
'KKPGHWMVRCKQGQKRVGLNRYMDNYSSPKNHMVRDHFHLWKWMPSENCPAECWADKLWYIMKSCP
ADQPFTALKQVIAQTEEQVNYNNVGAHMAADSCT'
>>> s2
'GGFMEGCCTPMYARTCVCDHCIGRVSERINKQGQKRVGLNLVRHGILIWHNFLVGNQVWPWLMECF
QAAGSTNKVYIREVPQIRKAIDYSLQYTINIVYL'
>>> subvals = FastSubValues( B50, PBET, s1, s2 )
>>> scmat, arrow = FastSW( subvals, s1, s2 )
>>> t1, t2 = SWBacktrace( scmat, arrow, s1, s2 )
>>> t1
'KQGQKRVGLNRYMDNYSSPKNHMVRDHFHLWKWMPSENC-PAECWADKLWYIMKSCP'
>>> t2
'KQGQKRVGLNLVRHGILIWHNFLVGNQ-V-WPWL-ME-CFQAAGSTNKV-YI-REVP'
```
Code 8-16

Figure 8-3 shows an image of the scoring matrix in which the darker pixels represent larger values in the matrix. The slanted streaks are *jets* that appear in the scoring matrix when alignment occurs. The main jet is quite obvious and starts at *scmat[10, 31]* because the first two aligning elements are *s1[10]* and *s2[31]*. The alignment should be only 10 elements long with the jet ending at *scmat[20, 41]*, but the jet does not end there. Recall that the goal is to have the largest value in the scoring matrix at the location where the two alignments end. This is a large number and is

FIGURE 8-3 Jets in the scoring matrix.

used to influence the subsequent elements in the scoring matrix via Equation (8-4). The third option in this equation will have two nonsimilar characters and the value returned from the BLOSUM matrix may be negative but not enough to return a 0 for this option. Alignments after that may return positive values from the BLOSUM matrix and thus increase the values in the cells of the scoring matrix after the alignment has ceased to exist. This is not a trivial problem, as can be seen in the figure. Smith-Waterman returned a large number of characters after the alignment, but this alignment was not terminated from the fading of a jet. It was terminated because one sequence had reached its end.

While this method did return the true alignment it also can return alignments of random characters. Thus, this is not the best alignment. It should also be noted that by viewing the scoring matrix in terms of its jets other possible alignments are seen. There are secondary jets that indicate other partial alignments between these two sequences.

8.7 Summary

Sequences can have bases added or deleted either through biology or errors in sequencing. The locations of these insertions or deletions are unknown, and their numbers are also unknown. A brute force computation that considers all possible combinations of alignments with insertions and deletions is computationally too expensive. Thus, the field has adopted dynamic programming as a method of finding a good alignment with gaps.

Creating a dynamic programming alignment can be accomplished by following the algorithm's equations. However, this creates a double-nested loop that may run slowly in Python. Thus, a modified approach is used to push one of the loops down into the compiled code and to leave only one loop up in the Python script. This makes the algorithm run fast by at least an order of magnitude.

Problems

1. Align two strings with some similarity, using gap penalties of 0, −2, −4, −8, and −16. Are there differences in the alignments?

2. Create a string with the form *XAXBX*, where *X* is a set of random letters and *A* and *B* are specific strings designed by the user. Each *X* can be a different length. Create a second string with the form *YAYBY*, where *Y* is a different set of random letters and each *Y* can have a different length. Align the sequences using Smith-Waterman. The scoring matrix will have two major maxima for the alignments of the *A* and *B* regions. Modify the program to extract both alignments.

3. Is it possible to repeat Problem 2 where the second string is of the form *YABY*?

4. Create a program that aligns two halves of strings. For example, the first string, *str1*, can be cut into two parts (*str1a* and *str1b*), where *str1a* is the first *K* characters and *str1b* is the rest of the string. The second string is similarly cut into two parts (*str2a* and *str2b*) at the same *K*. Align *str1a* with *str2a* (using Needleman-Wunsch) and *str1b* with *str2b*. For each alignment, compute the alignment score using **BlosumScore**. Is it the same value as the alignment of *str1* with *str2*?

5. Repeat Problem 4 for different values of *K*, where *K* ranges from 10 to *N* − 10 (*N* is the length of the strings). Did you find a case in which the alignment of the *a* and *b* parts performs better than the alignment of the original strings?

6. Align a set of proteins using different BLOSUM or PAM matrices. How different are the alignments?

7. Align two proteins using a BLOSUM matrix. Replace the substitution matrix with *M*, where

$$M_{i,j} = \begin{cases} 5 & i = j \\ -1 & i \neq j \end{cases}$$

Repeat the alignment using this substitution matrix. Does it make much of a difference?

9 Tandem Repeats

Some regions of the genome contain repeating regions of DNA. In some cases the repeats are quite simple—for example, GATGATGAT. In other cases repeats are much more complicated, with regions repeating minor variations or sets of nested repeating regions. This chapter will explore one method of finding these repeating regions.

9.1 Tandem Repeats

Repeats are consecutive repeating segments in a string. Two simple examples are

```
TCTCTCTC
ATTCATTCATTC
```

A compressed format for representing the repeat is to place a subscript for the number of repeats of a substring enclosed in parentheses—for example, TGTGTGTG, which can be written as $(TG)_4$.

Tandem repeats may become much more complex when a region repeats with a minor variation. In an example such as $(TGTA)_8$, the last letter in the alternating repeats may become a C, which will result in TGTA TGTC TGTA TGTC TGTA TGTC TGTA TGTC. In another complication, a set of nested repeats may consist of repeats inside of repeats, such as $(\,(TGTAT)_1\,(AC)_*\,)_3$ where the asterisk indicates a variable. This string thus becomes TGTAT ACACAC TGTAT ACACACAC TGTAT AC.

9.2 Hauth's Solution

The issue of finding tandem repeats in a string without prior knowledge of the type of repeat has been addressed by Hauth and Joseph (2002). Figure 9-1 shows the schematic for this algorithm. The system first selects a particular word size and then all of the words of this size are collected. Each of these words is considered, and their locations are collected and converted to a histogram. Detection of a repeat is performed from this histogram.

9.2.1 Foundation

Building this process starts with the extraction of words of a specific size from a string. Code 9-1 shows the function **WordExtract**, which collects all of the words into a list. The last line uses the **set** function, which removes the duplicates from this list.

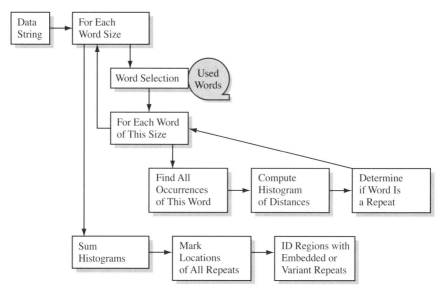

FIGURE 9-1 A schematic diagram of Hauth's algorithm.

```
# hauth.py
def WordExtract( seq, wsize ):
    words = []
    L = len( seq )
    for i in range( L-wsize+1 ):
        words.append( seq[i:i+wsize] )
    words = set( words )
    return list(words)
```

Code 9-1

When a single word is selected, it is necessary to find all occurrences of this word in the string. This is a simple operation performed by the function **FindOccurs** in Code 9-2. This simple function returns a list of the location of the specified word.

```
# hauth.py
def FindOccurs( seq, word ):
    cnt = seq.count( word )
    locs = []
    k = 0
    for i in range( cnt ):
        k = seq.find( word, k )
        locs.append( k )
        k +=1
    return locs
```

Code 9-2

The next step is to compute a histogram of the distances between locations of a specific word. If a word has a regular pattern, then there should be multiple instances in which the distance between two copies of this word is the same. It is easy to measure the distances between neighboring occurrences of a word but that may actually mask a repeat. Consider the string ATGACAACTGATGCATGA. The TG occurs every 8 locations but there is also a TG that is in between the repeating TGs. The locations of TG are [1, 8, 11, 15] and the distances between them are [7, 3, 4]. Even though TG occurs in every seventh position, the regular pattern is not evident. Therefore, it is necessary to consider distances between nonconsecutive locations. The function **AllDists** in Code 9-3 collects the distances between all possible pairs of numbers returned by **FindOccurs**.

```
# hauth.py
def AllDists( locs ):
    N = len( locs )
    D = []
    for i in range( N ):
        for j in range( i ):
            D.append( locs[i]-locs[j] )
    D = array( D )
    return D
```

Code 9-3

Code 9-4 shows the **Histogram** function, which generates a histogram from the results of **AllDists**. The user supplies an *upper* limit, which is the farthest that two pairs can be apart and still be counted. The histogram function uses the *if* statement to count only those pairs that are closer to each other than the value of *upper*.

```
# hauth.py
def Histogram( upper, D ):
    Nbins = upper + 1
    hist = zeros( Nbins, int )
    for i in range( len( D )):
        k = D[i]
        if 0< k < Nbins:  hist[k] +=1
    return hist
```

Code 9-4

In the example shown in Code 9-5, a random sequence is generated by converting 200 random integers to letters in the *abet*. The **take** function returns an array, and it is converted to a list using **list** and to a string using **join**. Then a repeat of $(ACTT)_{10}$ is inserted into the string. The **WordExtract** function finds all of the four-letter words in this string. Since there are only four letters in the alphabet, there is a maximum of 256 possible different words. In this string, 141 different words were found. The locations of a specified word "ACTT" are found and converted to a histogram. The value of 9 is found at *hist[4]*. This indicates that there are nine instances in which "ATCC" is followed by "ATCC." There is a bit of ringing in this method. There are also eight instances in which "ATCC" is followed by "xxxxatcc," where "x" indicates any letter. Of course, this occurs in the same run, as shown by the underlined letters atccatccatcc.

```
>>> from numpy import random, take
>>> random.seed( 414 )
>>> r = (random.rand( 200 )*4 ).astype(int)
>>> abet = 'ACGT'
>>> seq= take( list(abet), r )
```

```
>>> seq = ''.join( list( seq ))
>>> seq = seq[:45] + 'ACTT'*10 + seq[45:]
>>> words = WordExtract( seq, 4 )
>>> len( words )
136
>>> locs = FindOccurs( seq, 'ACTT' )
>>> len( locs )
10
>>> ad = AllDists( locs )
>>> hist = Histogram( 20, ad )
>>> hist
array([0, 0, 0, 0, 9, 0, 0, 0, 8, 0, 0, 0, 7, 0, 0, 0, 6, 0, 0, 0, 5])
```

Code 9-5

To determine whether a word is a repeating word, the number of instances in any of the histogram bins must be above a user-selected threshold. In Code 9-6, the function **FindRepeats** receives the histogram, the word in question, and the threshold *gamma*. In the example, the histogram needs to exceed a value of 7 in order to be considered a repeat. This function returns either an empty tuple or a tuple with the number of instances that this repeat was seen, the word, and the spacing between repeats.

```
# hauth.py
def FindRepeats( hist, word, gamma = 10 ):
    mx = hist.max()
    here = hist.argmax()
    if mx < gamma:
        ok = 0
        return ()
    else:
        return (mx,word, here)
>>> reps = FindRepeats( hist, 'actt', 7 )
>>> reps
(9, 'actt', 4)
```

Code 9-6

9.2.2 Multiple Words

In the previous example, we knew that the string being sought was "actt", but that type of information is usually not available. Therefore, all known words are used as a search query, and the histograms they generate are summed.

Code 9-7 shows this process in **MultipleWords**, in which each word is used to generate a histogram. If a word appears only once in the *seq*, then it will not produce any value in the histogram. The sequence "actt" is repeated ten times in this case. But that also repeats the sequence "ctta" nine times. Likewise, the sequences "ttac" and "tact" are also repeated nine times. Thus, the values at *hist[4]*, *hist[8]*, and so on, are increased. Clearly a pattern arises.

```
# hauth.py
def MultipleWords( words, seq ):
    NW = len( words )
    hist = zeros( 21, int )
```

```
    for i in range( NW ):
        locs = FindOccurs( seq, words[i] )
        D = AllDists( locs )
        h = Histogram( 20, D )
        hist += h
    return hist
>>> mw = MultipleWords( words, seq )
>>> mw
array([ 0, 0, 0, 0, 34, 0, 2, 0, 30, 0, 1, 0, 26, 1, 0, 1, 22, 0,
 0, 1, 18])
```

Code 9-7

9.2.3 Tandem Repeats

Recall the example ($(TGTAT)_1$ $(AC)_*$ $)_3$ that created the string TGTAT ACACAC TGTAT ACACACAC TGTAT AC. This string is more complicated than a simple repeat, but the system can still pick it up. Code 9-8 uses different word sizes to illustrate this example. There is a 5-letter element that repeats, and the two 7's in the example demonstrate its detection. There is also a 2-letter repeat and in the last example it is also picked up with the large values in *mw*.

```
>>> seq = 'TGTATACACACTGTATACACACACTGTATAC'
>>> words = WordExtract( seq, 5 )
>>> mw = MultipleWords( words, seq )
>>> mw
array([0, 0, 0, 0, 0, 0, 0, 0, 0, 0, 0, 7, 0, 7, 0, 0, 0, 0, 0, 0])
>>> words = WordExtract( seq, 7 )
>>> mw = MultipleWords( words, seq )
>>> mw
array([0, 0, 0, 0, 0, 0, 0, 0, 0, 0, 0, 5, 0, 7, 0, 0, 0, 0, 0, 0])
>>> words = WordExtract( seq, 2 )
>>> mw = MultipleWords( words, seq )
>>> mw
array([ 0, 0, 11, 0, 4, 0, 1, 2, 0, 5, 0, 12, 0, 13, 0, 4, 0, 1,
 0, 0, 1])
```

Code 9-8

9.3 Summary

DNA strings sometimes contain complicated repeating regions. The method considered here had the capability of finding complicated repeats as long as the user defined the length of the repeating segment. Thus, several word lengths were considered.

Bibliography

Hauth, A. H. and D. A. Joseph. (2002). Beyond tandem repeats: a complex pattern of structures and distant regions of similarity. *Bioinformatics*, 18, suppl. 1, S31–S37.

Problems

1. Write a program to find simple repeating regions.

2. Write a program that can find a repeating region in which the repeats are separated by a different number of characters. The separation of repeats has a finite range (for example, 0 to 6).

3. Write a program that can find repeats or their complements.

4. Get a Genbank file that contains repeating sections (NC_001133, NC_001136, NC_007646, NC_001147). Extract the DNA and use Hauth's method to find the repeating regions. Compare the known repeating regions described in the file.

10 Hidden Markov Models

The hidden Markov model (HMM) is a useful tool for computing probabilities of sequences. Since there are different types of sequences, there are different variations of the HMM. This chapter will review the basics of some of the most popular instantiations.

A typical HMM is a set of connected nodes, as shown in Figure 10-1. The information flows through the network from left to right. Each node computes the probability of emitting a given state, and each connection contains the probability of a transition from one node to the next. Commonly, the HMM uses only the emission or transition probabilities, but not both. Therefore, different HMMs are presented here using only one of these options.

10.1 The Emission HMM

The emission HMM consists of nodes that have emission probabilities, as shown in Figure 10-2. This node can emit one of three states (A, B, C), each with a different probability. Except in unusual circumstances, the sum of the probabilities equals 1.0, which is equivalent to stating that there is a 100 percent chance that the node will emit a state.

The emission HMM consists of a set of connected emission nodes, as shown in Figure 10-3. In this example the probability of the system emitting the sequence CAB is 0.5*0.1*0.6 = 0.03 or 3 percent. This HMM has a finite set of strings that it can

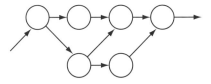

FIGURE 10-1 A simple hidden Markov model.

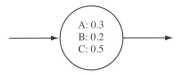

FIGURE 10-2 An emission HMM node.

125

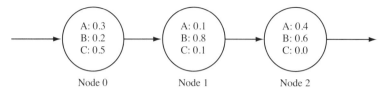

FIGURE 10-3 An emission HMM.

Table 10-1 The Emission Probabilities for All Possible States of Figure 10-3.

Emission State	Probability
AAA	0.012
AAB	0.018
AAC	0
ABA	0.096
ABB	0.144
ABC	0
ACA	0.012
ACB	0.018
ACC	0
BAA	0.008
BAB	0.012
BAC	0
BBA	0.064
BBB	0.096
BBC	0
BCA	0.008
BCB	0.012
BCC	0
CAA	0.02
CAB	0.03
CAC	0
CBA	0.16
CBB	0.24
CBC	0
CCA	0.02
CCB	0.03
CCC	0

produce. Table 10-1 shows the different emission states and their probabilities. The sum of all these states in this case must add up to 1.0 to indicate that there is a 100 percent chance that this system will emit a state.

To create the emission HMM in Python, it is necessary to represent the nodes and the connections between them. Since there are multiple choices for emissions, the emission information is contained in a list. Each item in the list is a tuple that contains

10.1 The Emission HMM

the emission state and the probability associated with that state. Thus, the emission information for the node in Figure 10-2 is represented by

[('A',0.3),('B',0.2),('C',0.5)]

This list contains only the emission information. The node needs to also have connection information. The philosophy adopted here is that each node will have an identifier and that the connections are denoted by these identifiers. In Figure 10-3 the three nodes are identified as 0, 1, and 2. Thus, the first node is represented by a tuple that contains the emission list and the transition to node 1,

([('A',0.3),('B',0.2),('C',0.5)], 1)

The HMM in Figure 10-3 contains three nodes and the collection of nodes (the HMM) is represented by a dictionary of nodes. The key is the node identification, and the value is the emission tuple. The creation of the HMM in Figure 10-3 is shown in Code 10-1.

```
hmm = {}
hmm[0] =( [('A',0.3),('B',0.2),('C',0.5)], 1 )
hmm[1] =( [('A',0.1),('B',0.8),('C',0.1)], 2 )
hmm[2] =( [('A',0.4),('B',0.6),('C',0.0)], -1 )
```

Code 10-1

Computing the probability of a given state is quite easy. Given as a string such as "BAB," the probabilities of each emission are computed in sequence. To accomplish this, two routines are written. Code 10-2 shows the function **EMatch**, which returns the emission for a single node given a specific letter. In the example, the first HMM node is searched for the letter "A," and **EMatch** returns a tuple with the identification and the emission probability. The function **EProb** in Code 10-3 computes the probability of a string given in an HMM model by sequentially considering each letter in the string. The example shows that the probability of this system emitting "BAB" is 1.2 percent.

```
# hmm.py
# find a match for a single node
def EMatch( hmmi, letter ):
    hit = (-1,-1)
    for i in hmmi[0]:
        if letter == i[0]:
            hit = i
            break
    return hit

>>> EMatch( hmm[0], 'A' )
('A', 0.3)
```

Code 10-2

```
# hmm.py
# compute the probability
def EProb( hmm, instr ):
    L = len( instr )
    k = 0
    pbs = 1.0
    for i in range( L ):
        # get emission
        emit = EMatch( hmm[k], instr[i] )
        pbs *= emit[1]
        # get transition
        k = hmm[k][1]
    return pbs

>>> EProb( hmm, 'BAB' )
0.012
```

Code 10-3

10.2 The Transition HMM

The transition HMM has nodes that emit only a single state, but they are connected to multiple nodes with individual probabilities. An example is shown in Figure 10-4. In this system the sum of the probabilities exiting a node must add up to 1.0 to indicate that there is a 100 percent chance that the node will transition to another node. The states realizable by this HMM are listed in Table 10-2. Again this HMM has a finite set of states, and so the total probabilities from this table add up to 1.0.

In Figure 10-4 there is only one entry node but two exit nodes. It is quite possible for a system to have multiple entry and exit nodes, although it can be cumbersome for

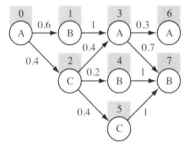

FIGURE 10-4 A transition HMM.

Table 10-2 The Emission Probabilities of All Possible States of Figure 10-4.

Emission State	Probability
ABAA	0.18
ABAB	0.42
ACAA	0.048
ACBB	0.08
ACAB	0.112
ACCB	0.16

programming. An easy solution is to create a BEGIN and END node at each end of the HMM. This provides a single entry and exit that sandwich the actual HMM.

Representing the HMM in Python follows a similar philosophy as in the emission case. A single node needs to represent both the emission and the transition. In this case, the emission is simply the state of the node (with an assumed probability of 1 for its emission), but the transition can have several choices. Thus, the transition is contained within a dictionary. The keys to the dictionary entries are the identifications of the nodes that can receive its transition, and the value for these is the probability of transition. Thus, the first node in Figure 10-4 is represented as

```
('A', {1:0.6,2:0.4} )
```

The transition HMM is thus represented by a dictionary of nodes. Figure 10-5 shows a simple network that can realize only two different states. The nodes that have a single transition have an assumed probability of 1.0. Code 10-4 creates this network. Note that since a dictionary is being used it is possible to identify the BEGIN and END nodes with strings while identifying the important nodes with integers.

```
# hmm.py
def SimpleTHMM( ):
    hmm = {}
    hmm['begin'] = ('',{0:1.0} )
    hmm[0] = ('A', {1:0.3,2:0.7} )
    hmm[1] = ('B', {3:1.0} )
    hmm[2] = ('C', {3:1.0} )
    hmm[3] = ('D', {'end':1.0} )
    hmm['end'] = ('',{} )
    return hmm
>>> hmm = SimpleTHMM()
```

Code 10-4

Computing the probability of the system is a bit more involved. In this case it is necessary to look at two characters in the input string to determine the correct transition. Consider the task of computing the probability of this HMM emitting the string ABD. The first node is A, but the program also needs to know that the second node will be the one with a B and not the C. The solution in function **NextNode** in Code 10-5 receives three arguments: (1) the entire HMM, (2) the identification of the node being considered, and (3) the state of the following node. In the example, the program

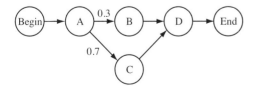

FIGURE 10-5 An HMM for Code 10-4.

is presented with the HMM, and it is asked to return information for node-0 that transitions to a node with a state "B." The program returns a tuple identifying the node and probability for the receiving HMM node.

```
# hmm.py
def NextNode( hmm, k, ask ):
    t = hmm[k][1].keys() # transition for this node
    hit = []
    for i in t:
        if hmm[i][0]==ask:
            hit = i, hmm[k][1][i]
            break
    return hit

>>> NextNode( hmm,0 , 'B' )
(1, 0.3)
```

Code 10-5

Computing the probability for the entire system then requires that each transition be discovered. In Code 10-6 the function **TProb** computes the probability that the given *hmm* will emit the state *instr*. It loops through **NextNode** to connect the *i*th node to the *i+1* node. The only two possible emissions for this system are shown as examples.

```
# hmm.py
def TProb( hmm, instr ):
    L = len( instr )
    pbs = 1.0
    k = 'begin'
    for i in range( L ):
        tran = NextNode( hmm,k,instr[i])
        k = tran[0]
        pbs *= tran[1]
    return pbs

>>> TProb( hmm, 'ABD' )
0.3
>>> TProb( hmm, 'ACD' )
0.7
```

Code 10-6

10.3 The Recurrent HMM

In both the emission and transition HMMs, the information flowed from the beginning to the end of the network only in a forward direction. In a *recurrent* HMM, the information can flow backward and a node can even feed itself, as shown in Figure 10-6.

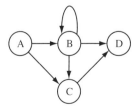

FIGURE 10-6 A recurrent HMM.

10.3 The Recurrent HMM

This system can emit strings such as ABD, ACD, ABCD, ABBBD, and ABBBBBCD. In fact, there is technically an infinite number of strings that this system can produce, and it is possible for this system to emit strings of differing lengths.

A program for a recurrent HMM can actually be quite easy to create. Consider the system shown in Figure 10-7, which can emit DNA strings. Every node is connected to every other node and to itself. This looks complicated but can be realized with a single 4×4 matrix. In this matrix the $M[i,j]$ element is the transition probability from node-i to node-j. The probability of the system is computed by considering consecutive letters in the string and finding the appropriate transition in the matrix M.

However, there is a problem with such networks because the probabilities are highly sensitive to the string length. Very long strings will produce a very long sequence of probabilities, each less than 1.0, that are multiplied together. This will create an extremely small number.

Consider a case in which the network has uniform probabilities. Since each node connects to four nodes, the transition probability for each connection is 0.25. Strangely, a network that has uniform connections is called a "random network" because it represents the network constructed from an infinite set of randomly generated strings. The problem is that a string of length 4 will have a probability of about 0.004. A string of length 10 will have a probability of 9e-7. The computation is 0.25^N, where N is the length of the string. Obviously, the computer will have problems with strings having a length of several hundred bases. Thus, the transition probabilities are replaced with log-odds. The log-odd is related to the probability of transition of the real data divided by the probability of transition for a random system. Consider a case in which the probability of seeing a "C" after an "A" is 30 percent. The ratio of the probability versus a random case is 0.3/0.25.

The odds ratio is greater than 1 if the transition is seen more often than the random case and less than 1 if the transition is rarely seen. By computing the log of this ratio, the ratios greater than 1 are converted to positive numbers and the ratios less than 1 are converted to negative values. Since the logs are being used, the transitions are added rather than multiplied.

To realize this network in code, a matrix containing the log-odds is created. The values from each transition in an input string are then summed. If the state string consists of very popular transitions, then there will be many log-odds values that are greater than 1 and thus the result will be a positive number. If the sequence contains many rare transitions, then the results will be negative. The resultant calculation is more dependent on the content of the string and less so on the length of the string.

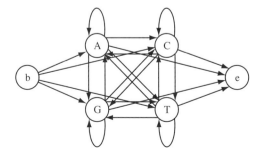

FIGURE 10-7 A recurrent HMM for DNA strings.

10.4 Constructing a Transition HMM

Constructing emission and recurrent HMMs can be rather straightforward and may not require any significant coding. The transition HMM though is a bit different, and thus it is presented here. Consider a case in which there are three data strings "GTATC," "GTTTC," and "GT-AC." The probability of getting a "G" as the first character is 1. The probability of getting a "T" as the fourth character is 2/3 since it is seen in 2 out of 3 strings. The premise of constructing a transition HMM is based upon measuring the probabilities of seeing characters at particular locations. It is common that each string has its own probability of being seen. Look again at the three strings discussed above, but this time note that "GTATC" and "GTTTC" are each seen 25 percent of the time while there is a 50 percent chance of seeing "GT-AC." Now, the probability of seeing a "T" as the fourth character is 50 percent. To build a transition HMM, the probabilities of each string are also taken into consideration.

The transition HMM is built in two stages because all of the nodes must have IDs before the connections can be made. Thus, it is necessary to examine the data once to create the nodes and a second time to create the connections. The function **NodeTable** in Code 10-7 computes a matrix that creates an HMM node for every necessary case. Basically, it assigns a node for each different character seen in the first position, then again for the second position, and so on. The table is $V \times H$, where V is the number of characters in the alphabet and H is the number of characters in a string.

```
from numpy import zeros
# hmm.py
def NodeTable( sts, abet ):
    # sts is a list of data strings
    L = len( sts )   # the number of strings
    D = len( sts[0] )   # length of string
    A = len( abet )
    NT = zeros( (A,D),int )-1
    nodecnt = 0
    for i in range( D ):
        for j in range( L ):
            ndx = abet.index( sts[j][i] )
            if NT[ndx,i] ==-1:
                NT[ndx,i] = nodecnt
                nodecnt +=1
    return NT

>>> data = ['GTATC', 'GTTTC', 'GT-AC']
>>> abet = 'ACGT-'
>>> nodetab = NodeTable( data, abet )
>>> nodetab
array([[-1, -1,  2,  6, -1],
       [-1, -1, -1, -1,  7],
       [ 0, -1, -1, -1, -1],
       [-1,  1,  3,  5, -1],
       [-1, -1,  4, -1, -1]])
```

Code 10-7

10.4 Constructing a Transition HMM

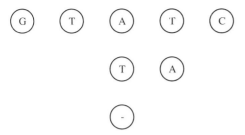

FIGURE 10-8 An unconnected HMM from Code 10-7.

In this example the table contains several −1 results indicating that no node is necessary. In the first column, there is a 0 in the third row. Since that corresponds to the third character in the alphabet, the *hmm[0]* node is now assigned to a "G" in the first position. The *hmm[1]* node is seen in the fourth row in the second column. Therefore it is now associated with the second position for the letter "T." This function has the ability to create an unconnected network, as shown in Figure 10-8. The emission states and the location of each node are known. At this point the transitions are not known.

The second part is quite detailed and requires a few functions. The function **MakeNodes** in Code 10-8 initializes the HMM by creating the dictionary *hmm*. The *sts* is a list of strings that are to be used to create the HMM. The *abet* is the alphabet of all possible letters, *weights* are the probabilities that each string will be seen, and the *nodet* is the node table. The *i*-loop considers the string *sts[i]*. The *j*-loop considers each pair of letters in the single string. If the transition has not been seen before, then it creates a node in the *hmm*. If it has been seen before, then the existing node is modified. This function collects the weights for every transition. Thus, the first node *hmm[0]* contains the emission state "G" and has a single transition to *hmm[1]*. This transition was seen in all three of the data strings (the first two letters are GT). Thus, it added the weights from all three strings to achieve a probability of 1.0.

```
# hmm.py
def MakeNodes( sts, abet, weights, nodet ):
    L = len( sts )
    D = len( sts[0] )   # length of string
    hmm = {}
    for j in range( D-1 ):
        for i in range( L ):
            # current letter
            clet = sts[i][j]
            # next letter
            nlet = sts[i][j+1]
            # node associated with current letter
            cnode = nodet[ abet.index(clet), j ]
            # node associated with next letter
            nnode = nodet[ abet.index(nlet), j+1]
```

```
            # connect the nodes
            if hmm.has_key( cnode ):
                # transition has been seen before
                # adjust the transition
                if hmm[cnode][1].has_key( nnode ):
                    hmm[cnode][1][nnode] += weights[i]
                else:
                    hmm[cnode][1][nnode] = weights[i]
            else:
                # transition has not been seen before - make new one
                hmm[cnode]= ( clet ,{ nnode: weights[i] })
    return hmm
>>> hmm = MakeNodes( data, abet, [0.25,0.25,0.5], nodetab )
>>> hmm
{0: ('G', {1: 1.0}), 1: ('T', {2: 0.25, 3: 0.25, 4: 0.5}), 2: ('A', {5: 0.25}),
 3: ('T', {5: 0.25}), 4: ('-', {6: 0.5}), 5: ('T', {7: 0.5}), 6: ('A', {7: 0.5})}
```

Code 10-8

At this point in the construction not all nodes will have a total transition probability of 1.0. For example, *hmm[2]* contains the letter "A," but only had a single incidence where it transitioned to *hmm[5]*, which contains the letter "T." This transition was seen in the first data string, and its weight was 0.25. The *hmm[2]* currently has a total transition probability of 0.25. As stated earlier, the total transition probabilities exiting a node must sum to 1.0. So, the second stage is to normalize all of the nodes to enforce this rule. Code 10-9 shows the **Normalization** function. This sums up the transition probabilities of each node and divides the probabilities by this sum. In the example, the *hmm[2]* now displays a 100 percent chance of transitioning to *hmm[5]*.

```
# hmm.py
def Normalization( hmm ):
    t = hmm.keys()
    for i in t:
        sm = 0
        for j in hmm[i][1].keys():
            sm += hmm[i][1][j]
        for j in hmm[i][1].keys():
            hmm[i][1][j] /= sm

>>> Normalization( hmm )
>>> hmm[2]
('A', {5: 1.0})
```

Code 10-9

The third step is to add the BEGIN and END nodes. Basically, the BEGIN node needs to transition to all of the entry nodes, and the END node needs to receive transition for all of the last nodes in the strings. This is accomplished by the function **Ends** in Code 10-10.

10.4 Constructing a Transition HMM

```
# hmm.py
def Ends( hmm, sts, abet, weights, nodet ):
    # add begin node
    T = {}
    L = len( sts )
    for i in range(L):
        clet = sts[i][0] # first letter in string i
        nlet = sts[i][1]
        idt = nodet[ abet.index(clet) ,0]
        # Build dictionary for the BEGIN node
        if idt != -1:
            if T.has_key( idt ):
                T[ idt ] += weights[i]
            else:
                T[ idt ] = weights[i]
    hmm['begin'] = ( '', T )
    # add end node
    hmm['end'] = ('',{} )
    for i in range( L ):
        clet = sts[i][-1]
        idt = nodet[ abet.index(clet) ,-1]
        hmm[idt] = (clet,{'end':1})

>>> Ends( hmm, data, abet, [0.25,0.25,0.5], nodetab )
>>> hmm['begin']
('', {0: 1.0})
```

Code 10-10

The final step is to write a program to drive these functions. The **BuildHMM** function in Code 10-11 will construct the HMM given a set of strings, the alphabet, and the weights of the strings. Confirmation is performed by computing the probabilities of the strings, which should match the original weights.

```
# hmm.py
def BuildHMM( sts, abet, weights ):
    nodet = NodeTable( sts, abet )
    hmm = MakeNodes( sts, abet, weights, nodet )
    Normalization( hmm )
    Ends( hmm, sts, abet, weights, nodet )
    return hmm

>>> hmm = BuildHMM( data, abet, [0.25,0.25,0.5] )
>>> TProb( hmm, data[0] )
0.25
>>> TProb( hmm, data[1] )
0.25
>>> TProb( hmm, data[2] )
0.5
```

Code 10-11

To construct the HMM, the BEGIN node is considered. In Code 10-12 the BEGIN node is printed, and it shows that it has no emission state and a 100 percent chance of transitioning to *hmm[0]*. After being printed, this node contains the emission "G" and

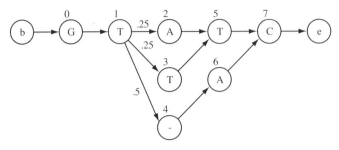

FIGURE 10-9 The HMM with the transition probabilities.

has a 100 percent chance of a transition to *hmm[1]*. The *hmm[1]* emits a "T" and connects to *hmm[2]*, *hmm[3]*, and *hmm[4]* with transitions 0.25, 0.25, and 0.5, respectively. Figure 10-9 shows the final product with small numbers above each node to identify it and transition probabilities printed for cases where it is less than 1.

```
>>> hmm['begin']
('', {0: 1.0})
>>> hmm[0]
('G', {1: 1.0})
>>> hmm[1]
('T', {2: 0.25, 3: 0.25, 4: 0.5})
```

Code 10-12

10.5 Considerations

The previous sections demonstrate the basic HMMs. However, there are problems that can quickly arise that should be noted or mitigated.

10.5.1 Assuming Data

In those cases having a large data set, the HMM is constructed from a sampling of the data set. In this scenario it is possible to miss some of the transitions. For example, a data set has 1,000 strings and 100 are used to train the HMM. All of the strings except for a few begin with "A," and none of these happen to be in the 100 subset. The HMM is constructed without the possibility of a "T" being in the first position.

An easy method of handling this situation is to assume one more string. In this case, if 100 strings start with "A" and it is known that there is a possibility of starting with a "T," then one more string starting with "T" is assumed. Thus, the emission probability of an "A" becomes 100/101 and the emission probability of a "T" becomes 1/101.

10.5.2 Spurious Strings

The methods discussed previously do build an HMM, but is it the correct HMM? Consider the data set "AGCTG," "ACCTA," and "CGGTA," which constructs the HMM shown in Figure 10-10. This network will indeed produce the correct probabilities for the strings as given, but it will also have probabilities for other strings as well,

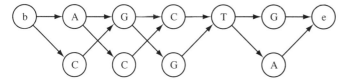

FIGURE 10-10 The HMM construct from the three strings "AGCTG," "ACCTA," and "CGGTA."

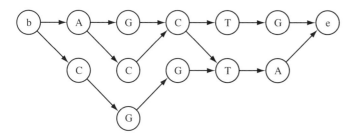

FIGURE 10-11 The HMM that prevents spurious states.

such as "CGCTA." If the goal is that the HMM produce *only* the strings used in training, then this method does not produce the correct HMM. The restrictive HMM is shown in Figure 10-11. Building this network is more involved. There are two instances where there are two nodes with the same emission at the same position. In the previous case, the fourth position "T" relied on only the immediate previous position. In this new construction, the "T" now relies on the entire string.

10.5.3 Recurrent Probabilities

In the recurrent network, a similar problem arises. The network shown in Figure 10-7 has a single transition probability for "C" feeding into itself. It will not be capable of handling cases in which it is more probable to see "CCC" and "CCCCCC" rather than "CCCC."

10.6 Summary

The hidden Markov model (HMM) is a system that uses probabilities of transition from one state to the next to compute the probabilities of a given output. There are two main types of HMMs: (1) the transition HMM, which uses probabilities between two consecutive states, and (2) the emission HMM, which computes probabilities of states at each node. More complicated HMMs are possible using both transition and emission states. Construction of the emission and recurrent HMMs are quite easy. Construction of a transition HMM requires two major steps, first to create the nodes and then to connect them.

Problems

1. Create an emission HMM for the following data: "AGTTACG," "TTGCTCAG," "TGACTCTAT," "TTTTGATCGCA." Validate the network by computing the probabilities of each of the training strings.

2. Create a transition HMM for the following data: "AGTTACG," "TTGCTCAG," "TGACTCTAT," "TTTTGATCGCA." Validate the network by computing the probabilities of each of the training strings.

3. Create a recurrent HMM for the following data: "AGTTACG," "TTGCTCAG," "TGACTCTAT," "TTTTGATCGCA." Validate the network by computing the probabilities of each of the training strings.

4. Compute the weights for Figure 10-10.

5. Design an HMM for the following case xxxx $(AT)_n$ xxxx. An x can be any of the ACGT bases and the n is a number between 10 and 20. Without further information you cannot assign values to the transitions but you do need to explain how you would compute the transition probabilities given a data set.

6. Describe a case in which you would need both emission and transition probabilities.

7. Consider a case in which an HMM is built from many DNA strings. Each column in the HMM can have four nodes (assuming no gaps). If position k in all strings is an A, what will be the transition probabilities between the nodes in columns k and $k+1$? What will be the transitions if the letter in position k is random (has an equal probability of being an A, T, G, or C?

8. The two scenarios given in Problem 7 show columns in an HMM that are not important. If all of the characters in a string are the same or are randomly chosen, then there is no importance to this knowledge. Design a metric that determines which columns in the HMM are important.

11 Genetic Algorithms

Cases arise in which abundant data is generated but the optimal function that could simulate the data is not known. For example, protein sequences as well as protein folding structures are known, but the missing part is the function that converts a protein sequence into a folded structure. This is a very difficult problem with no easy solution. However, it illustrates the idea that plenty of data can be available without knowing the exact function that associates them.

An approach to problems of this nature is to optimize a function through the use of an artificial genetic algorithm (GA). The idea of this system is that the GA contains several genes, each one encoding a solution to the problem. Some solutions are better than others. New genes are generated from the old ones with the better solutions being more likely to be used. The new generation of solutions should be better than the previous, and the process continues until a solution is reached or optimization has ceased.

Genetic algorithms are quite easy to employ and provide good solutions to tough problems. The downside is that they can be quite expensive to run on a computer. Before delving into GAs, it is first worth exploring simulated annealing, a simpler optimization scheme that naturally leads into GAs.

11.1 Simulated Annealing

Simulated annealing is a simple optimization process in which a solution is achieved through a perturbation process. In a single iteration the state vector is changed slightly and if the new vector provides a better solution then it becomes the state vector. The process iterates until a sufficient solution is found or optimization ceases.

The problem begins with a set of input data, $\{\vec{x}_i : i = 1, \ldots, N\}$ and output data $\{\vec{y}_i : i = 1, \ldots, N\}$. The goal is to generate a function that performs the following for all i's:

$$\vec{y}_i = f\{\vec{x}_i\}, \qquad \forall i \tag{11-1}$$

The user must define the solution $f\{\ \}$ but not its parameters. Quite often a perfect solution is not available, and so the problem is translated into minimizing the error between the calculated solution and the desired solution,

$$E = |f\{\vec{x}_i\} - \vec{y}_i|^2. \tag{11-2}$$

The process of *simulated annealing* changes the parameters of $f\{\}$ to decrease this error. This is best demonstrated through the following case. The task is to find a vector $\vec{v}$ that is perpendicular to every vector in a set of vectors. As long as the dimension of the vectors is greater than the number of vectors, then one or more perfect solutions exists.

The inner product of two perpendicular vectors is 0 and so the inner product is a function that can be used for optimization. Equation (11-1) becomes

$$y_i = \sum_j v_j x_{i,j} \qquad (11\text{-}3)$$

where $x_{i,j}$ is the jth element of the ith vector and the output is now a scalar. The elements v_j are not known, and it is the purpose of the simulated annealing process to find them.

This process begins with a random guess for $\vec{v}$. Each iteration changes the values of the vector; if it improves the results, then the changes are kept. The amount of change is controlled by the *temperature* of the system. Basically, in the beginning larger variations are allowed, and as the system cools smaller changes in v are allowed.

Code 11-1 shows the generation of random data vectors. The default generates nine vectors with length of 10. Since the number of vectors is just one less than the length of the vectors, there exists at least one vector that is perpendicular to all of the vectors in *vecs*. If any of the random vectors are parallel, then there is another solution as well.

```
# simann.py
def GenVectors( D=10,N=9 ):
    # D is the dimension of the vectors
    # N is the number of vectors
    vecs = random.ranf( (N,D) )
    return vecs

>>> random.seed( 1251 )
>>> vecs = GenVectors()
```

Code 11-1

The goal of the *target* vector is to evolve into the solution. Code 11-2 shows the function **SumInner**, which computes the inner product of the *target* to all of the vectors in *vecs*. Unit vectors that are parallel will produce an inner product of 1 and antiparallel unit vectors will produce an inner product of -1. In this application, the goal is to produce a vector that is perpendicular to all of the other vectors, which means that its inner product should be closer to 0. Thus, the **abs** function is used so that the perfect vector will produce a total sum of 0, which will also be the lowest possible score. The *for* loop computes the inner product of all vectors in *vecs* with *target*. However, the final answer is the sum of all of these inner products. The example shows that this sum is 21.7.

11.1 Simulated Annealing

```
# simann.py
def SumInner( vecs, target ):
    N = len( vecs )
    answ = zeros( N )
    for i in range( N ):
        answ[i] = abs(dot( vecs[i], target ))
    return answ.sum()
>>> smin = SumInner( vecs, vecs[0] )
>>> smin
21.7314954049
```

Code 11-2

The **SumInner** function is actually not necessary. The NumPy module can perform the entire process in a single command. The **dot** command in Code 11-2 receives two vectors that are the same length, which is normally how a dot product is considered. The **dot** command can also receive different dimensions under proper conditions. Consider the case with a matrix **A** that is $M \times N$ in size and a vector **B** of length N. Since the last dimension of **A** matches the dimension of **B** the **dot** function will compute the dot product of each row of **A** with **B** and return a vector of the results. Thus, the **SumInner** function can be replaced with Code 11-3.

```
>>> abs(dot( vecs, vecs[0])).sum()
21.7314954049
```

Code 11-3

The simulated annealing process starts with a random *target* and changes it by random numbers. The magnitude of the change is controlled by the temperature (*temp*), and *scaltmp* controls the rate at which the temperature is lowered. Code 11-4 shows the function **RunAnn**, which drives the simulated annealing process for this application. The *guess* is the new target, and it is scored using **dot**. If it produces a lower cost, then this proposed change is adopted and becomes the new *target*.

```
# simann.py
def RunAnn( vecs, D, temp=1.0, scaltmp=0.99 ):
    target = 2*random.rand( D )-1
    # compute the inner products for all vectors
    cost = abs(dot( vecs, target )).sum() # sum of inner prods
    ok = 1
    costs100,i = [],0
    while ok:
        # make a small change
        guess = target + temp*(2*random.rand(D)-1)
        # compute the inner products for all vectors
        gcost = abs(dot( vecs, guess )).sum()
        # if it is an improvement then keep the change
        if gcost < cost:
            target = guess + 0
            cost = gcost + 0
            #print cost, temp
        # lower the temperature
```

```
            if i % 100 ==0:
                costs100.append( cost )
            i+=1
            temp *= scaltmp
            if cost<0.1:
                ok= 0
        return target, array( costs100 )
```

Code 11-4

Code 11-5 runs the program and returns the *target* and the cost for each 100 iterations. The second command demonstrates that the *target* is fairly perpendicular to all of the vectors in *vecs* since the inner products are close to 0.

The process stops when the cost falls below 0.1. Certainly, by lowering this threshold the *target* will become closer to the perfect perpendicular vector, but the computational costs increase significantly. Figure 11-1 shows the costs measured for every 100 iterations. The decreasing values demonstrate that the system is optimizing the *target*, but as can be seen the amount of decrease fades as the system progresses. Better solutions come with a much higher cost.

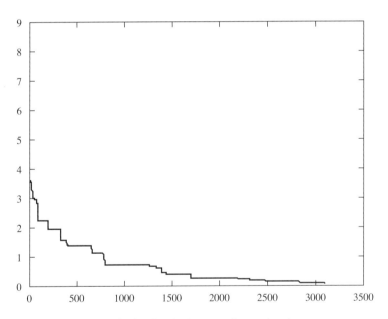

FIGURE 11-1 An example showing the decrease of a cost function.

```
>>> random.seed( 1252 )
>>> target, costs = RunAnn( vecs, 10, 2, 0.99999 )

>>> for i in range( 9 ):
      print "%1.3f" %dot(vecs[i], target )
0.005
-0.002
-0.012
0.005
0.004
-0.002
0.020
-0.010
0.029
```

Code 11-5

11.2 The Genetic Algorithm

The simulated annealing approach generates one single solution and attempts to optimize it. The genetic algorithm (GA) approach creates several solutions and attempts to optimize the process by creating new solutions from the old solutions. The advantage of this system is seen through a quick view of the optimization processes.

11.2.1 Energy Surfaces

Both optimization procedures attempt to find a minimum in an energy surface but in different ways. Figure 11-2 shows a simple view of an energy surface that can also be considered an error surface. The ball indicates the position of a solution and the error that accompanies this solution is the y-value. The purpose of optimization is to find the solution at the bottom of the deepest well.

In the case of simulated annealing, there is a single solution that is randomly located (since the initial target vector is random). Variations of this vector move this solution to a different location. Of course, large variations equate to large displacements of the solution. As the temperature is cooled, the solution cannot move around as much and eventually gets trapped in a well. Further optimization moves the solution down toward the bottom of the well. Of course, there is no guarantee that the solution will fall into the correct well. The term "caught in a local minimum" is used to describe a solution that is stuck in a well that is not the deepest.

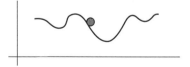

FIGURE 11-2 A simple rendition of an energy surface with a solution.

The GA approach is different in that there are several solutions. This is similar to placing several balls on the energy surface. The GA has a two-step optimization process. First, new solutions are created from old solutions, which is equivalent to replacing balls on the surface with a new set in which the likelihood is that the newer balls will be closer to the bottom of the wells. Second, the balls move slightly through a "mutation" step. The optimization occurs mostly by creating a new set of solutions that is better than the old set of solutions.

11.2.2 The Genetic Algorithm Approach

A simple GA iterates over the following sequence of steps:

1. Create a cost function.
2. Initialize the GA.
3. Score the current genes.
4. Iterate until the solution is good enough or stable, or until an iteration limit is reached.
   4.1 Create the next generation.
   4.2 Score the new generation.
   4.3 Replace the old generation.
   4.4 Mutate.
   4.5 Score the new generation.
   4.6 Check for a stop condition.

For the purposes of demonstration, the same problem of finding a perpendicular vector is used. Code 11-6 generates a set of random solutions as a list of vectors. Since this GA code may be used on data that is nonnumerical, the use of lists is preferred over the use of an array. The GA will keep track of two generations of solutions, and these are denoted as the parents (*folks*) and the offspring (*kids*). Code 11-6 generates the initial set of genes.

```
>>> random.seed( 1253 )
>>> folks = []
>>> for i in range( 10 ):
        folks.append( random.rand(10) )
```

Code 11-6

Each of these genes needs to be considered as a solution, and Code 11-7 thus presents the **CostFunction** that computes these values. It should be noted that there are two ways of approaching this part of the algorithm. The first is to compute the *score* of a gene in which a higher value is a better value. The second is to compute the *cost* of the gene in which the lower value is a better fit. There is no real advantage to either system, but one method has to be chosen. In the codes presented here, the cost of the genes are computed and thus a score of 0 is a perfect solution. In this example, the *fcost* is a vector that contains the cost of each of the *folks*.

11.2 The Genetic Algorithm

```
# ga.py
def CostFunction( data, genes ):
    ND = len( data ) # number of data points
    NG = len( genes ) # number of genes
    costs = zeros( NG, float )
    for i in range( NG ):
        # the next line performs the dot prod of all of the
        # data vectors with a single gene.
        costs[i] = (abs(dot( data, genes[i]))).sum()
    return costs

>>> fcost = CostFunction( vecs, folks )
>>> print fcost
[ 27.76374644  18.45164106  23.60291322  19.80241992  21.10378168
  27.2291017   25.47796998  25.12764945  12.06175036  22.94466166]
```

Code 11-7

Generating the next generation of solutions is a bit involved as shown in the **CrossOver** function in Code 11-8. The number of offspring is usually equal to the number of parents, and the offspring are generated in pairs. Thus for each iteration two parents are chosen along with a splice point, a random location in the vectors. The first child is created from the first part of the one parent and the second part of the other parent. The second child is created from the remaining parts, as shown in Figure 11-3. The parents are selected based upon their cost functions, which means that the parents with a lower cost have a better chance of being selected because they are the ones that will help generate the pair of children.

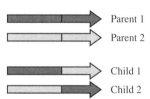

Parent 1
Parent 2

Child 1
Child 2

FIGURE 11-3 An iteration showing how two parents are spliced to create two children.

```
# ga.py
def CrossOver( folks, fcost ):
    # convert costs to probabilities
    dim = len( folks[0] )
    prob = cost + 0.0
    mx = prob.max()
    prob = mx - prob    # lowest cost is now highest number.
    mx = prob.max()
    prob = prob / mx    # makes sure numbers aren't too high
    prob = prob / prob.sum()  # normalized. sum(prob) = 1.0
    # make new kids
    kids = []
    NG = len( folks )
    for i in range( NG/2 ):
        rdad = random.rand()
```

```
            rmom = random.rand()
            # find which vectors to use
            sm = 0.0
            idad = 0
            while rdad > sm:
                sm = sm + prob[idad]
                idad = idad + 1
            sm = 0.0
            imom = 0
            while rmom > sm:
                sm = sm + prob[imom]
                imom = imom+1
            # these are the mom and dad
            idad,imom = idad-1,imom-1
            # make children
            x = int(random.rand()*dim)
# crossover
            kids.append(concatenate ((folks [idad][:x], folks [imom][x:])))
            kids.append(concatenate ((folks [imom][:x], folks [idad][x:])))
        return kids

>>> kids = CrossOver( folks, fcost )
>>> kcost = CostFunction( vecs, kids )
```

Code 11-8

The *prob* eventually becomes the probability of selecting each parent. Inside the *for* loop, the *rdad* and *rmom* are random numbers that are used to select the two parents based upon their probabilities. Figure 11-4 shows the method of selection of the parents. Each parent has a score (which is the opposite of a cost) that allocates space along a number line from 0 to 1. In this case parent *p5* performed better than the others and therefore is allocated more space on the number line. A random number *rdad* (or *imom*) is generated and *while* loops are used to determine which parent is to be used. Inside each *while* loop the *sm* sums up the score of each parent until it exceeds *sm*. The parent being considered when this condition is met is the parent that is to be used. In the figure, parent *p4* is selected as the "dad" for the new generation. The same is repeated to get the "mom" and then these two parents are used to generate new offspring.

The last two lines make two new children vectors, which are then appended to the *kids* list. The value *x* is the splice junction that is randomly selected. The result is that *kids* will become a list of new vectors created from the older vectors.

At this point in the algorithm the new vectors need to replace the old vectors. There are different theories on the best way to accomplish this. One option is to sim-

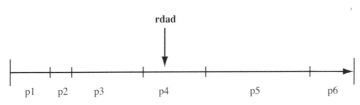

FIGURE 11-4 The number line shows the allocation of space for each parent and that the *dad* was selected by a random number that landed in the *p4* bin.

ply replace all parents with the children. The disadvantage of this system is that there is no guarantee that any of the children are better than the parents. It is possible that the solution could worsen after one iteration. Usually though, over the course of several iterations the system does optimize. The second option is that the children should replace parents only if the children are better. The disadvantage of this system is that *inbreeding* is more likely. This occurs when all of the children become the same vector. In either case it is necessary to compute the cost of the children vectors as shown at the end of Code 11-8. Code 11-9 shows a function **Feud** that replaces parents only if the children are better.

```
# ga.py
def Feud( folks, kids, fcost, kcost ):
    for i in range( 0, len(kids) ):
        if kcost[i] < fcost[i]:
            folks[i] = kids[i]
            fcost[i] = kcost[i]
>>> Feud( folks, kids, fcost, kcost )
```

Code 11-9

The final step is the mutation of the new parent vectors, which is needed to induce new values into the vectors. Otherwise the only values allowed would be those generated with the initial parents. The user selects a mutation rate that is a small value. In this case, a mutation rate of 0.05 indicates that 5 percent of the elements will be changed. Code 11-10 creates a vector of random values *r*, and it is compared to the *rate* and any value in *r* that is below *rate* is selected as a value to change. The vector *r* is then scaled within a range that is slightly bigger than the current range of the vector in order to include values that are slightly more or less than the values in the original *folks*. The changes are made in the last line of the function.

```
# ga.py
def Mutate( folks, rate ):
    cnt = 0
    NG = len( folks ) # number of genes
    for i in range( NG ):
        if sum( abs( folks[0]-folks[i]))< 0.01: cnt = cnt+1
    # if inbreeding occurs (the folks are very similar)
    # then change 90% of the elements
    if cnt >= NG/2 or random.rand()<0.01:
        rate = 0.9
        print 'Inbreeding warning'
    D = len( folks[0] ) # dim of vector
    for i in range( NG ):
        mx = folks[i].max()
        mn = folks[i].min()
        if mx > 0 :
            mx = mx * 1.05
        else:
```

```
            mx = mx * 0.95
        if mn>0:
            mn = mn * 0.95
        else:
            mn = mn * 1.05
        r = random.rand( D )
        hit = less(r,rate)
        r = (mx-mn)*r + mn
        folks[i] = (1.0-hit)*folks[i] + hit*r
>>> Mutate( folks, 0.05 )
```

Code 11-10

The final function drives the entire GA program as shown in Code 11-11. The **DriveGA** function performs iterations until the best vector scores below a minimal value (which is 0.1 in this case). In this program the best score of each iteration is printed to the console. For bigger jobs, this is too frequent, and the program will run faster if the printout occurs after several iterations. This program returns the vector that provided the best score that is considered as the solution. The last line computes the dot product with the original vectors and the best solution. They are printed to three decimal places using the **set_printoptions** command from NumPy. If the *best* vector were perpendicular to all of the vectors in *vecs*, then all of the values in the dot product would be 0. These are close to 0 and thus the *best* is not the perfect solution but perhaps a good solution. A better solution can be obtained by changing the *if* statement, but this will increase the number of iterations.

```
# ga.py
def DriveGA( vecs, NG=10, DM=10 ):
    # generate initial folks
    folks = []
    for i in range( NG ):
        folks.append( 2 * random.rand(DM) - 1 )
    # iterate
    fcost = CostFunction( vecs, folks )
    ok = 1
    while ok:
        kids = CrossOver( folks, fcost )
        kcost = CostFunction( vecs, kids )
        Feud( folks, kids, fcost, kcost )
        Mutate( folks, 0.05 )
        fcost = CostFunction( vecs, folks )
        # get the best vector
        best = fcost.min()
        besti = fcost.argmin()
        # test to see if the solution is good enough.
        if best < 0.1:
            ok = 0
        print best
    return folks[besti]

>>> random.seed( 1254 )
>>> best = DriveGA( vecs )
>>> set_printoptions( precision = 3) # to print out only 3 decimals
>>> dot( vecs, best )
array([-0.027,  0.027,  0.001, -0.003, -0.012, -0.  ,  0.002, -0.007,
        0.003])
```

Code 11-11

11.2.3 Checking the Solution

It is possible that a solution does not necessarily exist. For example, if the number of vectors equaled or exceeded that dimension of the vectors, then there is no guarantee that a vector exists that is perpendicular to the set. Yet Code 11-12 shows that the impossible occurs. In this case 100 vectors of length 10 are generated, and the GA is run. A solution is returned, and vector *a* contains the dot product of all of the vectors within the solution. The **greater** statement prints out all of the indices of *a* that are greater than 0.01, and there are none. This means that the dot product of *best* with any of the original vectors is less than 0.01. Does this indicate that *best* is perpendicular to all of the vectors?

```
>>> random.seed( 1255 )
>>> vecs = GenVectors( 10,100)
>>> best = DriveGA( vecs )
>>> a = abs(dot( vecs, best ))
>>> greater( a, 0.01 ).nonzero()[0]
array([], dtype=int32)
```
Code 11-12

If all of the vectors were perpendicular to *best*, then all of the values in *a* would be small. However, the converse is not necessarily true. Recall that perpendicular vectors have an inner product of 0, but there is another case that can produce the same result. Code 11-13 shows the *best* vector and that instead of being perpendicular to all of the vectors it was forced to have all of its elements close to 0. The lesson to be learned is that GAs are not magic and solutions should be checked to ensure that the application is properly solved.

```
>>> best
array([-0.001, -0.001,  0.002, -0.001, -0.002, -0.001,  0.001,  0.001,
        0.001,  0.  ])
```
Code 11-13

11.3 Nonnumerical Genetic Algorithms

In the previous example the genes in the GA were numerical vectors. There are cases, especially in bioinformatics, in which the information being manipulated is based on letters instead of numbers. The GA is a flexible approach and allows for adaptations to suit the idea of the GA to particular applications. To demonstrate this by example, the small problem of sorting data will be used. In this problem the goal of the GA is to sort letters of the alphabet. This will use a trivial cost function of matching a sequence from a gene to a target sequence. More complicated applications will require a more complicated cost function.

11.3.1 Notes on Copying

Before this GA is built, it is important to review how Python deals with copying lists. One method of building a list of shuffled letters is to create a list of letters and then iterate a shuffle with an append to the list. This, however, creates a problem. Code 11-14 performs these steps and shows the unfortunate result. The *abet* is the English alphabet in lowercase letters. The *ape* is this string converted to a list. The list *z* is created, and the *ape* is appended to it. Thus, the first item in the list is the letters in alphabetical order. Then the *ape* is randomly shuffled creating a new order for the letters and appended to *z*. Thus, *z[0]* has two lists, and the first was originally the letters in alphabetical order. However, Code 11-14 shows that *z[0]* was altered by shuffling *ape*.

```
>>> abet = string.lowercase[:26]
>>> ape = list( abet )
>>> z = []
>>> z.append( ape )
>>> random.shuffle( ape )
>>> z.append( ape )
>>> z[0]
['f', 'h', 'u', 'y', 'b', 'c', 'e', 'k', 's', 'x', 'a', 'm', 'i', 'j', 'v',
'g', 'q', 'w', 'z', 'o', 'd', 'p', 'l', 't', 'n', 'r']
>>> z[1]
['f', 'h', 'u', 'y', 'b', 'c', 'e', 'k', 's', 'x', 'a', 'm', 'i', 'j', 'v',
'g', 'q', 'w', 'z', 'o', 'd', 'p', 'l', 't', 'n', 'r']
```

Code 11-14

Python has actually appended the address of the list to *z*, not to the data. The result is that whenever *ape* is changed so are all entries in *z*. It is thus necessary to create a copy of the list before appending to *z*. Python contains a module named *copy*, which has functions useful in creating copies of data structures. However, the use of *numpy* may cause some problems. In many codes the line

```
from numpy import *
```

has been used to load many of the functions that will be needed. There is a function named **copy** inside of numeric. Code 11-15 demonstrates how the module *copy* is overridden by the function **numpy.copy**. There are two simple solutions. The first is to reverse the order of the imports. The second is to import only those functions from *numpy* that are necessary. This can be cumbersome since some applications import several functions from *numpy*.

```
# the imports are in the WRONG order
>>> import copy
>>> from numpy import *
```

Code 11-15

The *copy* module contains a few functions of which either of two is useful here. The function **copy.copy** creates a *shallow copy* of a data structure, and the function **copy.deepcopy** creates a *deep copy* of the data. Code 11-16 shows both. The variable *ape* contains the letters of the alphabet, but they are shuffled from Code 11-14. A list *a* is created that contains three numerals. A list *b* is created that contains two letters and then the list *a*. A shallow copy of *b* is created as the variable *c*, which is printed.

The first items of both *a* and *b* are then changed, and *c* is again printed. List *c* adopted the change to *a* but not to *b*. A shallow copy basically copies only the first layer of items. This includes the two letters and the pointer to list *a*. The **deepcopy** function is then called, and as shown list *c* is a completely separate copy of all items in *b*. The rule to follow is that if the item being copied has only one layer (does not have a list inside of a list, etc.) then **copy.copy** is sufficient. However, if the data structure is complicated, then **copy.deepcopy** is needed to create a duplicate of the entire data structure. Finally, it should be noted that this copy problem occurs only with data structures (lists, tuples, arrays, dictionaries, etc.), not for standard data types (float, int, etc.).

```
>>> a = [1,2,3]
>>> b = ['f','g', a ]
>>> c = copy.copy( b )
>>> c
['f', 'g', [1, 2, 3]]
>>> a[0] = 5
>>> b[0] = 'y'
>>> c
['f', 'g', [5, 2, 3]]
>>> c = copy.deepcopy( b )
>>> a[0] = 8
>>> c
['y', 'g', [5, 2, 3]]
```

Code 11-16

11.3.2 Creating Random Arrangements

To start this GA, a set of genes is created. Code 11-17 shows the function that receives an alphabet *abet* and the number of genes *ngenes*. It creates an *ape* that will be shuffled but does not alter the original data *abet*. Inside the loop, the *ape* is shuffled and copied into the list *folks*. The *abet* needs to be a list, and so the string *alpha* is converted to a list before it is sent to **Jumble**. Two examples are shown to placate fears expressed in Section 11.3.1. The **join** function is used to convert a list of characters into a more readable string.

```
from numpy import random
import copy, string

# gasort.py
def Jumble( abet, ngenes ):
    folks = []
    ape = copy.copy( abet )
    for i in range( ngenes ):
        random.shuffle( ape )
        folks.append( copy.copy( ape ))
    return folks

>>> random.seed( 1256 )
>>> alpha = string.lowercase[:26]
>>> folks = Jumble( list(alpha), 10 )
>>> ''.join( folks[0] )
'tcdesokupmyzvahrqgnjwxilfb'
>>> ''.join( folks[1] )
'nwicamvfxqdjterzplouhgkysb'
```

Code 11-17

11.3.3 The Genetic Algorithm

The genetic algorithm still uses the same basic philosophy in that each gene is scored and the **CrossOver** function is used to create new genes. These are scored and replace the previous generation. Finally, a mutation is applied. For this application, some of the previous functions need to be altered.

Certainly, the cost function needs to be altered, but this is true for each individual application. Code 11-18 shows a simple cost function that compares the gene to the perfect sorted alphabet. Again, this is a trivial cost function since the focus here is on a creation of the GA, not on its ability to perform a complicated function. The **CostFunction** receives the target sequence and a list of genes. The first loop will use *gene* to represent a single item in the list *genes*. The inner loop counts the number of dissimilarities between the target and an individual gene. The output is a list of costs. The example shows that the random genes were mostly dissimilar to the target.

```
# gasort.py
def CostFunction( target, genes ):
    NG = len( genes ) # number of genes
    cost = []
    for gene in genes:
        c = 0
        for i in range( len( target )):
            if target[i] != gene[i]:
                c += 1
        cost.append( c )
    return cost

>>> fcost = CostFunction( alpha, folks )
>>> fcost
[24, 26, 26, 25, 24, 26, 25, 25, 26, 26]
```

Code 11-18

The next step in the GA is to create the new genes. The **ga.CrossOver** function is sufficient to accomplish this as long as the input cost is a vector, not a list. In Code 11-19 the call to this function is performed. However, there is a problem here in that the new offspring are probably not legal. A new offspring received part of a list from two different parents, and there is no guarantee that all of the letters exist in the offspring. In fact, it would be highly unlikely that an offspring would receive exactly one letter of the alphabet from the two parents.

```
>>> import ga
>>> kids = ga.CrossOver( folks, array(fcost) )
```

Code 11-19

A solution for this is to make each offspring legal after the call to the **CrossOver** function. In this case, it is necessary to have each letter of the alphabet in the gene once and only once. In the function **Legalize** in Code 11-20 the *cnts* is a count of the number of times each letter in *valid* occurs in *gene* where *valid* is the string of valid characters. In this case *valid* is the alphabet. If the letter does not occur, then its count is 0 and the *mssg* (a list of letters that are missing) will collect it. For example, if the list *mssg* contains [0, 3] then both *valid[0]* and *valid[3]* are letters not found in *gene*. Likewise the list *dups* collects the locations in *valid* of the letters that are duplicated. The *dups* is shuffled to create a random replacement scheme. The replacement first finds the two locations of a duplicate and chooses one at random. This location is replaced with a letter that was missing. The result is that the gene will contain each letter only once. For example, the *ape* is a list of the alphabetical letters with the "k" replaced with another "a." Thus, one letter is absent, and one is duplicated. The **Legalize** function is called, and one instance of "a" is replaced with the missing "k." As shown in this example, either "a" may be replaced.

```
from numpy import equal, nonzero, zeros
# gasort.py
def Legalize( valid, gene ):
    # get the count for each letter
    LV = len( valid )
    LG = len( gene )     # length of this gene
    cnts = zeros( LV, int )
    for i in range( LV ):
        cnts[i] = gene.count( valid[i] )
    # get a list of the missing and duplicates
    mssg = nonzero( equal( cnts,0))[0]
    dups = nonzero( equal( cnts,2))[0]
    random.shuffle( dups )
    # replace
    for i in range(len(mssg)):
        # pick one of the dups
        k1 = gene.index( valid[dups[i]] )
        k2 = gene.index( valid[dups[i]], k1+1 )
        if random.rand() > 0.5:
            me = k1
        else:
            me = k2
        # replace
        gene[me] = valid[mssg[i]]
>>> ape = list( alpha[:10]+'a'+alpha[11:] )
>>> Legalize( alpha, ape )
>>> "".join(ape)
'kbcdefghijalmnopqrstuvwxyz'
```
Code 11-20

Now that all of the offspring are legal they can be scored, and they can replace their parents through a method of the user's choosing. Code 11-21 shows these two steps using the **ga.Feud** function as one method of replacing parents with their offspring.

```
>>> kcost = CostFunction( alpha, kids )
>>> ga.Feud( folks, kids, fcost, kcost )
>>> kcost
[23, 26, 24, 25, 25, 23, 26, 23, 25, 25]
```
Code 11-21

The final step is the mutation that needs to be radically altered from the previous version in Code 11-10, where the genes were composed of numerical vectors and random numbers were used to mutate the vectors. In this case, the data consists of characters. While it is possible to randomly replace a small number of letters, the result would make the genes illegal again. Another option is for the mutation step to switch two elements at random.

Code 11-22 shows the swapping **Mutate** function. The *for i* loop considers each gene. The array *r* will contain a few elements that were randomly selected. The variable *k* picks another element in the gene at random, and a swap occurs between the elements at positions *k* and *j*.

```
>>> from numpy import less

# gasort.py
def Mutate( genes, rate ):
    NG = len( genes )
    for i in range( NG ):
        DM = len( genes[i] )
        r = less( random.rand(DM), rate )[0]
        for j in r:
            k = int( random.rand()*DM)
            a = genes[i][k]
            genes[i][k] = genes[i][j]
            genes[i][j] = a

>>> Mutate( folks, 0.05 )
```

Code 11-22

The final step is to create a function to drive the GA. However, there needs to be a quick word on distinguishing between vectors (arrays) and lists. The **ga.CrossOver** function uses the **concatenate** function, which forces each gene to be an array. In the numerical version of the GA, this was acceptable. In this sorting version, however, each gene is a list.

There are a few solutions. The first is to create a new crossover function that is just like the last one except that the line

```
concatenate((folks[idad][:x],folks[imom][x:]))
```

is replaced with

```
folks[idad][:x] + folks[imom][x:]
```

In this new case the *folks* are a list, and the plus sign concatenates without converting to an array.

The second option would be to have the genes be an array of characters, but that would forfeit the use of the **count** and **index** functions that are used by the lists. The third option is to use **ga.CrossOver** and convert the offspring from vectors back to lists. Code 11-23 shows the driver function. The output of this function is the gene that lowers the cost function to 0.

```
# gasort.py
def DriveSortGA( ):
    target = list(string.lowercase[:26])
    alpha  = list(string.lowercase[:26])
    folks = Jumble( alpha, 10 )
    ok = 1
    fcost = CostFunction( target, folks )
    while ok:
        kids = ga.CrossOver( folks, array(fcost) )
        for k in range( len( kids )):
            kids[k] = list( kids[k] )
        for g in kids:
            Legalize( alpha, g )
        kcost = CostFunction( target, kids )
        ga.Feud( folks, kids, fcost, kcost )
        Mutate( folks, 0.01 )
        fcost = CostFunction( target, folks )
        if array(fcost).min() == 0:
            ok = 0
    me = array(fcost).argmin()
    return folks[me]

>>> random.seed( 1257 )
>>> genes = DriveSortGA()
>>> "".join( genes )
'abcdefghijklmnopqrstuvwxyz'
```

Code 11-23

11.4 Summary

There are several algorithms in the field of machine learning, which encompass programs that attempt to train on the data at hand. One approach that is widely used in bioinformatics is the genetic algorithm (GA). The GA contains a set of data genes (which can be vectors, strings, etc.), and through several iterations attempts to modify the genes to provide an optimal solution. This requires the user to define the metric for measuring the optimal solution. The unique quality of a GA is that new genes are constructed by mating old genes, and they are generated from copying parts of the older genes. GAs tend to use many iterations and can be quite costly to run. However, they can provide solutions that are more difficult to obtain using simpler methods.

Problems

1. Run a simulate annealing to minimize the cost *abs(vec − target)* where *target* is a user-defined target vector and *vec* is the vector being changed by simulated annealing. The algorithm should eventually change *vec* such that *vec=target*.

2. Repeat Problem 1 with the cost of *abs(vec − target1) + abs(vec − target2)*. Does *vec* become *target1*, *target2*, or a combination of the two?

3. Repeat Problem 2 with the cost of *abs(vec − target1) * abs(vec − target2)*.

4. Repeat Problem 1 using a genetic algorithm instead of simulated annealing. Which algorithm finds the solution faster?

5. Problem 3 should produce either *vec=target1* or *vec=target2*, depending on the initial value of *vec*. Repeat using a GA. Modify the GA to see if it is finding both solutions.

6. Create a GA that starts with random DNA strings of length N. Create a cost function such that the GA will compute the complement of a DNA target string.

12 Multiple Sequence Alignment

Aligning two sequences is a relatively straightforward process, but aligning multiple sequences adds a new complication. There are two types of approaches: (1) the *greedy approach*, which attempts to find the best pairs of sequences that align and to build on those alignments, and (2) the *nongreedy approach*, which attempts to find the best group of alignments. The advantages of the greedy approach are that the programming is not too complicated and the system runs fast. The advantage of the nongreedy system is that the answer is usually better.

12.1 The Greedy Approach

A common sequencing project begins with a very long strand of DNA. Because sequencing machines have a short limit on the number of consecutive bases that can accurately be identified, it is necessary to break a long strand of DNA up into many short strands. These short strands overlap each other, and their beginning and ending locations are not known. Thus it is necessary to align multiple sequences into an assembly to understand the nature of the original long strand. Figure 12-1 shows a typical assembly in which each vector represents a sequence. Vectors pointing in the opposite direction represent the complement of a sequence—convert C to G, G to C, A to T, T to A, and reverse the direction. In this diagram it should be noticed that the sequences are different lengths and some portions of the total sequence are sampled by more than two sequences. For this example, the complements of strings will not be considered since it is a trivial matter, and it is easy to add complements to an existing database. The only caveat is that if either the original or complement sequence is used then the other needs to be removed from further consideration.

FIGURE 12-1 Alignment of four sequences.

12.1.1 Sequence Comparison

The greedy approach starts with a comparison of all pairs of sequences. If there are four sequences, we then compute the following alignments: $(s1, s2), (s1, s3), (s1, s4), (s2, s3), (s2, s4)$, and $(s3, s4)$. This information can be represented in a triagonal matrix, **M**:

$$\mathbf{M} = \begin{Bmatrix} 0 & s1, s2 & s1, s3 & s1, s4 \\ 0 & 0 & s2, s3 & s2, s4 \\ 0 & 0 & 0 & s3, s4 \\ 0 & 0 & 0 & 0 \end{Bmatrix} \quad (12\text{-}1)$$

Each element of **M** keeps the score of the alignment of two sequences. Assuming that a large score indicates a better match, we can find the best of all possible pairings by finding the largest value in the matrix.

The best match aligns two sequences to form a *contig*. The second best match will either create a new contig or join with a previous one. For example, if the first contig consisted of $(s1, s2)$ and the second best match is $(s3, s4)$, then a second contig is created. However, if the second best match is $(s1, s4)$, then $s4$ is added to the contig that contains $s1$. The final possibility is that a match will be of two sequences that are contained in two other contigs. For this case, the two contigs are joined.

To demonstrate this approach, an example is employed in which a long string is separated into smaller sequences and then is sent to an assembler. Code 12-1 shows the function **ChopSeq**, which receives a long sequence and chops it up into overlapping subsequences. Complete coverage is not guaranteed because some portions of the original sequence may not be included in any of the subsequences. The variable *inseq* is the input sequence, the *nsegs* is the number of desired subsequences, and *length* is the length of each subsequence.

```
# greedy.py
def ChopSeq( inseq, nsegs, length ):
    segs = []
    # last possible starting location in inseq
    G = len( inseq ) - length
    for i in range( nsegs ):
        r = int( random.rand()*G )   # start the cut here
        segs.append( inseq[r:r+length] ) # extract the subsequence
    return segs
```

Code 12-1

Code 12-2 loads in the data from a Genbank file. This file contains a single protein that has 359 elements. The call to the **ChopSeq** function creates ten sequences that are each 50 elements long. In real applications the sequences are rarely the same length, but for the purpose of demonstrating the assembly process this anomaly is insignificant.

```
>>> import genbank
>>> data = genbank.ReadGenbank( 'genbank/XM_001326205.gb.txt')
>>> klocs = genbank.FindKeywordLocs( data )
>>> p1 = genbank.Amino( data, klocs[0] )
>>> len( p1 )
359
>>> chops = ChopSeq( p1, 10, 50 )
```

Code 12-2

Equation (12-1) shows the comparison matrix for all possible pairings. In this application the faster **BruteForceSlide** function is used since this data does not need to insert gaps into the alignments. It is possible to use dynamic programming to create this matrix, but this creates a difficult complication. The alignment of *s1* with *s2* may insert gaps into *s1*, and the alignment of *s1* with *s3* may insert a different set of gaps into *s1*. These gaps thus have to be tracked and collated. This complication will be considered in Chapter 13.

The function **FastMat** in Code 12-3 computes the matrix shown in Equation (12-1). Actually, it computes two matrices. The $M_{i,j}$ contains the score for the best alignment of *seq[i]* with *seq[j]*. The $L_{i,j}$ contains the relative shift between the two sequences that generated the alignment score. The element *M[1, 2]* indicates that the alignment of *chops[1]* with *chops[2]* has a large score of 184 and an alignment *L[1, 2]* of 72. Recall that the **BruteForceSlide** function returns a vector and that each element is a score for a particular relative position between the two sequences. A positive value for the position indicates that the beginning of the second sequence is aligning with some position in the first sequence. The final commands in Code 12-3 display the two subsequences with the first 22 elements of the first sequence removed. This shift is computed by *L[i,j] – len(seq[j])*.

```
# greedy.py
def FastMat( seqs, submat, abet ):
    # seqs is a list of strings
    # submat: substitution matrix. BLOSUM
    # abet: alphabet
    N = len( seqs ) # number of seqs
    M = zeros( (N,N), int )
    L = zeros( (N,N), int )
    for i in range( N ):
        for j in range( i+1,N ):
            bsl = easyalign.BruteForceSlide( submat, abet, seqs[i], seqs[j] )
            M[i,j] = bsl.max()    # peak value
            L[i,j] = bsl.argmax() # location
    return M,L

>>> M, L = FastMat( chops, blosum.BLOSUM50, blosum.PBET )
>>> M[:4,:4]
array([[  0,  10,   7,   6],
       [  0,   0, 184,  38],
       [  0,   0,   0, 170],
       [  0,   0,   0,   0]])
>>> L[:4,:4]
array([[  0, 78, 61,  2],
```

```
         [ 0,  0, 72, 94],
         [ 0,  0,  0, 72],
         [ 0,  0,  0,  0]])
>>> chops[1][72-50:]
'PTPLFFLNYFLRISGQTQESMLFARYIV'
>>> chops[2]
'PTPLFFLNYFLRISGQTQESMLFARYIVEMCLTSEKFNDVKASAIAATAV'
```

Code 12-3

Some of the alignments are good (such as *chops[1]* and *chops[2]*), but some are quite weak, with a low value in **M**. The threshold of an acceptable alignment is established by the user and is dependent upon the sequence lengths and the application. A score of 184 is high for sequences that are 50 amino acids long but small for sequences that are thousands long.

12.1.2 Assembly

The assembly of sequences is a complicated structure containing a number of contigs that are groups of aligned sequences. Thus the list *smb* represents the assembly, and each item in the list is a contig. Each contig contains several sequences that are aligned, which means that each contig is also a list. Each item in the contig list is a tuple that contains the sequence ID and the aligned sequence.

Each contig will have a number of sequences that will have relative shifts required for alignment, as shown in Figure 12-1. This alignment is achieved by inserting periods in front of a sequence to align it with the others. Code 12-4 shows the alignment of *chops[1]* with *chops[2]*, where 22 periods are inserted in front of *chops[2]*.

```
>>> chops[1]
'NREELVRKEIQLANITEFDFCFPTPLFFLNYFLRISGQTQESMLFARYIV'
>>> 22*'.' + chops[2]
'......................PTPLFFLNYFLRISGQTQESMLFARYIVEMCLTSEKFNDVKASAIAATAV'
```

Code 12-4

The alignment of two sequences is known from the location of the largest value in the vector returned from **BruteForceSlide**. Code 12-5 shows **ShiftedSeqs**, which receives the two sequences and the location of the peak in the vector. The variable *act* is the actual shift. If it is negative, the second sequence gets shifted; if it is positive, then the first sequence gets shifted.

```
# greedy.py
def ShiftedSeqs( seq1, seq2, loc ):
    L2 = len( seq2 )
    act = L2 - loc
    if act <=0:
        st1 = seq1
        st2 = (-act) *'.' + seq2
```

```
    else:
        st1 = act*'.' + seq1
        st2 = seq2
    return st1, st2
```

Code 12-5

A contig is represented by a list of tuples in which each tuple contains a sequence identification and a shifted version of the sequence. In this case, the sequences were just created randomly from a parent sequence, and so the identifiers are simple, as seen in Code 12-6. However, it is possible to use more complicated IDs such as gene names.

```
>>> ids = ['s0', 's1', 's2', 's3','s4','s5','s6','s7','s8','s9' ]
```

Code 12-6

The greedy algorithm starts with the best alignment and creates a new contig. It then considers the next best alignment and either creates a new contig or adds to the old one. If the best alignment is *s1* with *s2* and the second-best alignment is *s3* with *s4*, then the second alignment creates a new contig. If the second-best alignment is *s2* with *s3*, then *s3* is added to the existing contig that contains *s2*. The third possibility is that the alignment being considered has two sequences already belonging to two different contigs. In this case, the two contigs are joined.

Consider the development of an assembly in which several contigs have already been created. The next-best alignment is *sa* with *sb*. There are four possibilities that can occur:

1. Neither *sa* nor *sb* exist in any contig. Create a new contig.
2. The *sa* belongs in a contig but *sb* does not. Add the *sb* to the contig that contains *sa*. If *sb* belongs to a contig and *sa* does not then add *sa* to the contig with *sb*.
3. The *sa* and *sb* belong to different contigs. Join the two contigs.
4. The *sa* and *sb* belong to the same contig. Do nothing.

Initially there are no contigs and the assembly is an empty list. Code 12-7 shows the initialization of the assembly *smb*.

```
>>> smb = []
```

Code 12-7

The best alignment is the one with the largest score in matrix **M**. Code 12-8 shows that this maximum value is 317 and that it is located at *M[5, 6]*. The second argument of the **divmod** function is the number of sequences (or similarly the horizontal dimension of **M**).

```
>>> M.max()
317
>>> divmod( M.argmax(), len(M) )
(5, 6)
>>> M[5,6]
```

Code 12-8

Since no contigs currently exist, a new contig is to be created from the aligned sequences *chops[5]* and *chops[6]*. The location of the peak in **M** corresponds to the sequences involved in the alignment.

The function **NewContig** in Code 12-9 creates a new contig from two aligned sequences. This function does not return any variable but receives instead the assembly and appends the new contig to this assembly. The variables *fnams1* and *fnams2* are the identification of the two sequences involved. The final two lines print out the first two items of the first contig.

```
# greedy.py
def NewContig( smb, s1, s2, fnams1, fnams2 ):
    c = []  # the new contig
    c.append( (fnams1, s1) )
    c.append( (fnams2, s2) )
    smb.append( c )

>>> sa, sb = ShiftedSeqs( chops[5], chops[6], L[5,6] )
>>> NewContig( smb, sa, sb, ids[5], ids[6] )
>>> smb[0][0]
('s5', '..QDNDVQIKSEDVFVHTEMQIGDPTNIQDVIEYENIIYRSMRIRELQFPPV')
>>> smb[0][1]
('s6', 'KIQDNDVQIKSEDVFVHTEMQIGDPTNIQDVIEYENIIYRSMRIRELQFP')
```

Code 12-9

Displaying the results is actually an involved process. There will soon be several contigs and each one will have multiple sequences. Code 12-10 presents **ShowContigs**, which loops through each contig and then each member of the contig. It prints 50 characters for each aligned string. Because it is quite possible that some strings will have more than 50 dots in front of them, the optional variable *st* is used to start the printing at a different location in the strings.

```
# greedy.py
def ShowContigs(smb, st=0 ):
    lc = len(smb)
    for i in range( lc ):
        for j in smb[i]:
            print j[0],'\t', j[1][st:st+50]
        print ""
    print""

>>> ShowContigs( smb )
s5    ..QDNDVQIKSEDVFVHTEMQIGDPTNIQDVIEYENIIYRSMRIRELQFP
s6    KIQDNDVQIKSEDVFVHTEMQIGDPTNIQDVIEYENIIYRSMRIRELQFP
```

Code 12-10

The next best alignment in *M* is now ready to be considered. In order to find it, the current maximum value is destroyed. Now the second best alignment is the new maximum value, as shown in Code 12-11. The next best peak has a height of 269, which is still large, and the location is *M[1,8]*. The variables *v, h* are now used to represent the location in the matrix. Since neither of these is in the existing contig, a new contig is created.

```
>>> M[5,6] = 0
>>> M.max()
269
>>> v,h = divmod( M.argmax(), 10 )
>>> v,h
(1, 8)
>>> sa, sb = ShiftedSeqs( chops[v], chops[h], L[v,h] )
>>> NewContig( smb, sa, sb, ids[v], ids[h] )
>>> ShowContigs( smb )
s5    ..QDNDVQIKSEDVFVHTEMQIGDPTNIQDVIEYENIIYRSMRIRELQFP
s6    KIQDNDVQIKSEDVFVHTEMQIGDPTNIQDVIEYENIIYRSMRIRELQFP

s1    NREELVRKEIQLANITEFDFCFPTPLFFLNYFLRISGQTQESMLFARYIV
s8    ..........QLANITEFDFCFPTPLFFLNYFLRISGQTQESMLFARYIV
```
Code 12-11

The next step is to create a function that will search existing contigs for a sequence identification. Code 12-12 displays **Finder**, which receives the assembly and the sequence ID. It returns two numbers: (1) the index of the contig that contains *seqid* and (2) the location in that contig where *seqid* exists. If *answ=−1*, then *seqid* does not exist in any contig.

```
# greedy.py
def Finder( smb, seqid ):
    lc = len( smb )
    answ = -1
    ndx = -1
    for i in range( lc ):
        for j in range( len( smb[i] )):
            if seqid == smb[i][j][0]:
                answ = i
                ndx = j
                break
    return answ, ndx
```
Code 12-12

When the third alignment is considered, it indicates that *chops[2]* and *chops[8]* align with a score of 252. Using the **Finder** program in Code 12-13, we learn that *s2* does not belong to a contig and that *s8* belongs to the contig *smb[1]* and it is the second item in list (*hseqno=1*).

```
>>> M[v,h] = 0
>>> M.max()
252
>>> v,h = divmod( M.argmax(), 10 )
>>> v,h
(2, 8)
```

```
>>> vnum, vseqno = Finder( smb, ids[v] )
>>> hnum, hseqno = Finder( smb, ids[h] )
>>> vnum
-1
>>> hnum
1
>>> hseqno
1
```

Code 12-13

In this case it is necessary to add *s2* to the contig that contains *s8*. Adding a sequence to a contig requires a bit more alignment. In this case, *s8* is already shifted in the contig, but the call to **ShiftedSeqs** does not know of this shift. Thus once *sa* and *sb* are calculated, it is necessary to align them with the contig. In this example, the *s8* in the contig is shifted, and thus *sa* and *sb* need to be shifted by the same amount.

Code 12-14 shows **Add2Contig**, which adds a sequence to a contig. It receives six arguments: (1) the assembly *smb*, which is altered in the function; (2) the host sequence, *seqa* which is either *sa* or *sb* or whichever exists in the contig; (3) the host sequence *seqb*, which is the other choice to be added to the contig; (4) *seqbid*, which is the identification of the sequence that is being added; (5) *ctgn* is the contig number; and (6) *ctgndx* is the location where the previous sequence was found. These are the results of the **Finder** program. There are three conditions in this function. Condition 1 considers the case in which the incoming sequence is shifted (has leading periods). The *ndots* is the shift of the host, which is then added to the front of the incoming sequence (*temp*) so that it properly aligns in the contig, as shown in Figure 12-2. Condition 2 considers the case in which the incoming sequence is not shifted and the other sequence is shifted. Even though this is the same sequence that is in the contig,

```
Condition 1

Contig
   ABCD
   ..CDEF    (host)

   CDEF      (seqa)
   ..EFG     (seqb)
ndots = 2

Final contig
   ABCD
   ..CDEF
   ....EFG
```

FIGURE 12-2 Building a contig with Condition 1.

12.1 The Greedy Approach

```
Condition 2                    Condition 3

Contig                         Contig
   ABCD                           ABCD
   ..CDEF    (host)               ..CDEF    (host)

   ...CDEF   (seqa)               .CDEF     (seqa)
   ZABCD     (seqb)               BCDE      (seqb)
diff = -1                      diff = 1

Final contig                   Final contig
   .ABCD                          ABCD
   ...CDEF                        ..CDEF
   ZABCD                          .BCDE

       (a)                            (b)
```

FIGURE 12-3 Building a contig with either (a) Condition 2 or (b) Condition 3.

there are two different shifts. In the example in Figure 12-3(a), the *diff=1* and *seqb* need to be shifted by *diff* in order to align. If *diff* is positive, then there are more dots in *seqa* than in the sequence in the contig. In this case the entire contig needs to be shifted by *diff*. If *diff* is negative (Figure 12-3b), then only the incoming sequence is shifted by the amount *–diff*.

The calls in Code 12-14 align the two sequences that were the best match and then adds them to the appropriate contig. In this case it was *chops[h]* that was found in a contig. Had it been *chops[v]*, then the call would have been *Add2Contig(smb, sa, sb, ids[h], hnum, hseqno)*.

```python
# greedy.py
def Add2Contig( smb, seqa, seqb, seqbid, ctgn, ctgndx ):
    # seq1a seq2a from ShiftedSeqs
    # ctgn, ctgndx from Finder
    if seqb[0] == '.':
        # CONDITION 1
        ndots = smb[ctgn][ctgndx][1].count('.')
        temp = '.'*ndots + seqb
        smb[ctgn].append( (seqbid,temp) )
    else:
        nd1 = seqa.count('.')
        # number of dots in ctg seq
        ndc = smb[ctgn][ctgndx][1].count('.')
        diff = nd1 - ndc
        if diff >0:
            # CONDITION 2
            for i in range( len( smb[ctgn] )):
                smb[ctgn][i] = smb[ctgn][i][0], diff*'.' + smb[ctgn][i][1]
        else:
            # CONDITION 3
            seqb = (-diff)*'.' + seqb
        smb[ctgn].append( (seqbid,seqb) )
```

```
>>> sa, sb = ShiftedSeqs( chops[v], chops[h], L[v,h] )
>>> Add2Contig( smb, sb, sa, ids[v], hnum, hseqno )
>>> ShowContigs( smb )
s5    ..QDNDVQIKSEDVFVHTEMQIGDPTNIQDVIEYENIIYRSMRIRELQFP
s6    KIQDNDVQIKSEDVFVHTEMQIGDPTNIQDVIEYENIIYRSMRIRELQFP

s1    NREELVRKEIQLANITEFDFCFPTPLFFLNYFLRISGQTQESMLFARYIV
s8    ..........QLANITEFDFCFPTPLFFLNYFLRISGQTQESMLFARYIV
s2    ...................PTPLFFLNYFLRISGQTQESMLFARYIV
```

Code 12-14

Code 12-15 shows the further progression of the system. Sequences are added to contigs as necessary. Code 12-16 shows the state of the assembly at the end of Code 12-15. In Code 12-15 the decision as to which contig to be used is performed manually. An automated approach using **Finder** will be developed later.

```
>>> M[v,h] = 0
>>> M.max()
200
>>> v,h = divmod( M.argmax(), 10 )
>>> v,h
(3, 9)
>>> sa, sb = ShiftedSeqs( chops[v], chops[h], L[v,h] )
>>> NewContig( smb, sa, sb, ids[v], ids[h] )

>>> M[v,h] = 0
>>> M.max()
194
>>> v,h = divmod( M.argmax(), 10 )
>>> v,h
(5, 7)
>>> sa, sb = ShiftedSeqs( chops[v], chops[h], L[v,h] )
>>> Add2Contig( smb, sa, sb, ids[h], 0,0 )

>>> M[v,h] = 0
>>> M.max()
184
>>> v,h = divmod( M.argmax(), 10 )
>>> v,h
(1, 2)
# do nothing because (1,2) already exist in same contig

>>> M[v,h] = 0
>>> M.max()
184
>>> v,h = divmod( M.argmax(), 10 )
>>> v,h
(4, 6)
>>> sa, sb = ShiftedSeqs( chops[v], chops[h], L[v,h] )
>>> Add2Contig( smb, sb, sa, ids[v], 0,1 )

# next two iterations give seqs already in same contig
```

Code 12-15

There are three contigs in this assembly. The next alignment considers pairs $s2$ with $s3$. Both of these are already in contigs, and so these two contigs need to be joined. Code 12-17 shows the **JoinContigs** function. It needs to receive the assembly, the contig numbers, the locations in the contigs, and the two aligned sequences. One

```
>>> ShowContigs( smb )
s5                      ......................QDNDVQIKSEDVFVHTEMQIGDPTNI
s6                      .....................KIQDNDVQIKSEDVFVHTEMQIGDPTNI
s7                      ............................................DPTNI
s4                      DSDDEDDVIPDEIDLQILTSPKKIQDNDVQIKSEDVFVHTEMQIGDPTNI

s1                      NREELVRKEIQLANITEFDFCFPTPLFFLNYFLRISGQTQESMLFARYIV
s8                      ..........QLANITEFDFCFPTPLFFLNYFLRISGQTQESMLFARYIV
s2                      .....................PTPLFFLNYFLRISGQTQESMLFARYIV

s3                      FARYIVEMCLTSEKFNDVKASAIAATAVVIMRVVYSETPWTEDLMMFSRY
s9                      ..................KASAIAATAVVIMRVVYSETPWTEDLMMFSRY
```

Code 12-16

of the aligned sequences will have dots in front of it, and this will cause the members of its contigs to shift in order to properly align. This function will actually create a new contig that collects members from the old contigs and then will destroy the two older contigs.

```
# greedy.py
def JoinContigs( smb, cnumb1, cnumb2, cseq1, cseq2, sa, sb ):
    # create a new contig from two old ones.
    c = []
    # how many dots should ctg1 be shifted?
    sh1 = sa.count('.') - smb[cnumb1][cseq1][1].count('.')
    # how many dots for ctg2
    sh2 = sb.count('.') - smb[cnumb2][cseq2][1].count('.')
    if sh1 < 0:
        sh2 -= sh1
        sh1 = 0
    if sh2 < 0 :
        sh1 -= sh2
        sh2 = 0
    # create new contig
    for i in range( len( smb[cnumb1] )):
        temp = ( smb[cnumb1][i][0], sh1*'.'+smb[cnumb1][i][1] )
        c.append( temp )
    for i in range( len( smb[cnumb2] )):
        temp = ( smb[cnumb2][i][0], sh2*'.'+smb[cnumb2][i][1] )
        c.append( temp )
    # destroy old contigs
    if cnumb1 > cnumb2:
        a = smb.pop( cnumb1 )
        a = smb.pop( cnumb2 )
    else:
        a = smb.pop( cnumb2 )
        a = smb.pop( cnumb1 )
    smb.append( c )
```

Code 12-17

Code 12-18 shows the call to the functions and the alignment when the contigs are joined. In this assembly *s9* has more than 50 dots at the beginning of the sequence, and so the display from **ShowContigs** is moved to start at position 30 to show some of *s9*.

```
>>> M[v,h] = 0
>>> M.max()
170
>>> v,h = divmod( M.argmax(), 10 )
>>> v,h
(2, 3)
>>> sa, sb = ShiftedSeqs( chops[v], chops[h], L[v,h] )
>>> JoinContigs( smb, 1,2,2,0,sa,sb )

>>> ShowContigs( smb,30 )
s5   IKSEDVFVHTEMQIGDPTNIQDVIEYENIIYRSMRIRELQFPPV
s6   IKSEDVFVHTEMQIGDPTNIQDVIEYENIIYRSMRIRELQFP
s7   ..............DPTNIQDVIEYENIIYRSMRIRELQFPPVIFKQAI
s4   IKSEDVFVHTEMQIGDPTNI

s1   YFLRISGQTQESMLFARYIV
s8   YFLRISGQTQESMLFARYIVEMCLTSEKFN
s2   YFLRISGQTQESMLFARYIVEMCLTSEKFNDVKASAIAATAV
s3   ..............FARYIVEMCLTSEKFNDVKASAIAATAVVIMRVVYS
s9   ..............................KASAIAATAVVIMRVVYS
```

Code 12-18

All of the components are now in place. Each sequence pair from **M** is considered until the best remaining score falls below a threshold. For each pair, a new contig is created, a contig is increased, or contigs are joined. When the threshold is reached, no other alignment pairs are considered but there may still be some sequences that remain. It is possible that a sequence did not align well with any of the other sequences, and therefore does not belong to any contig. For each of these, a new contig is created with just a single member and added to the assembly.

Code 12-19 shows the **Assemble** program, which drives all of the functions. It receives the sequence identifications, sequences, substitution matrix, alphabet, and threshold. It creates the **M** and **L** matrices as well as an empty assembly *smb*. The *while* loop processes the best alignment pair and decides if a contig is created, added to, or joined with another contig. At the end, the sequences that did not join any contig will create new solo contigs and join the assembly.

```
# greedy.py
def Assemble( fnms, seqs, submat, abet, gamma = 500 ):
    used = zeros( len( fnms ))  # set these elements to 1 when a seq is used
    M,L = FastMat( seqs, submat, abet )
    #print M
    ok = 1
    smb = []
    nseqs = len( seqs )
    while ok:
        v,h = divmod( M.argmax(), nseqs )
        if M[v,h] >= gamma:
            vnum, vseqno = Finder( smb, fnms[v] )
            hnum, hseqno = Finder( smb, fnms[h] )
            s1, s2 = ShiftedSeqs( seqs[v], seqs[h], L[v,h] )
            if vnum == -1 and hnum == -1:
                # create a new contig
                NewContig( smb, s1, s2, fnms[v], fnms[h] )
            if vnum != -1 and hnum == -1:
                Add2Contig( smb, s1, fnms[h], vnum, vseqno )
            if vnum == -1 and hnum != -1:
                Add2Contig( smb, s2, s1, fnms[v], hnum, hseqno )
            if vnum != -1 and hnum != -1 and vnum != hnum:
```

```
                    JoinContigs( smb, vnum, hnum, vseqno, hseqno, s1, s2 )
            M[v,h] = 0
            used[v] = used[h] = 1
        else:
            ok = 0
# make single contigs for all sequences not used
notused = nonzero( 1-used)[0]
for i in notused:
    smb.append( [(fnms[i],seqs[i])] )
return smb
```

Code 12-19

Code 12-20 shows the call to the function and the results for two tests. Recall that the **ChopSeq** function chopped up the sequence in a random fashion, which means that a new trial was created by simply chopping up the same sequence in a different manner.

```
>>> smb = Assemble( ids, chops, blosum.BLOSUM50, blosum.PBET, 50 )
>>> ShowContigs( smb )
s5      ...............................................Q
s6      .............................................KIQ
s7      .................................................
s4      ........................DSDDEDDVIPDEIDLQILTSPKKIQ
s0      RRPLAFVSNQLIQPQTLISKTTTIFDSDDEDDVIPDEIDLQILTSPKKIQ

s1      NREELVRKEIQLANITEFDFCFPTPLFFLNYFLRISGQTQESMLFARYIV
s8      .........QLANITEFDFCFPTPLFFLNYFLRISGQTQESMLFARYIV
s2      ...................PTPLFFLNYFLRISGQTQESMLFARYIV
s3      .............................................FARYIV
s9      .................................................

# Test 2
>>> chops = ChopSeq( p1, 10, 50 )
>>> smb = Assemble( ids, chops, blosum.BLOSUM50, blosum.PBET, 50 )
notused []
>>> ShowContigs( smb )
s8      IYRSMRIRELQFPPVIFKQAITNSQKGQMIDWIDRLHYKSQCCTTSLYRA
s9      .........LQFPPVIFKQAITNSQKGQMIDWIDRLHYKSQCCTTSLYRA

s3      .VSNQLIQPQTLISKTTTIFDSDDEDDVIPDEIDLQILTSPKKIQDNDVQ
s7      .VSNQLIQPQTLISKTTTIFDSDDEDDVIPDEIDLQILTSPKKIQDNDVQ
s2      FVSNQLIQPQTLISKTTTIFDSDDEDDVIPDEIDLQILTSPKKIQDNDVQ
s5      ...NQLIQPQTLISKTTTIFDSDDEDDVIPDEIDLQILTSPKKIQDNDVQ
s1      ....................DEDDVIPDEIDLQILTSPKKIQDNDVQ
s6      ................TTIFDSDDEDDVIPDEIDLQILTSPKKIQDNDVQ

s0      ITPDSMRQFAAASLLIASKMEDLQPVSIDILIQCSKNTLNREELVRKEIQ
s4      .........................................REELVRKEIQ
```

Code 12-20

12.2 The Nongreedy Approach

The greedy approach is based on finding the best pairs of alignments. While there is some logic to this approach, it does not necessarily find the best alignment. The nongreedy approach only scores the total alignment and does not attempt to find the best pairs of alignments. Only one of the many different nongreedy approaches is presented here.

The example approach uses a genetic algorithm (GA) to create several sample assemblies and then optimizes the situation by creating new assemblies from the best of the older assemblies. Each gene creates an assembly, and each assembly contains multiple contigs. Each contig is used to generate a consensus sequence. The assembly is converted to a *catsequence*, which is the concatenation of the consensus sequences. The goal in this case is to find the assembly that creates the shortest catsequence, and thus the cost of the gene is the length of the catsequence that it eventually generates.

The data for this system is generated as in the greedy case. Code 12-21 reviews the commands needed to generate the data for this section.

```
>>> data = genbank.ReadGenbank( 'genbank/XM_001326205.gb.txt')
>>> klocs = genbank.FindKeywordLocs( data )
>>> p1 = genbank.Amino( data, klocs[0] )
>>> chops = greedy.ChopSeq( p1, 15, 50 )
>>> M,L = greedy.FastMat( chops, blosum.BLOSUM50, blosum.PBET )
```

Code 12-21

12.2.1 Creating Genes

The gene for the GA needs to encode a method by which an assembly is created. In the greedy case the assembly was created by considering pairs of sequence alignments in order of their alignment score. In the nongreedy case the use of alignment scores for pairs of sequences is not used. Rather an assembly is created by a random sequence of alignment pairs. The matrix **M** contains the scores for the alignments, and in this case its sole purpose is to provide a list of possible alignment pairs, which are those elements in **M** that are above a small threshold. Code 12-22 presents the function **BestPairs**, which creates a list of all elements in **M** that are above a threshold *gamma*. Each entry in the list is the v, h from the $M[v, h]$ locations that qualify. In this case the data generated 99 elements in **M** that were above the threshold of 5. The first ten of these are shown.

```
# nongreedy.py
def BestPairs( M, gamma ):
    # M from greedy.FastMat
    work = M+0
    hits = []
    ok = 1
    V,H = work.shape
    while ok:
        mx = work.max()
        if mx > gamma:
            v,h = divmod( work.argmax(), H )
            hits.append( (v,h) )
            work[v,h] = gamma-1
        else:
            ok = 0
    return hits
```

12.2 The Nongreedy Approach

```
>>> hits = BestPairs( M, 5 )
>>> len( hits )
99
>>> hits[:10]
[(9, 13), (2, 6), (4, 5), (11, 12), (7, 11), (1, 3), (5, 10), (4, 14), (7, 12), (1, 6)]
```

Code 12-22

This particular list is ordered according to the magnitude of the values in **M**. Basically, these would be the base pairs that would be extracted in each loop inside of Code 12-19. However, a rearrangement of these pairs can also be used to create an assembly. Code 12-23 is similar to Code 12-19 except that the list of pairs is fed into the function rather than generated by it. The argument *pairs* is the list of gene *hits* from Code 12-22. The argument *gene* is the ordering in which these pairs are considered. In this case *gene* is thus an ordered list of numbers from 0 to 98. Code 12-24 shows the greedy assembly using this method. Basically, this function creates an assembly by considering each alignment pair in a prescribed order. A different order of the same alignment pairs creates a different assembly.

```
# nongreedy.py
def Gene2Assembly( gene, pairs, seqs, seqnames, L ):
    # L from greedy.FastMat
    smb = []
    used = zeros( len( seqs ), int )
    for g in gene:
        i1, i2 = pairs[g] # the indices of the sequences
        # find the locations of the sequences in the assembly
        b1, c1 = greedy.Finder( smb, seqnames[i1] )
        b2, c2 = greedy.Finder( smb, seqnames[i2] )
        # decide what to do with the information
        s1, s2 = greedy.ShiftedSeqs( seqs[i1], seqs[i2], L[i1,i2] )
        if b1==-1 and b2==-1:
            greedy.NewContig( smb, s1, s2, seqnames[i1], seqnames[i2] )
        if b1!=-1 and b2==-1:
            greedy.Add2Contig( smb, s1, s2, seqnames[i2], b1, c1 )
        if b1==-1 and b2!=-1:
            greedy.Add2Contig( smb, s2, s1, seqnames[i1], b2, c2 )
        if b1!=-1 and b2!=-1 and b1!=b2:
            greedy.JoinContigs( smb, b1, b2, c1, c2, s1, s2 )
        used[i1] = used[i2] = 1
    # unused
    for i in nonzero( 1-used)[0]:
        smb.append( [(seqnames[i], seqs[i])] )
    return smb
```

Code 12-23

Thus, the *gene* in this case is merely the order in which the pairs of sequences are considered in building an assembly. The assembly, however, still needs to be converted to a catsequence. This is accomplished by converting each contig to a consensus sequence, as shown in Figure 12-4. The letters in column k of a contig are used to create an element of the consensus sequence $cs[k]$.

```
>>> ids = []
>>> for i in range( 15 ):
    ids.append( 's'+str(i))
>>> smb = Gene2Assembly( range(99), hits, chops, ids, L )
>>> greedy.ShowContigs( smb )
s9      ................................................
s13     ................................................
s0      ................................................
s8      ................................................
s1      ..............................................EI
s3      ........................................DDEDDVIPDEI
s2      ................................................
s6      ................................................
s11     ........KRVQRRPLAFVSNQLIQPQTLISKTTTIFDSDDEDDVIPDEI
s12     .............RPLAFVSNQLIQPQTLISKTTTIFDSDDEDDVIPDEI
s7      QNINPNASKRVQRRPLAFVSNQLIQPQTLISKTTTIFDSDDEDDVIPDEI
s4      ................................................
s5      ................................................
s10     ................................................
s14     ................................................

>>> greedy.ShowContigs( smb,50 )
s9      ................................................
s13     ................................................
s0      ................................................
s8      ................................................
s1      DLQILTSPKKIQDNDVQIKSEDVFVHTEMQIGDPTNIQDVIEYENIIY
s3      DLQILTSPKKIQDNDVQIKSEDVFVHTEMQIGDPTNIQD
s2      ....................SEDVFVHTEMQIGDPTNIQDVIEYENIIYRS
s6      ...............VQIKSEDVFVHTEMQIGDPTNIQDVIEYENIIYRS
s11     DLQILTSP
s12     DLQILTSPKKIQD
s7
s4      ................................................
s5      ................................................
s10     ................................................
s14     ................................................

>>> greedy.ShowContigs( smb,100 )
s9      ................................................
s13     ................................................
s0      ................................................
s8      ................................................
s1
s3
s2      MRIRELQFPPVIFKQAITN
s6      MRIRELQFPPVIFKQ
s11
s12
s7
s4      .......................MIDWIDRLHYKSQCCTTSLYRAIGIF
s5      ................ITNSQKGQMIDWIDRLHYKSQCCTTSLYRAIGIF
s10     ..IRELQFPPVIFKQAITNSQKGQMIDWIDRLHYKSQCCTTSLYRAIGIF
s14     ...................................QCCTTSLYRAIGIF
```

Code 12-24

In real applications there is not a complete agreement in each column, as there can be more than one letter in a column. Often there is a *consensus letter*—one letter that

12.2 The Nongreedy Approach

```
ABCDEF
..CDEFGH
.....FGHIJ

ABCDEFGHIJ (consensus)
```

FIGURE 12-4 Building a consensus sequence.

is seen considerably more often than the others. Code 12-25 shows the **ConsensusCol** function, which receives a list of characters from a single column of contig *stg*. It will extract the consensus character from this list excluding the periods.

```
# nongreedy.py
def ConsensusCol( stg ):
    chrs = []
    cnts = []
    ape = copy.copy( stg )
    while len( ape )>0:
        C = ape[0]
        chrs.append( C )
        N = ape.count( ape[0] )
        if C!='.':
            cnts.append( N )
        else:
            cnts.append(0)
        for i in range( N ):
            ape.remove( C )
    # find most common
    vec = array( cnts )
    ndx = vec.argmax( )
    return chrs[ndx]

>>> a = ['a','b','c','a','d','b','b']
>>> ConsensusCol( a )
'b'
```

Code 12-25

The consensus sequence is created for all columns by the function **CatSeq** in Code 12-26. The *for i* loop considers each contig inside of the assembly. The second *for j* loop considers each column in the contig. The *for k* loop considers each string. Basically, for each position a letter is extracted from each string, excluding those strings that do not have a character at this position. These are collected in the list *y* and sent to **ConsensusCol**, which returns a single character that is added to the string *sq*. The second step in this process is to realize that an assembly contains several contigs and that these do not overlap. For the purposes of scoring the assembly, a single long string is created from all of the contigs. The nonoverlapping contigs are concatenated into *sq*. The example creates a single string from the assembly generated in Code 12-24.

```
# nongreedy.py
def CatSeq( smb ):
    NC = len( smb ) # number of contigs
```

```
        sq = ''
        for i in range( NC ):
            NS = len( smb[i] )
            # get the length of the strings
            lgs = zeros( NS, int )
            for j in range( NS ):
                lgs[j] = len( smb[i][j][1] )
            # find the max length
            mxlg = lgs.max()
            # for each column
            for j in range( mxlg ):
                # grab all characters
                y = []
                for k in range( NS ):
                    if lgs[k] > j:
                        y.append( smb[i][k][1][j] )
                sq += ConsensusCol( y )
        return sq

>>> sq = CatSeq( smb )
>>> sq
'QNINPNASKRVQRRPLAFVSNQLIQPQTLISKTTTIFDSDDEDDVIPDEIDLQILTSPKKIQDNDV
QIKSEDVFVHTEMQIGDPTNIQDVIEYENIIYRSMRIRELQFPPVIFKQAITNSQKGQMIDWIDRLH
YKSQCCTTSLYRAIGIFNRAINLTNITPDSMRQYIVEMCLTSEKFNDVKASAIAATAVVIMRVVYSE
TPWTEDLMMFSRYSLKDLSSNIRDAYEILTDLEREESTF'
```

Code 12-26

A gene is merely an ordering of the sequence pairs used to create an assembly. Code 12-27 creates an instance of a *GA* class and uses the function **InitGA** to create random arrangements of the sequence pairs. In this example each gene is a list of numbers from 0 to 98 in a random arrangement.

```
>>> import ga
# nongreedy.py
def InitGA( pairs, Ngenes ):
    # pairs from BestPairs
    # Ngenes = desired number of GA genes
    work = arange( len(pairs) )
    genes = []
    for i in range( Ngenes ):
        random.shuffle( work )
        genes.append( copy.deepcopy(work) )
    return genes

>>> folks = InitGA( hits, 10 )
```

Code 12-27

12.2.2 Steps in the Genetic Algorithm

The cost of a gene is the length of the consensus sequence that it creates. Code 12-28 shows the function **CostAllGenes**, which considers each gene in the *for* loop. Each gene creates an assembly *smb* that in turns creates a catsequence *cseq*. The cost of this sequence is its length. In this example there are ten genes, and the costs they generated are shown.

12.2 The Nongreedy Approach

```
# nongreedy.py
def CostAllGenes( genes, pairs, seqs, seqnames, L ):
    NG = len( genes )
    cost = zeros( NG )
    for i in range( NG ):
        smb = Gene2Assembly( genes[i], pairs, seqs, seqnames, L )
        cseq = CatSeq( smb )
        cost[i] = len( cseq )
    return cost

>>> fcost = CostAllGenes( folks, hits, chops, ids, L )
>>> fcost
array([ 136.,  174.,  187.,  178.,  249.,  171.,  230.,  160.,  177.,  176.])
```

Code 12-28

The crossover function is not changed from the original GA program. Code 12-29 shows the calls to create new genes and to compute their cost. The problem with the new genes is that they may not contain all of the pairings and that they may contain two copies of other pairings. The two **nonzero** calls in Code 12-29 show that the first new gene, *kids[0]*, does not contain a 0 but does contain two 1's.

```
>>> kids = net.CrossOver( fcost )
>>> kcost = CostAllGenes( kids, hits, chops, ids, L )
>>> kcost
array([ 160.,  187.,  178.,  176.,  158.,  189.,  177.,  174.,  176.,  177.])
>>> nonzero( equal(kids[0], 0) )
(array([], dtype=int32),)
>>> nonzero( equal(kids[0], 1) )
(array([14, 98]),)
```

Code 12-29

A gene should contain each pair of sequences from the original list, and these new genes are not correct. The function **FixGene** in Code 12-30 considers a gene and finds those elements that are missing (*missg*) and those that are duplicated (*dups*). It then randomly chooses one of the duplicates and replaces it with one of the missing elements. The **random.shuffle** command changes the order of the duplicates so that the first duplicate is not always replaced by the first missing element. In the example, each new gene is sent to this function for repair.

```
# nongreedy.py
def FixGene( gene, pairs ):
    # count the number of times each pairing is used
    NP = len( pairs )
    gene = list( gene )
    cnts = zeros( NP, int )
    for i in range( NP ):
        cnts[i] = gene.count( pairs[i] )
    # find the missing and the duplicates
    missg = nonzero( equal( cnts, 0 ) )[0]
    dups = nonzero( equal( cnts, 2 ))[0]
    # rearrange the dups
    random.shuffle( dups )
```

```
            # replace a duplicate with a missing
            for i in range( len( dups )):
                # locate the two duplicates
                d1 = gene.index( pairs[dups[i]] )
                d2 = gene.index( pairs[dups[i]], d1+1 )
                # select one
                if random.rand()>0.5:
                    choose = d1
                else:
                    choose = d2
                # replace
                gene[choose] = pairs[missg[i]]
            gene = array( gene )
            return gene
>>> for i in range(len(kids)):
            kids[i] = FixGene( kids[i], arange(len(hits)) )
>>> nonzero( equal(kids[0], 0) )
(array([36]),)
>>> nonzero( equal(kids[0], 1) )
(array([14]),)
```

Code 12-30

The cost of the *kids* can now be computed by calling the **CostAllGenes** function as shown in Code 12-31.

```
>>> kcost = CostAllGenes( kids, hits, chops, ids, L )
>>> kcost
array([ 227., 221., 201., 177., 165., 270., 187., 170., 172., 176.])
```

Code 12-31

The mutation stage of the GA is to randomly swap a small number of elements in the gene. The swap will preserve the requirement that all numbers from 0 to 98 exist only once in each gene. Code 12-32 shows the function **SwapMutate**, which performs this swapping. The *for i* loop considers each gene and the *for j* loop considers each position in the gene.

```
# nongreedy.py
def SwapMutate( genes, rate ):
    for i in range( len(genes) ):
        dm = len( genes[i] )
        for j in range( dm ):
            if random.rand() < rate:
                pick = int( random.rand() * dm )
                a = genes[i][pick]+0
                genes[i][pick] = genes[i][j] + 0
                genes[i][pick] = a + 0
```

Code 12-32

12.2.3 The Test Run

All of the parts are now in place to perform the GA. Code 12-33 shows the function **RunGA**, which drives the GA process. It settles rather quickly on an assembly that creates a consensus sequence that has a length of 120.

12.2 The Nongreedy Approach

```
# nongreedy.py
def RunGA( hits, seqs, seqnames, L ):
    NH = len( hits )
    folks = InitGA( hits, 10 )
    fcost = CostAllGenes( folks, hits, seqs, seqnames, L )
    print fcost.min(), fcost.argmin()
    for i in range( 10 ):
        kids = ga.CrossOver( folks, fcost )
        for i in range(len(kids)):
            kids[i] = FixGene( kids[i], arange(NH) )
        kcost = CostAllGenes( kids, hits, seqs, seqnames, L )
        ga.Feud( folks, kids, fcost, kcost )
        SwapMutate( folks, 0.03 )
        fcost = CostAllGenes( folks, hits, seqs, seqnames, L )
        print fcost.min(), fcost.argmin()
    return folks[fcost.argmin() ]

>>> g = RunGA( hits, chops, ids, L )
141.0 2
126.0 0
126.0 0
120.0 0
120.0 0
120.0 0
120.0 0
120.0 0
120.0 0
120.0 0
120.0 0
```

Code 12-33

The results of the nongreedy test are compared to the greedy approach. Code 12-34 shows the steps used to create a greedy consensus. The length of the greedy consensus is 267 while the length of the nongreedy approach is only 120. Obviously, the nongreedy approach significantly outperformed the greedy approach. The cost of this improvement though is that the nongreedy approaches are usually computationally expensive.

```
>>> smb = greedy.Assemble( ids, chops, blosum.BLOSUM50, blosum.PBET, 20 )
>>> cseq = CatSeq( smb )
>>> len( cseq )
267
```

Code 12-34

12.2.4 Improvements

The nongreedy approach presented is still not the best system and does have a flaw. Consider the following sequences:

$S1$ = abcdef
$S2$ = defghi
$S3$ = jkldef

It is quite possible to align *S1* with *S2* and then *S2* with *S3*. In doing so the following assembly is created:

```
abcdef
...defghi
jkldef
```

In this assembly the *S1* and *S3* do not align all that well. Such problems are likely to occur when building an assembly from pairs of sequences. An improvement to the GA program would be to prevent such poor secondary alignments from occurring or to increase the cost of the assembly if there is a poor consensus.

It is important to note that there is no set method of creating a nongreedy algorithm. The GA is only one method, and as we saw it could be modified to behave differently. The main purpose of the nongreedy approach is to create a system that scores the entire assembly rather than finding the best matches within it.

12.3 Summary

As we saw in the previous chapter, aligning two sequences is relatively straightforward, but many applications require the alignment of more than two sequences. Multiple sequence alignment can be performed through two methods. The first is the greedy approach in which the assembly is constructed by adding pairs of sequences according to their pair alignment scores. The second is the nongreedy approach, which attempts to find the best overall assembly by using machine learning techniques. This approach does not consider the alignments according to their pairing scores, but rather it attempts to optimize the entire alignment. The latter approach is much more expensive but can provide better results.

Problems

1. Apply the greedy algorithm to English text. Chop written text up into many subsequences and then assemble using the greedy approach. Is this assembly similar to the original?

2. Use different matrices (BLOSUM, PAM, etc.) in computing **BruteForceSlide**. Does the use of a different matrix change the assembly?

3. Measure the scale-up effect on computation time. For strings of different sizes, compute the assembly and measure the time of computation. Plot the computational time versus the size of the original data string.

4. Modify the greedy algorithm to handle sequences and their complements. The program should note that if a string is used in making a contig then it and its complement should be removed from further consideration.

5. For strings of different sizes, compare the assemblies created by the greedy and nongreedy methods. Which algorithm provides the best alignment?

6. Modify the best algorithm in the previous problem to handle the situation addressed in Section 12.2.4.

13 Gapped Alignments

In Chapter 8 sequences were aligned using dynamic programming and in Chapter 12 an assembly was created using nongapped alignments. This chapter will combine the two approaches to create an assembly of gapped alignments. The procedures presented here follow the logic of programs such as CLUSTALW (Thompson, 1999).

13.1 Theory of Gapped Alignments

When considering multiple sequence alignments with gapped sequences, there is an added level of complexity. Consider the alignments of ABCE, ACDE, and AABE. Aligning the first two sequences would create the contig [ABC-E, A-CDE]. Aligning the first and third sequences provides [A-BCE, AAB-E]. Both are valid alignments and both contigs include the first sequence. The two contigs can therefore be joined into a single contig. However, the representations of the first sequence in the two contigs are now different since gaps have been added.

The general rule is that gaps are added and not removed. Thus, in order to align the two contigs through a similar sequence, it is necessary to add gaps to make the similar sequences identical. Furthermore, if one sequence in a contig receives a gap during the joining process, then all of the sequences in that contig receive the same gap. This is illustrated in Figure 13-1. Both contigs gain gaps that make the first sequence of each equivalent, and they are then joined to create a single contig.

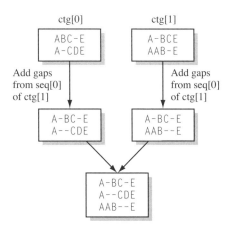

FIGURE 13-1 Joining gapped sequences.

The process for creating a gapped assembly is similar to the ungapped assembly in the previous chapter. A matrix is created that scores the individual alignment pairs, but in this case it is necessary to maintain information about the gaps. The best alignments are considered in a greedy fashion to create contigs, add to contigs, and join contigs. However, each of these operations must be modified to manage the gaps as in Figure 13-1.

13.2 Chopping the Data

To illustrate an example a sequence is chopped up into smaller segments. In this case the overlapping segments must have some differences in order to infuse the gaps. Function **ChopFunny** (Code 13-1) receives an input sequence (*inseq*), a number of segments (*nsegs*), and the length of the segments (*length*). Because Python does not allow elements to be removed from strings, in the second *for i* loop each string is converted to a list. Then 5% of the elements chosen at random are removed. The **join** statement then converts the list back to a string. The example illustrates 15 sequences of length 100 at the bottom of Code 13-1. The **random.seed** argument is used to produce a particular set of random numbers so that the reader's replication of these codes may match those printed here. By removing this line or changing the seed number, the **ChopFunny** function will produce a different set of substrings.

```
# gapalign.py
def ChopFunny( inseq, nsegs, length ):
    segs = []
    locs = []
    G = len( inseq ) - length    # last possible starting location
    for i in range( nsegs ):
        r = int( random.rand()*G )    # start the cut here
        locs.append( r )
        segs.append( inseq[r:r+length] )
    # remove random elements
    for i in range( len( segs) ):
        a = list( segs[i] )
        r = less( random.rand(len(a)), 0.05 )
        nz = nonzero( r )[0][::-1]
        for j in nz:
            dump = a.pop( j )
        segs[i] = ''.join( a )
    return segs, locs

# use the random seed only to replicate results in this text
>>> random.seed( 1059 )
>>> data = file( 'data/choppedseq.txt' ).read()
>>> segs, locs = ChopFunny( data, 15, 100 )
```

Code 13-1

In this example, 15 strings of length 100 were generated. The original string was 576 characters in length, and 1,500 characters were generated in the string segments, which means that on average each element in the original string should be in two or three substrings. A quick visual method of displaying the overlapping segments is in the function **ShowOverlap** in Code 13-2, and the output is printed in Code 13-3. The left column is the segment identification. Because a string of 576 characters will not fit on a single line on the page, the data is compressed by a scale factor (in this case a factor of 10). Each space or X in the horizontal line represents 10 characters in the string, thus making the information on each line 57 characters long. The first substring was extracted from position 167 to 267. In positions 16 to 26 there are X's to indicate the location of this string. Strings that overlap will have X's in the same vertical position. This chart should be compared with the final results of the assembly to ensure validity of the code.

```
# gapalign.py
def ShowOverlap( locs, L, W, scale=10 ):
    # L = length of initial sequence
    # W = length of segs[0]
    N = len( locs )
    for i in range( N ):
        st = ''
        for j in range( 0, L, scale ):
            if locs[i]<j<locs[i]+W:
                st += 'X'
            else:
                st += ' '
        print i,'\t', st
```

Code 13-2

```
>>> ShowOverlap( locs, len(data), 100 )
0                 XXXXXXXXXX
1                                               XXXXXXXXXX
2           XXXXXXXXXX
3           XXXXXXXXXX
4          XXXXXXXXXX
5                                          XXXXXXXXXX
6                      XXXXXXXXXX
7                                            XXXXXXXXXX
8        XXXXXXXXXX
9                             XXXXXXXXXX
10                             XXXXXXXXXX
11                              XXXXXXXXXX
12       XXXXXXXXXX
13                        XXXXXXXXXX
14            XXXXXXXXXX
```

Code 13-3

Each string segment can contain five letters ($ACGTN$), a gap ($-$), or a space ($.$). Thus, the substitution matrix and the associated alphabet need to handle all possibilities. Code 13-4 creates the matrix and alphabet.

```
# gapalign.py
>>> GDNA = array( [[3,-1,-1,-1,0,0,0],\
                   [-1,3,-1,-1,0,0,0],\
                   [-1,-1,3,-1,0,0,0],\
                   [-1,-1,-1,3,0,0,0],
                   [0,0,0,0,0,0,0],\
                   [0,0,0,0,0,0,0],\
                   [0,0,0,0,0,0,0]] )
>>> GBET = 'ACGTN-.'
```

Code 13-4

13.3 Pairwise Alignments

Chapter 12 contained functions to make an assembly from ungapped sequences. By adding gaps to the process, many changes are required. One such change is that it is sometimes necessary to distinguish between gaps and alignment spacing. After an alignment, a string may have aligning gaps at the beginning or end of the sequence. These need to be converted to spacing symbols. The **string.replace** function can convert all gaps to spaces, but this would also include gaps that are inside the string. Therefore a different approach is used. **LeadTrailGaps** in Code 13-5 uses **string.lstrip** and **string.rstrip** to count the number of leading and trailing gaps. These are converted to periods without changing the gaps in the middle.

```
# gapalign.py
def LeadTrailGaps( st ):
    w = st.lstrip('-')
    d = len( st ) - len(w)
    st = '.'*d + st[d:]
    w = st.rstrip('-')
    d = len( st ) - len( w )
    if d>0:
        st = st[:-d] + '.'*d
    return st
```

Code 13-5

Another change is that the alignments need to return more information, as in the Smith-Waterman backtrace. **SWBacktrace2** in Code 13-6 is the same as **dynprog.SWBacktrace** except that it performs a few extra tasks. In the Smith-Waterman backtrace, the trace begins at the location of the largest value in *scormat* and ends at a different location, perhaps in the interior of *scormat*. The beginning and ending of the trace are stored in *(v2, h2)* and *(v1, h1)*, respectively. The Smith-Waterman algorithm returns only the aligning portions of the strings, but in the assembly process it is necessary to return all the strings aligned. Toward the end of the function, the *st1* and *st2* are the aligning string segments. The portions of the strings that were not involved in the alignment are attached at the end of this function.

This function also tracks the locations where the gaps were inserted and stores these into two lists, *vskips* and *hskips*. During the assembly process, this information is needed.

```
# gapalign.py
def SWBacktrace2( scormat, arrow, seq1, seq2 ):
    st1, st2 = ".","
    v,h = arrow.shape
    ok = 1
    v,h = divmod( scormat.argmax(), len(seq2)+1 )
    v2,h2 = v,h
    vskips, hskips = [], []
    while ok:
        if arrow[v,h] == 0:
            st1 += seq1[v-1]
            st2 += '-'
            vskips.append( v-1 )
            v -= 1
        elif arrow[v,h] == 1:
            st1 += '-'
            st2 += seq2[h-1]
            hskips.append( h-1 )
            h -= 1
        elif arrow[v,h] == 2:
            st1 += seq1[v-1]
            st2 += seq2[h-1]
            v -= 1
            h -= 1
        elif arrow[v,h] == 3:
            ok = 0
        if v==0 and h==0:
            ok = 0
    v1,h1 = v,h
    # reverse the strings
    st1 = st1[::-1]
    st2 = st2[::-1]
    # replace leading and trailing gaps
    st1 = LeadTrailGaps( st1 )
    st2 = LeadTrailGaps( st2 )
    # append portions that weren't aligned
    st1 = seq1[:v1] + st1 + seq1[v2:]
    st2 = seq2[:h1] + st2 + seq2[h2:]
    return st1, st2,vskips, hskips
```

Code 13-6

The function **greedy.FastMat** computed two matrices that contained information on the score of each possible alignment pair and the relative shift that was needed to achieve this score. In this case the idea of comparing each possible pair of strings is still used but performed in a very different manner. The alignments must now use the Smith-Waterman process instead of **easyalign.BruteForceSlide**, and each alignment must retain a gapped alignment. In the previous case the value $L[a, b]$ was the relative shift between $seg[a]$ and $seg[b]$. In this case it will be necessary to keep the gapped alignments between the two strings.

These initial alignments are performed in **InitAligns** in Code 13-7. This function receives the names of the segments (*ids*), the segments (*segs*), and the substitution matrix and alphabet. Instead of returning two matrices as did **FastMat**, it will return a list. Each item in the list is a tuple that contains four items. These are the names of the two sequences and the gapped and aligned strings. The example at the end of Code 13-7 shows the alignment of the first two strings.

```
import blosum, dynprog

# gapalign.py
def InitAligns( ids, seqs, submat, abet ):
    lines = []
    N = len( seqs )
    for i in range( N ):
        for j in range( i ):
            if i!=j:
                bv = dynprog.FastSubMatrix( submat, abet, seqs[i], seqs[j] )
                sc, ar = dynprog.FastSW( bv, seqs[i], seqs[j] )
                t1, t2, vs, hs = SWBacktrace2( sc, ar, seqs[i], seqs[j] )
                lines.append( (ids[i], ids[j],t1,t2) )
    return lines

>>> ids = ['cut0', 'cut1', 'cut2', 'cut3', 'cut4', 'cut5', 'cut6', 'cut7',
'cut8', 'cut9', 'cut10', 'cut11', 'cut12', 'cut13', 'cut14']
>>> ilines = InitAligns( ids, segs, blosum.NDNA, blosum.NBET )
>>> ilines[0][:2]
('cut1', 'cut0')
>>> ilines[0][2]
'GTAATACGGGTTATCCAAGGAATCAGGGGATAACGCAGGAAGACATG--TG-CAAA-
GGGCAGCAAAGGGCAGGAACCCTAAAAAGG-CGC-GTTGGTG'
>>> ilines[0][3]
'---AATTCCACACAACTACGAGCCGGAAGCATAAAGTGTAAAGCCTGGGGTGCCTAAT
GAGGAGCTACTCACATTAATTGCGTTGCGCTCACTGCCCGCT'
```

Code 13-7

The **greedy.FastMat** also returned the scores of the alignments in matrix **M**. The alignments in **InitAligns** were not scored. Because this assembly is more complicated, the scoring is performed in a different function. The list *ilines* is sent to the function **InitScores** (Code 13-8). Each item in the list contains two aligned, gapped sequences, and so the scoring is quite easily performed by **easyalign.BlosumScore**.

The example shows the first five scores and then computes the sort order *ag*. The top five scores are shown starting with a value of 226 down to 182. The user will soon need to determine what is a sufficient score for aligning sequences in the assembly, and this is, of course, dependent on the length of the sequences. The best alignment has a score of 226 and is from *ilines[60]*. As shown, this is the alignment of *cut11* with *cut5*. The sequences are shown, and they are indeed similar. The list *locs* from **ChopFunny** contains the locations where the segment was extracted from the original string. It is clear that these two sequences were extracted from nearly the same region that is expected for the best aligning pairs.

```
import easyalign

# gapalign.py
def InitScores( lines, submat, abet ):
    N = len( lines )
    sc = zeros( N, int )
    for i in range( N ):
        sc[i] = easyalign.BlosumScore( submat, abet, lines[i][2], lines[i][3] )
    return sc

>>> sc = InitScores( ilines, blosum.NDNA, blosum.NBET )
>>> sc[:5]
# first five scores
array([-25, 121,  37,  93,   7])
>>> ag = argsort( sc )[::-1]
>>> ag[:5]
    # sort order of the scores
array([60, 91,  5, 62, 26])
>>> sc[ag[:5]]
# top five scores
array([226, 199, 192, 191, 182])
>>> ilines[60][:2]
('cut11', 'cut5')
>>> ilines[60][2]
'AGGAATCAGGGG-TAACGCAGGAAAGACATGTGAGCAAAAGGG-AGCAAAAGGGCAGGAACCCTAAAAAGGCGCGT-
GGTGGGNTTTTC-ATAGGGTCC'
>>> ilines[60][3]
'...AATCAGGGGATAACGCAGGAAAGACATGTGAGCAAAAGGGCAGCAAAAGGGCAGGAACCC-
AAAAAGCCGCGTTGGTGGGNTTTTCCATAGGGTCCCCC'
>>> locs[11]
445
>>> locs[5]
448
```

Code 13-8

At this point the *ilines* contains the alignments of all possible pairings and the *sc* contains the score of these pairings. The *ag* is the sort order from highest to lowest of these alignments. The data is now ready to be assembled.

13.4 Building the Assembly

Before the assembly is constructed, the best alignments are printed in Code 13-9. Each row contains the index number, the score, and the two participating sequences. The logic of Chapter 12 will again be used. Two sequences not yet seen in any contig will form a new contig. An alignment containing a sequence already in a contig and one that is not will add the latter to the contig. Two contigs will be joined if the two sequences are contained in different contigs. If the two aligning sequences are in the same contig, then nothing new will occur. The first alignment is from *ilines[60]*, and it obviously needs to create a new contig.

```
>>> for i in range( 20 ):
      print ag[i], sc[ag[i]], ilines[ag[i]][:2]

60 226 ('cut11', 'cut5')
91 199 ('cut14', 'cut0')
5 192 ('cut3', 'cut2')
62 191 ('cut11', 'cut7')
26 182 ('cut7', 'cut5')
84 174 ('cut13', 'cut6')
32 168 ('cut8', 'cut4')
56 166 ('cut11', 'cut1')
11 164 ('cut5', 'cut1')
94 162 ('cut14', 'cut3')
9 160 ('cut4', 'cut3')
50 151 ('cut10', 'cut5')
65 149 ('cut11', 'cut10')
54 143 ('cut10', 'cut9')
46 137 ('cut10', 'cut1')
74 132 ('cut12', 'cut8')
1 121 ('cut2', 'cut0')
8 117 ('cut4', 'cut2')
37 94 ('cut9', 'cut1')
3 93 ('cut3', 'cut0')
```

Code 13-9

13.4.1 Creating New Contigs

The best alignment is *cut11* with *cut5*, and since this is the first alignment a new contig is created. This is accomplished by **NewCtg**, as shown in Code 13-10. Each contig is a dictionary in which the *key* is the sequence ID and the data is the gapped aligned sequence. In this code a new assembly *smb* is created as an empty list, and this first contig is appended into it. The first alignment is from *ilines[60]*, which is also *ilines [ag[0]]*. Eventually, the *ag* notation will have to be adopted to automate the assembly, but for the next few codes both will be shown.

The second and third best alignments include sequences that are not in previous contigs and so these create two new contigs, as shown in Code 13-11.

```
# gapalign.py
def NewCtg( linei ):
    ctg = { }
    id1, id2 = linei[:2]
    ctg[id1] = linei[2]
    ctg[id2] = linei[3]
    return ctg

>>> ctg = NewCtg( ilines[60] )   # ctg = NewCtg( ilines[ag[0]] )
>>> smb = []
>>> smb.append( ctg )
```

Code 13-10

```
>>> ctg = NewCtg( ilines[91] )   # ag[1]
>>> smb.append( ctg )
>>> ctg = NewCtg( ilines[5] )    # ag[2]
>>> smb.append( ctg )
```

Code 13-11

At this point there are three contigs in the assembly, and it is prudent to display them. Since the contigs are designed differently, a new display function is needed. The function **ShowMe** in Code 13-12 replaces **greedy.ShowContigs**. It allows for sequence names that are up to 10 spaces long. The *spcs* variable is the number of spaces needed to start the aligning sequences at the same place in the print.

```
# gapalign.py
def ShowMe( smb, start=0 ):
    N = len( smb )
    for i in range( N ):
        K = smb[i].keys()
        for k in K:
            spcs = 10-len(k)
            print k,' '*spcs, smb[i][k][start:start+50]
        print ''

>>> ShowMe( smb )
cut5         ...AATCAGGGGATAACGCAGGAAAGACATGTGAGCAAAAGGGCAGCAAA
cut11        AGGAATCAGGGG-TAACGCAGGAAAGACATGTGAGCAAAAGGG-AGCAAA

cut14        AAATTGTTATCCGCTCACAATTCCACACAACATACGAGCCGGAAGCATAA
cut0         ................AATTCCACACAAC-TACGAGCCGGAAGCATAA

cut2         ............ATCATGGTCATAGCGTTTCCTGTGTGAA-TTGTTATC
cut3         ATAGCTTGGCGTAATCATG-TCATAGCGTTTCCTGTGTGAAATTGTTATC
```

Code 13-12

13.4.2 Adding to a Contig

Code 13-13 shows the next case in which one of the sequences already exists in a contig. In this case it will be necessary to add *cut7* to the first contig, which already contains *cut11*. As in the ungapped case, it is necessary to identify in which contig the sequences reside. The **Finder** program in Code 13-14 will identify in which contig a sequence resides. The *f1* is −1, which means that *id1* does not exist in any contig, but *f2* is 1, which indicates that *smb[1]* contains the sequence of *id2*. In the example, *cut11* is found in the contig *smb[0]*.

```
>>> ilines[62][:2]    # ilines[ag[3]}[:2]
('cut11', 'cut7')
```

Code 13-13

```
# gapalign.py
def Finder( smb, askid ):
    N = len( smb )
    hit = -1
    for i in range( N ):
        if askid in smb[i].keys():
            hit = i
            break
    return hit
```

```
>>> id1, id2 = ilines[ag[3]][:2]
>>> f1 = Finder( smb, id1 )
>>> f2 = Finder( smb, id2 )
>>> f1, f2
(0, -1)
```

Code 13-14

In Chapter 12 the addition of a sequence to a contig was straightforward. It was aligned and inserted. The major complication of the gapped assembly occurs at this point. Figure 13-1 shows the problem in which the sequence already in the contig and the sequence that comes from *ilines* have different sets of gaps, and these need to be reconciled. Because there are two sequences involved, the sequence that is also contained in the contig is identified as *matseq* in order to alleviate confusion. This is not the sequence that is in the contig but rather the one in *ilines* that has the same name as the one in the contig. The *umatseq* is the sequence that does not match with a contig. This example uses *cut11* as the match sequence since it is in contig *smb[0]*, as seen in Code 13-15.

```
>>> matseq = ilines[ag[3]][2]  # cut11
>>> umatseq = ilines[ag[3]][3] # cut7
```

Code 13-15

It is now necessary to align the *matseq* with *smb[f1][id1]*, the sequence that exists in the contig. Code 13-16 performs the first part of the dynamic programming. This aligns the *cut11* sequence in the contig with the *cut11* sequence from *ilines*. Code 13-17 calls the backtrace function to create aligned sequences and also to return the positions of the gaps that were inserted.

```
>>> bv = dynprog.FastSubMatrix( submat, abet, smb[fm][idm], matseq )
>>> scm, ar = dynprog.FastSW(bv, smb[fm][idm], matseq )
```

Code 13-16

```
>>> t1,t2,vgap,hgap = SWBacktrace2( scm, ar, smb[f1][id1], matseq )
>>> vgap
[89, 76, 43]
>>> hgap
[87, 16, 15, 14, 13, 12, 11, 10, 9, 8, 7, 6, 5, 4, 3, 2, 1, 0]
```

Code 13-17

Figure 13-2 shows the contig in the assembly, which contains *cut11*, and the *ilines[62]*, which contains *cut11*. These two versions of *cut11* have different gaps. Because the alignment between the two differs only in gaps, it creates *t1* and *t2*, which are identical versions of *cut11* with both sets of gaps.

13.4 Building the Assembly

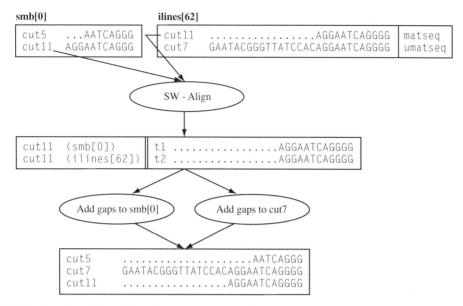

FIGURE 13-2 Flow of the addition of a sequence to a contig.

If *hgap* contains elements, then it is necessary to insert gaps at those positions for all sequences in the contig. The first step is to create the function **GapSeqByList** (Code 13-18), which receives a single sequence and inserts gaps according to the list *gaplist*. Since Python does not allow the insertions of characters into a string, the string is converted to a list. The gaps are inserted into the list—a process that must be done in order—and then the list is converted back to a string. This function is repeatedly called by **GapCtgByList** (Code 13-19) to insert gaps into all of the sequences in a contig.

```
# gapalign.py
def GapSeqByList( seq, gaplist ):
    glist.sort()
    temp = list( seq ) # convert to list
    for i in gaplist:
        temp.insert( i, '-' )
    nseq = ''.join( temp ) # convert to string
    return nseq
```
Code 13-18

```
# gapalign.py
def GapCtgByList( ctg, gaplist ):
    k = ctg.keys()
    for i in k:
        ctg[i] = GapSeqByList( ctg[i], gaplist )
        ctg[i] = LeadTrailGaps( ctg[i] )

>>> GapCtgByList( smb[f1], hgap )
```

```
>>> ShowMe( smb, 0 )
cut5                  ....................AATCAGGGGATAACGCAGGAAAGACATGTG
cut11                 ..................AGGAATCAGGGG-TAACGCAGGAAAGACATGTG

cut14                 AAATTGTTATCCGCTCACAATTCCACACAACATACGAGCCGGAAGCATAA
cut0                  ..................AATTCCACACAAC-TACGAGCCGGAAGCATAA

cut2                  ..............ATCATGGTCATAGCGTTTCCTGTGTGAA-TTGTTATC
cut3                  ATAGCTTGGCGTAATCATG-TCATAGCGTTTCCTGTGTGAAATTGTTATC
```

Code 13-19

The sequence to be added to the contig is *umatseg*, or in this case *cut7*. From *ilines[62]* it contains gaps. However, it will need new gaps defined in *vgap* in order to align with the contig. Code 13-20 gaps this sequence and then converts the leading and trailing gaps to spaces. Finally, this new sequence is added to the proper contig in the assembly.

```
>>> newseq = GapSeqByList( umatseq, vgap )
>>> newseq = LeadTrailGaps( newseq )
>>> smb[f1][id2] = newseq

>>> ShowMe( smb, 0 )
cut5                  ....................AATCAGGGGATAACGCAGGAAAGACATGTG
cut7                  GAATACGGGTTATCCACAGGAATCAGGGGATAACGCAGGAAAG-ACAT-T
cut11                 ..................AGGAATCAGGGG-TAACGCAGGAAAGACATGTG

cut14                 AAATTGTTATCCGCTCACAATTCCACACAACATACGAGCCGGAAGCATAA
cut0                  ..................AATTCCACACAAC-TACGAGCCGGAAGCATAA

cut2                  ..............ATCATGGTCATAGCGTTTCCTGTGTGAA-TTGTTATC
cut3                  ATAGCTTGGCGTAATCATG-TCATAGCGTTTCCTGTGTGAAATTGTTATC

>>> ShowMe( smb, 30 )
cut5                  TAACGCAGGAAAGACATGTGAGCAAAAGGGCAGCAAAAGGGCAGGAACCC
cut7                  TAACGCAGGAAAG-ACAT-TGACAAAAGGGCAGCAAAAGGGCAGGA-ACC
cut11                 TAACGCAGGAAAGACATGTGAGCAAAAGGG-AGCAAAAGGGCAGGAACCC

cut14                 CATACGAGCCGGAAGCATAAAGTG-AAAGCCTGGGGTGCCTAATGAGTA-
cut0                  C-TACGAGCCGGAAGCATAAAGTGTAAAGCCTGGGGTGCCTAATGAGGAG

cut2                  TCCTGTGTGAA-TTGTTATCCGCTCACAATTCCACACAACATACGAGCCG
cut3                  TCCTGTGTGAAATTGTTATCCGCTCACAATTCC-CACAACATACGAGCCG
```

Code 13-20

These steps are combined into a single function, **AddToCtg**, as shown in Code 13-21. This function must receive the assembly (*smb*), the identity of the contig that is to be altered (*fm*), the identities of the matching and unmatching sequences (*idm* and *idum*), the tuple from *ilines* (*ilinei*), and the substitution matrix and alphabet.

```
# gapalign.py
def AddToCtg(smb, fm, idm, idum, ilinei, submat, abet ):
    print 'AddToCtg', idum, 'added to ', fm
    if ilinei[0]==idm:
        matseq = ilinei[2] # the seq that is also in the smb
        umatseq = ilinei[3]
```

```
        else:
            matseq = ilinei[3] # the seq that is also in the smb
            umatseq = ilinei[2]
        # align match seq with the same seq in smb
        bv = dynprog.FastSubMatrix( submat, abet, smb[fm][idm], matseq )
        scm, ar = dynprog.FastSW(bv, smb[fm][idm], matseq )
        t1,t2,vgap,hgap = SWBacktrace2( scm, ar, smb[fm][idm], matseq )
        # gap contig
        GapCtgByList( smb[fm], hgap )
        # create and add new sequence
        newseq = GapSeqByList( umatseq, vgap )
        newseq = LeadTrailGaps( newseq )
        smb[fm][idum] = newseq
```

Code 13-21

Code 13-22 brings the assembly up to date. The string *cut7* has been added to the assembly from Code 13-20. The next alignment pair is from *ilines[ag[4]]*, which contains *cut7* and *cut5*. Since these are already in *smb[0]*, no action is taken. The contigs *ag[5]* and *ag[6]* present two alignments that require two new contigs. The next two present alignments having pairs that already exist in a contig. It is *ag[9]* that requires that new steps be taken.

```
>>> # ag[4] is cut7 and cut5 already together in smb[0]
>>> smb.append( NewCtg( ilines[ag[5]] ) )
NewCtg ('cut13', 'cut6')
>>> smb.append( NewCtg( ilines[ag[6]] ) )
NewCtg ('cut8', 'cut4')
>>> # ag[7] and ag[8] already in smb[0]
```

Code 13-22

13.4.3 Joining Contigs

The assembly at this point is shown in Code 13-23. There are five contigs. The alignment *ag[9]*, which is shown in Code 13-24, contains *cut14* and *cut3*, which exist in different contigs. Therefore, it is necessary to join the two contigs.

```
>>> ShowMe( smb, 30 )
cut5        TAACGCAGGAAAGACATGTGAGCAAAAGGGCAGCAAAAGGGCAGGAACCC
cut7        TAACGCAGGAAAG-ACAT-TGACAAAAGGGCAGCAAAAGGGCAGGA-ACC
cut11       TAACGCAGGAAAGACATGTGAGCAAAAGGG-AGCAAAAGGGCAGGAACCC

cut14       CATACGAGCCGGAAGCATAAAGTG-AAAGCCTGGGGTGCCTAATGAGTA-
cut0        C-TACGAGCCGGAAGCATAAAGTGTAAAGCCTGGGGTGCCTAATGAGGAG

cut2        TCCTGTGTGAA-TTGTTATCCGCTCACAATTCCACACAACATACGAGCCG
cut3        TCCTGTGTGAAATTGTTATCCGCTCACAATTCC-CACAACATACGAGCCG

cut6        ATCGGCCAACGCGCGGGGAGAGGCGGTTTGCGTATTGGG-GCTCTCCGCT
cut13       .TCGGCCAACGCGCGGGGA-AGGCGGTTTGCGTATTGGGCGCTCTTCCCT

cut4        AAGCTTGAGTATTCTATAGTGTCACCTAAATAGCTTGGCGTAAT-ATGGT
cut8        AAGC-TGAGTATTCTATAGTGTCACCTAAA-AGCTTGG-GTAATCATGGT
```

Code 13-23

```
>>> ilines[ag[9]][:2]
('cut14', 'cut3')
>>> id1, id2 = ilines[ag[9]][:2]
>>> f1 = Finder( smb, id1 )
>>> f2 = Finder( smb, id2 )
>>> f1, f2
(1, 2)
```

Code 13-24

The alignment of two contigs is more complicated than in Chapter 12 but not for a new reason. In this case *ilines[ag[9]]* contains two sequences that have been aligned by gaps. These sequences exist in contigs with a different set of gaps. In order to join two contigs, it will be necessary to reconcile the gaps in the contig sequences with the gaps in the *ilines* sequences. Fortunately, this does not require any new tools. The function **JoinCtgs** in Code 13-25 receives the assembly, the identities of the two contigs, the identities of the two sequences, and the substitution matrix and alphabet. It aligns the two sequences that exist in the assemblies, which creates the lists of gaps necessary to align the contigs. Each contig is gapped, and the second contig is added to the first in the *for* loop. The final command destroys the second contig.

```
# gapalign.py
def JoinCtgs( smb, f1, f2, id1, id2, submat, abet):
    print 'JoinCtgs', f1, f2
    bv = dynprog.FastSubMatrix( submat, abet, smb[f1][id1], smb[f2][id2] )
    scm, ar = dynprog.FastSW(bv, smb[f1][id1], smb[f2][id2] )
    t1,t2,vgap,hgap = SWBacktrace2( scm, ar,smb[f1][id1], smb[f2][id2] )
    GapCtgByList( smb[f1], hgap )
    GapCtgByList( smb[f2], vgap )
    for i in smb[f2].keys():
        smb[f1][i] = smb[f2][i]
    del( smb[f2] )
```

Code 13-25

The example in Code 13-26 joins the two contigs that contained *cut14* and *cut3*. The results are shown.

```
>>> JoinCtgs( smb, f1, f2, id1, id2, GDNA, GBET )
>>> ShowMe( smb, 30 )
cut5        TAACGCAGGAAAGACATGTGAGCAAAAGGGCAGCAAAAGGGCAGGAACCC
cut7        TAACGCAGGAAAG-ACAT-TGACAAAAGGGCAGCAAAAGGGCAGGA-ACC
cut11       TAACGCAGGAAAGACATGTGAGCAAAAGGG-AGCAAAAGGGCAGGAACCC

cut14       .........AAATTGTTATCCGCTCACAATTCCACACAACATACGAGCCG
cut0        .......................AATTCCACACAAC-TACGAGCCG
cut2        TCCTGTGTGAA-TTGTTATCCGCTCACAATTCCACACAACATACGAGCCG
cut3        TCCTGTGTGAAATTGTTATCCGCTCACAATTCC-CACAACATACGAGCCG

cut6        ATCGGCCAACGCGCGGGGAGAGGCGGTTTGCGTATTGGG-GCTCTCCGCT
cut13       .TCGGCCAACGCGCGGGGA-AGGCGGTTTGCGTATTGGGCGCTCTTCCCT

cut4        AAGCTTGAGTATTCTATAGTGTCACCTAAATAGCTTGGCGTAAT-ATGGT
cut8        AAGC-TGAGTATTCTATAGTGTCACCTAAA-AGCTTGG-GTAATCATGGT
```

Code 13-26

13.4.4 Performing the Assembly

All of the tools are now in place. The alignment pairs are considered in order according to the alignment score. Each consideration creates a new contig, adds to a contig, joins two contigs, or does nothing. Similar to **greedy.Assemble**, the **Assemble** function in Code 13-27 performs these tasks. It receives the segments, scores, *ilines*, minimum alignment score, and substitution information. The minimum alignment score is the lowest score allowed for two sequences to be considered sufficiently aligned. The list *notused* keeps track of those sequences not used in any contig. At the end of the function, each of these is used to create a contig with a single sequence, and it is appended to the assembly.

```
# gapalign.py
def Assemble( seqs, sc, ids, lines, mnsc, submat, abet ):
    smb = []
    ok = 1
    ag = argsort( sc )[::-1]
    k = 0
    notused = copy.copy( ids )
    while ok:
        # get the best score
        i = ag[k]
        if sc[i]> mnsc:
            # find the elements in contigs
            id1, id2 = lines[i][:2]
            f1 = Finder( smb, id1 )
            f2 = Finder( smb, id2 )
            # decide to create, add, or join
            if f1==-1 and f2 == -1:
                # create
                smb.append( NewCtg( lines[i] ))
            if f1 != -1 and f2 == -1:
                # the id1 is in ctg (smb[f1] )
                AddToCtg( smb, f1, id1, id2, lines[i], submat, abet)
            if f1==-1 and f2 != -1:
                AddToCtg( smb, f2, id2, id1, lines[i], submat, abet )
            if f1 != -1 and f2 != -1 and f1 != f2:
                # Join 2 contigs
                JoinCtgs( smb, f1, f2, id1, id2, submat, abet )
            if id1 in notused:
                notused.remove( id1 )
            if id2 in notused:
                notused.remove( id2 )
        k += 1
        if k>= len( ag ): ok = 0
    # add unused sequences
    for i in notused:
        ndx = ids.index( i )
        ctg = {i: seqs[ndx]}
        smb.append( ctg )
    return smb
```

Code 13-27

Code 13-28 shows the use of this command with the *mnsc=160*. The assembly creates seven contigs, and these should be compared to Code 13-3.

```
>>> smb = Assemble( segs, sc, ids, ilines, 160, GDNA, GBET )
>>> ShowMe( smb, 30 )
cut5        TAACGCAGGAAAGACATGTGAGCAAAAGGGCAGCAAAAGGGCAGGAACCC
cut7        TAACGCAGGAAAG-ACAT-TGACAAAAGGGCAGCAAAAGGGCAGGA-ACC
cut1        TAACGCAGGA-AGACATGTG--CAAAGGGC-AGC-AAAGGGCAGGAACCC
cut11       TAACGCAGGAAAGACATGTGAGCAAAAGGG-AGCAAAAGGGCAGGAACCC

cut14       .........AAATTGTTATCCGCTCACAATTCCACACAACATACGAGCCG
cut0        ........................AATTCCACACAAC-TACGAGCCG
cut2        TCCTGTGTGAA-TTGTTATCCGCTCACAATTCCACACAACATACGAGCCG
cut3        TCCTGTGTGAAATTGTTATCCGCTCACAATTCC-CACAACATACGAGCCG

cut6        ATCGGCCAACGCGCGGGGAGAGGCGGTTTGCGTATTGGG-GCTCTCCGCT
cut13       .TCGGCCAACGCGCGGGGA-AGGCGGTTTGCGTATTGGGCGCTCTTCCCT

cut4        AAGCTTGAGTATTCTATAGTGTCACCTAAATAGCTTGGCGTAAT-ATGGT
cut8        AAGC-TGAGTATTCTATAGTGTCACCTAAA-AGCTTGG-GTAATCATGGT

cut9        CGGCGAGCGGTATCAGCTCACTCAAAGGCGGGTAATACGGGTTATCACAG

cut10       CGGGTATCCACAGGAATCGGGGATAACGCAGGAAAGACATGTGAGCAAAA

cut12       CGGGGATCCTCTAGAGTCGACCTCAGGATGCAAGCTTGAGTATTCTATAG
```

Code 13-28

13.5 Summary

The previous chapter created a multiple sequence alignment, but the sequences did not contain gaps. This chapter considered the case of such an alignment with the addition of gaps. The problem that must be abated is that the gaps inserted by alignments of sequences *s1* and *s2* will affect the alignment of *s2* and *s3*. This chapter addresses these issues and constructs Python code for multiple, gapped alignments.

Bibliography

Thompson, J. D., D. G. Higgins, T. J. Gibson. (1994). CLUSTALW: Improving the sensitivity of progressive multiple sequence alignment position through sequence weighting, position-specific gap penalties and weight matrix choice. *Nucleic Acids Res.* 22(22), 4673–4680.

Problems

1. Given a string *st* that is sent to the **ChopFunny** program, write a program that computes the number of times that each character in the original string appears in the *segs* returned by the function.

2. Using Problem 1, create two plots. The first fixes the value of *length* and changes the value of *nsegs*. Plot the average number of times each character is used (Problem 1 program) versus *nsegs*. The second plot fixes the value of *nsegs* and plots the average number of times each character is used versus *length*.

3. Perform alignments using different scoring matrices (BLOSUM, PAM, etc.). Do the different matrices affect the alignment?

4. Perform alignments using multiple values of the gap penalty. Describe how the gap penalty affects the alignments.

5. Write a program that limits the density of gaps during the assembly. In other words, sequences do not join a contig if the density of gaps is greater than a user-defined threshold. Likewise, contigs are not joined if the gap density is greater than the same threshold.

14 Trees

Some applications require the knowledge of how items can be linked together given specific scoring functions. These applications and many others use trees to sort and organize the data in a very efficient manner. The tree obtains this efficiency by linking data such that searches through the data follow short paths. At each node in the tree the search makes a simple decision to create this path. Tree searches are highly efficient means of organizing and searching data.

14.1 Basic Tree Theory

Tree theory starts with the simpler concept of *linked lists*. Consider a case in which data arrives in no particular order but must be sorted while it arrives. In Python it is easy to put data into a list and sort it. However, if sorting is required each time that data arrives and/or the list becomes large, this approach becomes very slow. Lists are indeed a wonderful concept but they can be quite cumbersome for large problems.

Figure 14-1 shows two steps in this process. In the first, three data items each have a measurement T and are sorted by T. In the next step, a fourth data item arrives and needs to be placed between Items 2 and 3. It is possible to move Item 3 over to the right and insert Item 4 between them. In the computer this would require moving data for Item 3 to make room for Item 4 and moving data (especially large chunks of data) is a time-consuming process.

The solution is to use a linked list. Each item has an arrow that points to the next item according to T. The items are stored in memory in the order that they arrived, and only these arrows are changed. In this manner it is possible to sort the list without moving data items about.

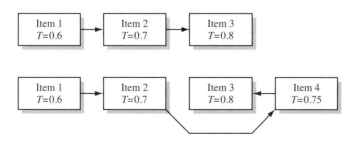

FIGURE 14-1 Two steps in a sorting process. The first has three items, and the second shows how a new item arrives and the arrows change to realize the proper sorting order.

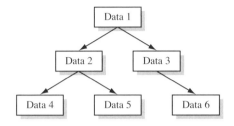

FIGURE 14-2 A typical binary tree.

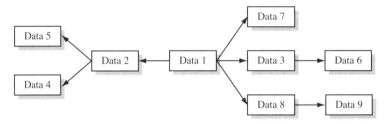

FIGURE 14-3 An unrooted tree.

A tree is not much different than a linked list except that each item has two (or more) arrows. Figure 14-2 shows a typical *binary tree* in which each item can have up to two arrows connecting it to other items. Furthermore, items only receive one arrow. The tree shown in Figure 14-2 shows a *rooted tree* with a block that is the starting point. Figure 14-3 shows an *unrooted tree*.

14.2 Python and Trees

There are several ways that a tree can be created in Python. For trees in which the connections have a single direction, a dictionary is a powerful tool to use. Each item in the dictionary has a key and data. The key is the node's identification, and the data includes the links to the other items and information such as T-values in Figure 14-1. The tree in Figure 14-4 contains six different nodes and each one contains some data (the T value) and possible links to two other nodes.

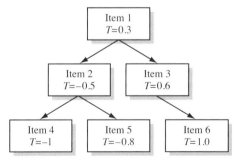

FIGURE 14-4 A tree with six nodes having data associated with Code 14-1.

The information of a single node is contained in a list that has three items: *[T, left-child, right-child]*. The children are the identifications of the subsequent nodes. The first node would then have the list *[0.3, 2, 3]*, as shown in Code 14-1. Many of the nodes do not have children, and in this case a -1 is used to show that there is no connection.

```
>>> tree = { }
>>> tree[1] = [0.3, 2, 3]
>>> tree[2] = [-0.5, 4, 5 ]
>>> tree[3] = [0.6, 6, -1]
>>> tree[4] = [-1, -1, -1]
>>> tree[5] = [-0.8, -1, -1]
>>> tree[6] = [1, -1, -1]
```

Code 14-1

This type of code works for a binary tree. Other trees will require modifications. For example, some trees will have an unlimited number of children. In this case a list can be used to store all of the children's identifications by a single data structure. Other cases may require travel in both directions inside the tree. In these cases it is important to have each node keep the identification of its parent.

14.3 An Example Using UPGMA

The UPGMA (Unweighted Pair Group Method with Arithmetic Mean) algorithm builds a simple tree in a greedy fashion. Each pair of data are compared and scored. Those with the best scores will be joined to either form a new tree or to join an existing tree.

A simple example is used to demonstrate this system. In Code 14-2 six data vectors are created at random. They are compared using a simple absolute subtraction, and the matrix **M** contains the scores of the comparisons. The highest score indicates better similarity.

```
>>> from numpy import random, zeros
>>> data = random.ranf( (6,10) )
>>> M = zeros( (6,6), float )
>>> for i in range( 6 ):
        for j in range( i ):
            M[i,j] = 10 - (abs( data[i]-data[j])).sum()

>>> M
array([[ 0. ,  0. ,  0. ,  0. ,  0. ,  0.   ],
       [ 7.2,  0. ,  0. ,  0. ,  0. ,  0.   ],
       [ 6.1,  5.4,  0. ,  0. ,  0. ,  0.   ],
       [ 5.6,  5.5,  6.9,  0. ,  0. ,  0.   ],
       [ 6.0,  6.1,  7.0,  7.4,  0. ,  0.   ],
       [ 6.8,  8.5,  6.3,  6.3,  7.0,  0.   ]])
```

Code 14-2

In this example the best score is 8.5 and belongs to column 1 and row 5. Therefore, the best matching data vectors are *data[1]* and *data[5]*. The UPGMA creates a small tree from these two data vectors, as shown in Figure 14-5, where each *data* vector is represented by *V*.

FIGURE 14-5 A small tree in which the first two nodes are connected.

At the top of this tree is V7, which is not part of the original data. This is an artificial data vector created from the average of V1 and V5. This new data vector is added to the other data vectors, and V1 and V5 are both removed from further consideration.

This maneuver will require that **M** be of a bigger size. In fact, in the UPGMA algorithm the size of M is $(2N - 1) \times (2N - 1)$ where N is the number of original data vectors. Code 14-3 initializes M to this new size and fills it with the scores shown in Code 14-2. The maximum value is located and returned as location v, h. Furthermore, there is going to be a need for $(2N - 1)$ data vectors, and these are established as *vecs*.

```
>>> M = zeros( (11,11), float )
>>> for i in range( 6 ):
        for j in range( i ):
            M[i,j] = 10 - (abs( data[i]-data[j])).sum()
>>> v,h = divmod( M.argmax(), 11 )
>>> v,h
(5, 1)
>>> vecs = zeros( (11,10), float )
>>> for i in range( 6 ):
        vecs[i] = data[i] + 0
```

Code 14-3

Figure 14-5 illustrates that a new vector, *vecs[7]*, is created, which is shown in Code 14-4. The *i*-loop computes the score of this new vector compared to the others. The final four commands eliminate all rows and columns that are associated with *vecs[1]* and *vecs[5]*. The variable *last* keeps track of the last known vector in the list, and it increments with each new vector.

```
>>> last = 7
>>> vecs[last] = (vecs[v] + vecs[h])/2.0
>>> for i in range( last ):
        M[last,i] = 10 - (abs( vecs[i]-vecs[last])).sum()
>>> M[v] = zeros(11)
>>> M[h] = zeros(11)
>>> M[:,v] = zeros(11)
>>> M[:,h] = zeros(11)
>>> last += 1
```

Code 14-4

The next iteration finds the new largest value in **M** and repeats the process. In this example *vecs[3]* and *vecs[4]* generate the best match, and so a new tree is created with these two. This is shown in Figure 14-6. On the third iteration the best match is between *vecs[2]* and *vecs[8]*, however, *vecs[8]* is already in a tree. Thus, *vecs[2]* is

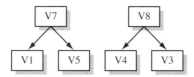

FIGURE 14-6 The second iteration.

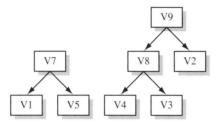

FIGURE 14-7 The third iteration.

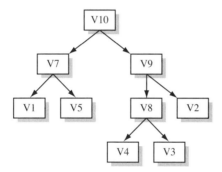

FIGURE 14-8 The fourth iteration.

attached to the existing tree creating *vecs[9]*. This is shown in Figure 14-7. The final type of iteration is one in which both of the vectors exist in different trees. In this case the two trees are joined together as shown in Figure 14-8.

The UPGMA algorithm is shown in Code 14-5. The input data *indata* is a list of the data vectors (not a matrix). The *scmat* is the matrix that contains the pairwise scores (similar to the **M** matrix in previous examples). The list *net* collects the nodes as they are computed. The list *used* collects the names of data vectors after they have been used to prevent their reuse. In the major *i*-loop the best match is found in the **divmod** statement, which returns the location in **M** where the best match occurs. It is appended to the *net*, and the average of the two constituent vectors is computed as *avg*. The *j*-loop computes the similarity of the new vector with the existing unused vectors. The final command removes the comparison scores for the two vectors that are being removed from further consideration.

In the example at the end of Code 14-5 six random data vectors of length 10 are used as inputs. The *tree* is computed and printed. Recall that the tree is a dictionary

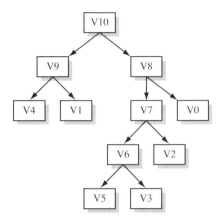

FIGURE 14-9 A tree illustrating the results in Code 14-5.

and that the data of the dictionary contains the two children and the score. The tree produced by this system is shown in Figure 14-9.

```
# upgma.py
def UPGMA( indata ):
    data = copy.deepcopy( indata )
    # data is a list of vectors
    N = len( data )  # number of data vectors
    N2 = 2*N-1
    BIG = 999999
    scmat = zeros( (N2,N2), float ) + BIG
    # initial pairwise comparisons
    for i in range( N ):
        for j in range( i ):
            scmat[i,j] = (abs( data[i]-data[j] )).sum()
    #
    tree, used = {}, []
    for i in range( N-1 ):
        v,h = divmod( scmat.argmin(), N2 )
        tree[N+i] = (v, h, scmat.min() )
        used.append( v )
        used.append( h )
        avg = ( data[v] + data[h])/2.
        data.append( avg )
        for j in range( N+i ):
            if j not in used:
                scmat[N+i,j] = (abs( avg-data[j] )).sum()
        scmat[v] = zeros( N2 ) +BIG
        scmat[h] = zeros( N2 ) +BIG
        scmat[:,v] = zeros( N2 ) +BIG
        scmat[:,h] = zeros( N2 ) +BIG
    return tree

>>> from numpy import set_printoptions
>>> set_printoptions( precision = 3 )
>>> data = []
>>> for i in range( 6 ):
      data.append( random.rand( 10 ))

>>> net = UPGMA( d )
>>> for i in net.keys():
      print i,net[i]
```

```
8 (7, 0, 2.380)
9 (4, 1, 3.257)
10 (9, 8, 2.728)
6 (5, 3, 2.260)
7 (6, 2, 2.270)
```

Code 14-5

14.4 Examples of Trees

Trees are used for a variety of applications and therefore come in a variety of forms. This section will review a few of the most popular trees.

14.4.1 Sorting Trees

Python (*numpy*) offers the **argsort** command, which will sort data from lowest to highest and generally does so faster than any program that is scripted in Python. In the case where all data is present before any sorting is required, the **argsort** command will be the best approach. There are cases, however, where building the code for the tree will be better. Consider a scenario in which an event occurs (name it E0) and triggers a reaction, and then three other events (E1, E2, E3) occur at later times. These four events create a small tree. Each of these new events will trigger other events that will occur at various times. In this application it is necessary to consider each event in order of the time that it occurs. Now, consider a case where E1 triggers events E4 and E5, and in this case, E4 will occur before E3. The events are added to the tree, and each time an event occurs it is removed from the tree. The tree will keep the data sorted according to the time that it should occur. The function **argsort** will sort the entire list each time that it is called. In this example, it is only necessary to properly place the new events into the tree of sorted events. In cases where there is a very large number of events, the tree will be faster than calling **argsort**.

In this example the task of the tree is to sort the data according to a parameter T. The data is added to the tree sequentially but not in a sorted order. The process starts the tree with the first data, and the second data is attached as the left child if it has a lower T value and as the right child if it has a larger T value. In Code 14-6 a set of random values is generated, and these are to be put into a tree. The top of the tree (root node) is *data[0]*. *data[1]* is a higher value and so it is attached as the right child, and *data[2]* is lower and so it is attached as the left child. *data[3]* is also lower, but there is already a left child and so it considers it for attachment. Because *data[3]* is less than *data[1]*, it is attached as the left child. Progress at this point is shown in Figure 14-10.

```
>>> data = random.rand( 13 )
>>> data[0]
0.66
>>> data[1]
0.84
>>> data[2]
0.64
>>> data[3]
0.58
```

Code 14-6

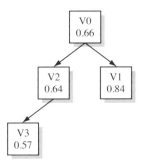

FIGURE 14-10 The tree constructed from the data shown in Code 14-6.

The process continues. Each new node moves down the tree to the left or right, depending on whether it is less than or greater than the node's value. When it comes to a node that doesn't have an appropriate child then the new node is attached. Once a node in this tree is attached, it is not moved. The final tree for this randomly generated data is shown in Figure 14-11. The removal (or removal and reattachment) process continues until all of the nodes in the tree have been removed. This is the sorted order for the data.

Code 14-7 displays the function **AddNode**, which will find the correct location in a tree to add the node. It receives the tree that will be modified, the ID of the first node, the ID of the incoming node, and the incoming data. The integer k keeps track of the node being considered. Inside the loop it sets *go* to indicate to which side the new node goes. If there is no child already there, then the new node is created with $[-1, -1,$

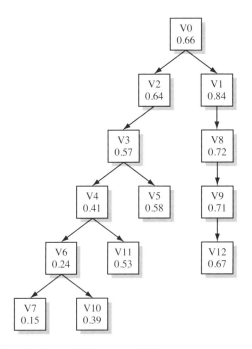

FIGURE 14-11 The final tree for the randomly generated data.

newdata] and attached to the tree. Otherwise, the *k* is changed to the ID of the child node, and the instructions inside the *while* loop are repeated. To run this program, an empty dictionary is created. The first node is manually installed, and then each piece of data is added sequentially.

```
# sorttree.py
def AddNode( tree, top, newid, newdata ):
    # top is the ID of the first node
    # newid and newdata are the ID and data for new node
    k = top
    ok = 1
    while ok:
        # decide if the new node goes to the left or right
        go = 0 # left is the default
        if newdata > tree[k][2]:
            go = 1 # instead go to the right
        # is there a child hanging here
        if tree[k][go] == -1:
            # no child exists
            tree[k][go] = newid
            tree[newid] = [-1, -1, newdata ]
            ok = 0
        else:
            # there is a child. move down the tree
            k = tree[k][go]

>>> tree = {}
>>> tree[0] = [-1,-1,data[0]]
>>> for i in range( 4, 13 ):
        AddNode( tree, 0, i, data[i] )
```

Code 14-7

From this tree it is easy to sort the data. The lowest value is the left-most end node. To find the second smallest value, the smallest node is removed from the tree and the process repeats. In this case the next smallest is V6. Although V6 does have a child, it is the right-child and therefore is not considered in this search. However, removing V6 would also remove V10, and so in this case V6 is removed and V10 is attached to the parent of V6, as shown in Figure 14-12.

Two functions are needed in order to read the tree. The first is to find the lowest node, and the second is to remove this node. Code 14-8 shows the function **FindLefty**, which finds the left-most node. Basically, *k* is the top node and the loop iterates to the left child until a node having no left child is found. This function returns two items. The first is the ID of the leftmost node and the second is the ID of the parent of that node.

```
# sorttree.py
def FindLefty( tree, top ):
    k, mom = top, -1
    ok = 1
    while ok:
        if tree[k][0] == -1:
            ok = 0
            return k, mom
```

```
        else:
            mom = k
            k = tree[k][0]
>>> FindLefty( tree, 0 )
7, 6
```

Code 14-8

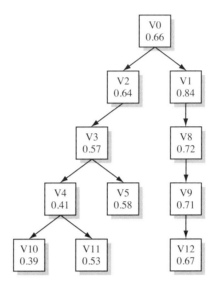

FIGURE 14-12   The tree after removing V7 and V6.

Code 14-9 shows **RemoveNode**, a function that removes the specified node and replaces it with a right-child if one exists. This function assumes that the node being removed does not have a left-child. The function does need to consider differently the case in which the top node is being replaced (*mom* = −1). If the top node is the one being removed, then its right-child becomes the new top node. This requires that *top* be returned from **RemoveNode** to catch this change. The call to the function sets the top node and iteratively calls **FindLefty** and **RemoveNode**. It should be noted that the tree is being altered, and at the end of this *while* loop the tree is reduced to an empty dictionary. It may be prudent to keep a copy of the tree (using **copy.deepcopy**) if its use is required for further calculations. The example prints out a list of node IDs and the *T* values in the sorted order.

```
# sorttree.py
def RemoveNode( tree, loc, mom, top ):
    # if the node has a right child then perform a transplant
    rchild = tree[loc][1]   # the right child
    if rchild != -1 :
        if mom != -1:
            tree[mom][0] = tree[loc][1]
        else:
            top = tree[loc][1]
```

```
        elif mom != -1:
            tree[mom][0] = -1   # left child no longer exists
        # remove the node
        trash = tree.pop( loc )
        return top

>>> top = 0
>>> while len( tree )>0:
        me, mom = FindLefty( tree, top )
        print me, "%.2f"%tree[me][2]
        top = RemoveNode( tree, me, mom, top )

7 0.15
6 0.25
10 0.40
4 0.41
11 0.53
3 0.58
5 0.59
2 0.64
0 0.66
12 0.67
9 0.71
8 0.72
1 0.84
```

Code 14-9

14.4.2 Dictionary Trees

Spell-checking in a word processor is a common function. Because the number of words in a language is very large, it is not prudent to search this list for each word in a document to determine if the spelling is correct. To facilitate this search, a *dictionary tree* can be used.

Consider a language of only four words: CAT, CART, COB, and COBBLE. The dictionary tree for these words is shown in Figure 14-13. In this case valid spellings connect nodes in the tree. A valid end to a word is a shaded node. In the case of COB and COBBLE, a node may be both the end of a word and a continuation onto other words.

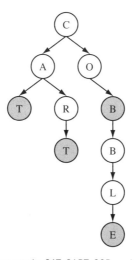

FIGURE 14-13 A dictionary tree for four words: CAT, CART, COB, and COBBLE.

Nodes in this tree are different than in the previous section. A node may have several children and may also be an end node. Thus the node is defined as *[listOfKidLetters, listOfKidIDs, letter, endFlag]*. The *listOfKidLetters* is a list of all of the letters of its kids. The *listOfKidIDs* is a list of the node IDs for these same children. The *letter* is the letter of the current node and the *endFlag* is True if this node represents the end of a word.

In this example all words start with the same letter, which of course is not a real case. It is possible to have 26 different trees each starting with a different letter of the alphabet. It is also possible to have a START node that has no letter but does have 26 children.

Code 14-10 shows the function **AddWord**, which will add a word to the dictionary tree. The input word is compared to the tree. The first *if* statement determines whether the new word has letters that are already in the tree. Inside of the function there is a second *if* statement that determines whether the *word* has reached its last letter. In this case the *endFlag* is set to *True*. If the word has a sequence of letters not in the tree, then the first *if* statement is *False* and nodes are added to the tree to finish out the word.

```
# spelltree.py
def AddWord( tree, word ):
    treeloc = 0
    wi = 0  # location in word
    NW = len( word )
    ok = 1
    while ok:
        if word[wi] in tree[treeloc][0]:
            # this letter is already in the tree
            if wi==NW-1:
                # it is the last letter of the word
                ndx = tree[treeloc][0].index( word[wi] )
                treeloc = tree[treeloc][1][ndx]
                tree[treeloc][3] = True # end of word
                ok = 0
            else:
                # move down the tree
                ndx = tree[treeloc][0].index( word[wi] )
                treeloc = tree[treeloc][1][ndx]
                wi += 1
        else:
            # the rest of the word is not in the tree
            TL = len( tree ) # current tree length
            last = TL
            tree[treeloc][0].append( word[wi] )
            tree[treeloc][1].append( TL )
            for i in range( wi, NW-1 ):
                newnode = [ [word[i+1]], [last+1], word[i], False]
                tree.append( newnode )
                last +=1
            # last letter
            TL = len( tree )
            newnode = [ [], [], word[-1], True ]
            tree.append( newnode )
            ok = 0
```

Code 14-10

To operate this function, it is first necessary to create a tree and establish the first node as the start node. Code 14-11 starts the tree with the first two commands. The list *words* is a list of words to be used in the first example. The *for* loop considers each word and adds it to the tree. There are 47 nodes to this tree of which a few are printed.

```
>>> tree = []
>>> tree.append( [ [], [], '', False] )
>>> words =
['python','pylon','nylon','rayon','crayon','crayola','payola','payment'\
,'pavement','pave','paves','paving']
>>> for i in words:
        AddWord( tree, i )
>>> tree[0]
{['p', 'n', 'r', 'c'], [1, 10, 15, 20], '', False]
>>> tree[1]
{['y', 'a'], [2, 28], 'p', False]
>>> tree[7]
{['o'], [8], 'l', False]
>>> tree[29]
{['o', 'm'], [30, 33], 'y', False]
```

Code 14-11

A part of the tree is shown in Figure 14-14. Not all of the nodes are shown here. For example, the node *tree[3]* has a single chain of nodes (*4, 5, 6*) that have the letters (*h, o, n*). Instead of showing this long chain, the letters of a single chain are listed just below the node. Otherwise, the figure gets more cluttered.

The final example is shown in Code 14-12. In this case Shakespearean sonnets are loaded. The program calls **miner.Hoover**, which removes punctuation and converts all letters to lowercase. A total of almost 30,000 words is reduced to a tree of only 15,000 nodes.

```
>>> import miner
>>> b = miner.Hoover( 'data/sonnets.txt')
>>> words = b.split()
>>> len( words )
29641
>>> tree = []
>>> tree.append( [ [], [], '', False] )
>>> for i in words:
      AddWord( tree, i )

>>> len( tree )
15143
```

Code 14-12

14.4.3 Percolation Trees

One of the problems with a sorting tree is that it can be quite lopsided, with some branches being much longer than others. In the worst case, a sorting tree would be just one long branch, offering no advantage over brute force sorting. The percolation tree offers the advantage of keeping the branches somewhat the same length, but the cost is a slightly more complicated tree.

210 CHAPTER 14 TREES

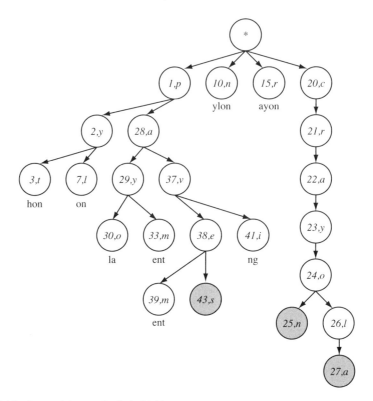

FIGURE 14-14 A part of the tree for Code 14-11.

The philosophy of this tree is best explained through an example. In this case the goal is to sort the data according to the lowest data value. Each node should therefore have a lower data value than its children. When a node is added to the end of a branch, it percolates upward as long as it has a lower value than its current mother. The top node of the tree is the one with the smallest value. To obtain the sorted data, a list appends the top node. This node is then removed, and the tree is adjusted, which then places the node with the next smallest value at the top of the tree.

Before the tree is constructed, a data representation must be solidified. In this case it is necessary to maintain the node's identification, the data value, and the identification of the node's mother as well as the children. It is also necessary to keep a list of the nodes that do not have children, noting that it is possible that a node can be listed twice in this list if it has no children. In Code 14-13 the function **StarterTree** creates the tree shown in Figure 14-15. This tree is not perfect in that V1 and V4 are not in the correct positions.

14.4 Examples of Trees

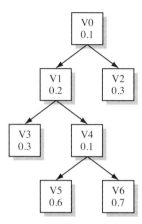

FIGURE 14-15 A starter tree. The node identification starts with V and the data values are shown.

```
# perctree.py
def StarterTree( ):
    tree = {}
    tree[0] = [0.1, -1, 1, 2]
    tree[1] = [0.4, 0, 3, 4]
    tree[2] = [0.3, 0, -1, -1]
    tree[3] = [0.3, 1, -1, -1]
    tree[4] = [0.1, 1, 5, 6]
    tree[5] = [0.6, 4, -1, -1]
    tree[6] = [0.7, 4, -1, -1]
    avail = [3, 2, 2, 5, 5, 6, 6]
    return tree, avail
>>> tree, avail = StarterTree()
```

Code 14-13

The corrected tree is shown in Figure 14-16. To obtain this tree, the nodes V1 and V4 are swapped. However, in order to accomplish this several nodes are affected. The

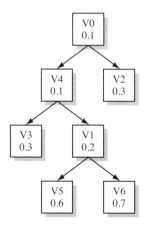

FIGURE 14-16 The corrected tree.

Table 14-1 The Connections of the Affected Nodes Both Before and After Swapping V1 and V4.

Node	Before			After		
	Mom	Kid1	Kid2	Mom	Kid1	Kid2
V0	—	V1	V2	—	V4	V2
V1	V0	V3	V4	V4	V5	V6
V3	V1	—	—	V4	—	—
V4	V1	V5	V6	V0	V1	V3
V5	V4	—	—	V1	—	—
V6	V4	—	—	V1	—	—

V0 node needs to adjust one of its children. The V3, V5, and V6 nodes need to adjust their parents, and nodes V1 and V4 need to adjust the links to both their parents and children. Table 14-1 shows the old links and the new links for this swap.

Before following the several steps needed to accomplish this swap, we can first consider a more whimsical method of denoting the nodes. In this scenario node V1 is considered the mom and V4 is the kid. Thus, V0 becomes the grandma, V5 and V6 are the grandkids, and V3 is the aunt. All of these nodes need to be adjusted. Code 14-14 shows the function **Swap**, which performs the necessary movements. The grandma can have two kids and so it must be determined which one is the mom. The grandkids need to change their moms. The *if* statements for them are necessary since there is no guarantee that the grandkids will exist. If an aunt exists, then it needs to change its mom. Finally, the mom and kid need to exchange their moms and one of their kids. In the example, nodes V1 and V4 are declared to be the mom and kid and are swapped. The adjusted nodes are then printed showing the new moms and kids, which agree with the right side of Table 14-1.

```
# perctree.py
def Swap( tree, mom, kid ):
    # swap tree[mom] with tree[kid]
    grandma = tree[mom][1]
    if grandma != -1:
        if tree[grandma][2] == mom:
            tree[grandma][2] = kid
        else:
            tree[grandma][3] = kid
    # grandkids
    gkid1, gkid2 = tree[kid][2:]
    if gkid1 != -1:
        tree[gkid1][1] = mom
    if gkid2 != -1:
        tree[gkid2][1] = mom
    # aunts
    if tree[mom][2] == kid:
        aunt = tree[mom][3]
    else:
        aunt = tree[mom][2]
    # swap
    tree[mom][1] = kid
```

```
        tree[kid][1] = grandma
        if aunt != -1: tree[aunt][1] = kid
        temp = tree[kid][2:]
        if tree[mom][2] == kid:
            tree[kid][2:] = mom,tree[mom][3]
        else:
            tree[kid][2:] = tree[mom][2], mom
        tree[mom][2:] = temp

>>> Swap( tree, 1, 4 )
>>> tree[0]
[0.1, -1, 4, 2]
>>> tree[1]
[0.4, 4, 5, 6]
>>> tree[3]
[0.3, 4, -1, -1]
>>> tree[4]
[0.1, 0, 3, 1]
>>> tree[5]
[0.6, 1, -1, -1]
>>> tree[6]
[0.7, 1, -1, -1]
```

Code 14-14

In this example node V4 only needed to move up one position. For larger trees it is possible that a node may need to move up several positions. It percolates upward, which is the same as performing several swaps. To demonstrate this process, a new node is created and placed as a kid to V5. This is node V7 with a data value of 0.12, which should require two swaps. In Code 14-15 the function **AddNode** creates a new node and appends it to the specified location *newmom*. The identity of the new node is *ident*.

```
# perctree.py
def AddNode( tree, newmom, ident, data ):
    att = 2 # attach the kid here unless...
    if tree[newmom][2]!=-1:
        att = 3
    # new node
    tree[ident] = [data, newmom, -1, -1 ]
    tree[newmom][att] = ident

>>> AddNode( tree, 5, 7, 0.12 )
>>> tree[5]
[0.6, 1, -1, 7]
>>> tree[7]
[0.12, 5, -1, -1]
```

Code 14-15

In Code 14-13 the function returned a list named *avail*, which is all of the nodes that are available for adding a kid. A node that has no kids is listed twice. When one is chosen, it is then removed from this list. When a node is created, it must be added twice to this list. If a node in this list is percolated upward, then it must be removed from this list and its swapping partner must be added.

Once a node is added, it must be tested for percolation until it finds its proper location in the tree. The function **PercUp** in Code 14-16 performs this function. The identity of the node being considered is *me*, and it is tested inside of the *while* loop with its *mom*. If the data of *me* is lower than the data of *mom*, a swap occurs. If *me* reaches the top of the tree, the percolation stops (*ok=0*). In the example, the node V7 is moved up two positions so that it is a child of V4 and adopts V1 and V6 as its children. The affected nodes are shown.

```
# perctree.py
def PercUp( tree, me ):
    ok = 1
    while ok:
        mom = tree[me][1]
        if tree[mom] < tree[me]:
            ok = 0
        else:
            # perculate one step
            Swap( tree, mom, me )
        if tree[me][1] == -1:
            ok = 0   # reached the top

>>> PercUp( tree, 7 )
>>> tree[7]
[0.12, 4, 1, 6]
>>> tree[4]
[0.1, 0, 3, 7]
>>> tree[1]
[0.4, 7, -1, 5]
>>> tree[6]
[0.7, 7, -1, -1]
```

Code 14-16

The node with the smallest data value is at the top of the tree. When sorting the data, this node is noted as the next one in the sorting order and should be removed from the tree. When it is removed, it needs to be replaced by its child with the lowest value. This, however, will require the child to adjust its children and so on. So, the effect of removing a node sends a ripple effect down the tree until an end node is reached. At the bottom of the tree a new available position will be created and needs to be added to *avail*.

Instead of removing the node, a slightly different approach is adopted. The data of the top node is set to a very large number, and this node is percolated downward. This involves the same amount of work as the ripple effect. This node is swapped with its child with the lowest data value until it reaches the bottom of the tree. It is then removed and its final mom now has an available kid link and is added to *avail*. The percolation downward function **PercDown** in Code 14-17 is quite similar to **PercUp** except that it is necessary to decide which child to swap. A node can have no kids, one kid, or two kids and a single child can either have a left or right child. All of these options need to be considered, which creates the *if – elif – else* loop.

In the example, the data of V0 is set to 9 (a value much larger than any node). When it is percolated downward, it is replaced by V4. Then it swaps with V1 and V5.

It can now be removed from the tree, and V5 can be added to *avail* since it was the final mom.

```
# perctree.py
def PercDown( tree, me ):
    ok = 1
    while ok:
        # data from the kids
        kid1, kid2 = tree[me][2:]
        # if both kids are -1 then stop
        if kid1 == -1 and kid2 == -1:
            ok = 0
        else:
            # find kid with lowest value
            if kid1 == -1:
                kid = kid2
            elif kid2 == -1:
                kid = kid1
            else:
                # tree[me] has two kids
                if tree[kid1][0] < tree[kid2][0]:
                    kid = kid1
                else:
                    kid = kid2
            # kid is the kid to swap with. He is valid.
            # if mom value > kid value : swap
            if tree[me][0] > tree[kid][0]:
                Swap( tree, me, kid )
>>> tree[0][0] = 9
>>> PercDown( tree, 0 )
>>> for i in range( len(tree) ):
        print i,tree[i]

0 [9, 5, -1, -1]
1 [0.4, 7, 5, 6]
2 [0.3, 4, -1, -1]
3 [0.3, 7, -1, -1]
4 [0.1, -1, 7, 2]
5 [0.6, 1, -1, 0]
6 [0.7, 1, -1, -1]
7 [0.12, 4, 3, 1]
```

Code 14-17

Now, consider a reaction problem such as the simulation of a neural network. When a neuron fires it communicates with other neurons, which will fire at a later time with a delay that is dependent upon the distance between the neurons. As an example, the first node (V0) fires at $t=0$ and communicates with nodes V1 and V2 with delays of 0.1 and 0.3. Node V1 will fire next and communicate with V3 and V4 with delays of 0.1 and 0.4. Thus V3 and V4 are scheduled to fire at $t=0.2$ and $t=0.5$, and node V2 is still scheduled to fire at $t=0.3$. This percolation would start with a single node V0 and add nodes V1 and V2. When V0 fires, it is percolated downward and then removed from the tree. V1 is next and it creates V3 and V4, which are each added and percolated upward. V1 is percolated downward and removed, leaving V3 as the new top node. The process continues.

Code 14-18 shows these steps. The *cnnx* matrix contains the delay times from one node to another. In this case it is created from a random array, but only those values less than 0.05 are kept. On average each neuron will be connected to five other neurons. Also, neuron V0 was randomly selected to be the first to fire at time $t=0$. A tree is created with only the node V0. The list *used* will keep track of the neurons that have fired so that they will not be used twice (which does not replicate some neural systems but keeps the problem small). V0 is connected to neurons V18, V22, V27, V40, V49, V53, V59, and V71. So, each of these are added to the network. The first two added are V18 and V22, which are attached to V0 and do not need to be percolated since V0 is the only node that fires at $t=0$. The delay for V18 and V22 are the values from the *cnnx* matrix. The third node that is added is V27 but it cannot be added to V0 since it already has two children. It is added to V18 and then percolated. The process continues adding the rest of the nodes to nodes that have available slots and then it is percolated upward.

When the nodes are added, the top node V0 has its time set to a very large value and it is percolated downward. In this example, its final mom is V53. V0 is removed (the *pop* statement) and the appropriate child of V53 is changed from 0 to -1. The top node in the tree is now V22, and it is connected to several nodes (the list from 17 to 90). The time t is advanced to *cnnx[0,22]* because this is the amount of delay from the time that V0 fired until V22 fires. Before these nodes are added, they are checked against the *used* list to make sure they have not been used before. The first node is V17 and it is added to the tree using the delay *cnnx[22,17]* plus the current time t. It is percolated, and the rest of the nodes connected to V22 are considered. Once this is complete V22 is percolated downward and removed. The new top node is the next neuron to fire, and the time t is adjusted to the new firing time.

The process in this case would continue until all nodes connected to the system fire. In other networks the neurons listed in *used* could be removed after a certain amount of time to replicate a system in which the neurons reset and are available for firing again. The list *avail* was not employed in the example, but for a fully automated system it will also have to be managed. Each time a node is added to the tree, it is added twice to *avail* to indicate that is has two children available. This node's mom is removed from *avail*. When a node is percolated, *avail* must also be adjusted.

```
>>> from numpy import random, less, nonzero
>>> N = 100
>>> cnnx = random.ranf( (N,N) )
>>> mask = less( cnnx, 0.05 )
>>> cnnx *= mask
>>> for i in range( N ):
        cnnx[i,i] = 0
>>> t = 0
>>> tree = {}
>>> used = [0]
>>> nonzero( cnnx[0] )[0]
array([18, 22, 27, 40, 49, 53, 59, 71])
>>> tree[0] = [0,-1, -1, -1 ]
>>> AddNode( tree, 0, 18, cnnx[0,18] )
>>> used.append( 18 )
>>> AddNode( tree, 0, 22, cnnx[0,22] )
>>> used.append( 22 )
```

```
>>> AddNode( tree, 18, 27, cnnx[0,27] )
>>> used. Append( 27 )
>>> PercUp( tree, 27 )
# repeat for other nodes
>>> tree[0][0] = 999
>>> PercDown( tree, 0 )
# last mom of V0 is 53
>>> a = tree.pop( 0 )
>>> t += cnnx[0,22]
>>> tree[53][3] = -1
# top node is now V22
>>> nonzero( cnnx[22] )[0]
array([17, 20, 31, 54, 65, 73, 78, 81, 84, 90])
>>> 17 in used
False
>>> AddNode( tree, 49, 17, cnnx[22,17]+t )
>>> used.append( 17 )
>>> PercUp( tree, 17 )
# repeat for all nonzeros not in used
# PercDown V22
```

Code 14-18

14.4.4 Suffix Trees

A suffix tree captures the different complexities contained in a string. The first step is to create a list of strings using each letter as the beginning. A tree is then constructed from subsequent combinations. As an example the string *'ramada armada'* is used as a test case. The suffix tree is shown in Figure 14-17. Consider the string *'mada'*, which appears twice in the string. One of them leads to *'_armada'* and the other is at the end of the string. (The underscore indicates that the blank space is included.) In this chart node 3 is *'mada'*, which can either terminate or lead on to node 11.

In order to build this tree, the original sequence is used to create a set of strings. These new strings start with each character in the string. Code 14-19 shows the routine that creates these strings, which are sorted alphabetically.

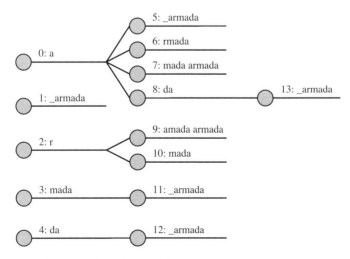

FIGURE 14-17 The suffix tree for 'ramada armada.'

```
# suffixtree.py
def CutUp( instr ):
    cuts = []
    for i in range( len(instr) ):
        cuts.append( instr[i:] )
    cuts.sort()
    return cuts

>>> cuts = CutUp( 'ramada armada' )
>>> cuts
[' armada', 'a', 'a armada', 'ada', 'ada armada', 'amada armada', 'armada',
'da', 'da armada', 'mada', 'mada armada', 'ramada armada', 'rmada']
```

Code 14-19

In Code 14-20 the function **Divide** will separate the strings according to their first letters. It will return a list in which each item is a list of words that begin with a particular letter. A couple of examples are shown.

```
# suffixtree.py
def Divide( cuts ):
    # get first letters
    flett = []
    for i in cuts:
        flett.append( i[0] )
    flett = set( flett )
    # divide into groups by first letters
    divd = []
    for i in flett:
        temp = []
        for j in cuts:
            if j[0]==i: temp.append( j )
        divd.append( temp )
    return divd

>>> divd = Divide( cuts )
>>> divd[0]
['a', 'a armada', 'ada', 'ada armada', 'amada armada', 'armada']
>>> divd[2]
['ramada armada', 'rmada']
>>> divd[3]
['mada', 'mada armada']
```

Code 14-20

It is necessary to isolate the groups of letters that are in common to all strings in a list. As seen in the suffix tree, many of the nodes contain several letters. These are obtained by stripping off all of the letters that a set of strings have in common at their beginnings. The function **CommonStart** in Code 14-21 performs these tasks. The example uses *divd[3]*, which contains two strings that both start with *'mada'* but are different after that.

```
# suffixtree.py
def CommonStart( alist ):
    st = alist[0][0]   # first letter
    ok = 1
    k = 1
    while ok:
        for j in alist:
```

```
                if len(j)<= len(st): ok =0
        if ok:
            t = st + alist[0][k]
            for j in alist:
                if not(j.startswith(t)):
                    ok = 0
                    break
            if ok:
                st = t
        k+=1
    return st

>>> CommonStart( divd[3] )
'mada'
```

Code 14-21

The next function needed will strip the common string from the beginning of all strings in a list. This is accomplished in Code 14-22 by the function **Strip**. This completes all of the tools necessary to build the suffix tree.

```
# suffixtree.py
def Strip( alist, cs ):
    a = []
    N = len( cs )
    for i in alist:
        if len(i) > N:
            a.append( i[N:] )
    return a
```

Code 14-22

The construction of the suffix tree will first cut up the original string in the set of strings starting with each letter. These are divided into groups, and each group defines the need for a node in the tree. At first all the nodes that will be attached will not have a parent node. In Code 14-23 the function **SuffixTree** starts with calls to **CutUp** and **Divide** to create these initial nodes. The list *wait* will contain items that will be considered in a subsequent loop. Each node will be assigned the set of strings from the *divd* list. For example, the first node in this case will have the letter *'a'* and then be assigned all of the stings that start with *'a'*. These are all defined in *divd[0]*. The *wait* list contains two items in a tuple. The first is the ID of the parent node, and a −1 indicates that the particular node has no parent. The second is the list of sequences that belong to the particular node.

In the *while* loop each of the items in the *wait* list are considered. The first item in the *wait* list is: (−1, [*'a'*, *'a armada'*, *'ada'*, *'ada armada'*, *'amada armada'*, *'armada'*]). The common starting string is extracted, which in this case is *'a'*. The *node* to the tree is created as a tuple containing three items: the node ID, the node's parent, and the common starting string. This common start is stripped from the strings and the resulting strings are then divided. Code 14-24 shows this process for the first node.

```
# suffixtree.py
def SuffixTree( instr ):
    cuts = CutUp( instr )
```

```
divd = Divide( cuts )
nodes = []
wait = []  # parts waiting to be considered
k = 0  # keeps track of current node
for i in range( len( divd )):
    wait.append( (-1,divd[i]))
while len(wait) >0:
    a = wait.pop(0)
    cs = CommonStart( a[1] )
    nodes.append( (k,a[0],cs) )
    sp = Strip( a[1], cs )
    divd = Divide( sp )
    for j in divd:
        wait.append( (k,j) )
    k +=1
return nodes
```

Code 14-23

```
>>> sp = Strip( divd[0], 'a' )
>>> sp
[' armada', 'da', 'da armada', 'mada armada', 'rmada']
>>> Divide( sp )
{[' armada'], ['rmada'], ['mada armada'], ['da', 'da armada']}
```

Code 14-24

This new list of lists is added to the *wait* list and each one is assigned the current node as its parent. This process continues until the *wait* list is empty. The entire process is driven in Code 14-25 and the result is printed. The construction of Figure 14-17 follows directly from these results. Each tuple provides the node ID and its parent and data.

```
>>> tree = SuffixTree( 'ramada armada' )
>>> tree
[(0, -1, 'a'), (1, -1, ' armada'), (2, -1, 'r'), (3, -1, 'mada'), (4, -1,
'da'), (5, 0, ' armada'), (6, 0, 'rmada'), (7, 0, 'mada armada'), (8, 0, 'da'),
(9, 2, 'amada armada'), (10, 2, 'mada'), (11, 3, ' armada'), (12, 4, '
armada'), (13, 8, ' armada')]
```

Code 14-25

14.5 Decision Trees and Random Forests

In the previous examples each node is sorted by a single parameter, but many applications have more than one parameter. In a medical example the database consists of several patients, some of whom have a specific illness and the others do not. Each patient has several parameters (age, smoker versus nonsmoker, hereditary illnesses, etc.). The goal of a decision tree is to sort the data such that it becomes apparent which parameters are important. One important difference between this tree and the previous one is that each data entry has a truth (the patient is ill or sick), which is the key to sorting the data.

In this case each node attempts to divide the data samples (patients) by a single parameter. For example, a node may divide the patients into two groups (those younger than 45 and those that are not). Some of the patients in each group will be sick and some well. The score of the node is its ability to separate the sick patients from the healthy ones.

In order to demonstrate this tree, a fake data set is created. In Code 14-26 a function is created that will produce *N* patients with *M* parameters for each. The *prms* is a set of coefficients that indicate how each parameter affects the patient's health. If *prms[k]* is large, then parameter *k* tends to make a person sick. *mylife* is established for each person, indicating how much of each parameter a person has. If *mylife[k]* is small, then this person has very little of condition *k* (which is good news if *prms[k]* tends to make a person sick). The *data* is a list, and each item is a single patient with a patient ID, the classification of sick (*True*) or healthy (*False*), and the *mylife* data. The *prms* are not returned since this is usually the information that is being sought by the tree. In this example, 20 patients are created with five measurements each.

```
# decidetree.py
def FakeIt( N, M ):
    # N is the number of patients
    # M is the number of parameters
    # each parameter affects the chance of being sick
    prms = random.rand( M )
    data = []
    for i in range( N ):
        mylife = random.rand(M)
        temp = (mylife * prms).sum()/sqrt(M)
        sick = temp > 0.7
        data.append( (i, sick, mylife) )
    return data
>>> data = FakeIt( 20, 5 )
```

Code 14-26

The first node in the tree will attempt to separate the patients according to one of the parameters. The parameter chosen is the one that best separates the individuals. Thus each parameter must be tested. Code 14-27 presents the function **ScoreParam**, which scores a single parameter for a set of data. The patients are divided into two groups (sick and healthy). The average and standard deviation, which defines a normal distribution for a class, is computed for each group. The next step is to find the location *x* between the two averages where the two distributions cross. At this location the following condition exists:

$$\mu_1 + \alpha\sigma_1 = x = \mu_2 - \alpha\sigma_2. \tag{14-1}$$

Here the μ_1 and μ_2 are the averages of the two classes and $\mu_2 > \mu_1$. The location *x* is where the two distributions overlap and the parameter α is the percentage of standard deviations that separates *x* from the two averages. In the program the parameter α is found and that is used to compute *x*. Finally, this function will use *x* as the dividing threshold. In the previous example this could be the age of 45. The patients are then divided according to this threshold. If the threshold is perfect, then one group will have all the sick patients and the other group will have all the healthy patients. The *score* will be 1 if one of these is achieved and 0.5 if there is total confusion.

```python
# decidetree.py
def ScoreParam( data, prm ):
    # prm is the index of the parameter to score
    # separate the parameters according to sickness
    well, sick = [], []
    for d in data:
        if d[1] == False:
            well.append( d[2][prm] )
        else:
            sick.append( d[2][prm] )
    # convert to vectors and get stats
    well = array( well );   sick = array( sick )
    wellavg = well.mean();  wellstd = well.std()
    sickavg = sick.mean();  sickstd = sick.std()
    # find crossover
    if wellavg > sickavg:
        alpha = (wellavg-sickavg)/(wellstd+sickstd)
        x = sickavg + alpha * sickstd
    else:
        alpha = (sickavg-wellavg)/(wellstd+sickstd)
        x = sickavg - alpha * sickstd
    # score
    # count the sicks on the left side
    cnt = 0 # sick on left side
    tot = 0. # total sick
    for d in data:
        if d[1]==True and d[2][prm]<x:
            cnt += 1
        if d[1]==True:
            tot += 1
    # if prm is perfect then cnt=0 or cnt=tot
    score = max( (cnt/tot, (tot-cnt)/tot) )
    return score, x
```

Code 14-27

Code 14-28 scores the five parameters for all of the patients. The best performing parameter is the last one with a score of 0.85, and its threshold is 0.52. Thus the first node in the tree separates the data according to the last parameter, as is shown in Figure 14-18. The items in the left column are all of the patients that have their last

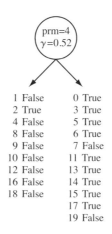

FIGURE 14-18 The first node and its separations.

14.5 Decision Trees and Random Forests

parameter less than 0.52, and the nodes in the right column are all of those that are greater. A perfect separation would place all of the sick (*True*) patients on one side and the healthy patients on the other. This node does separate most of them.

```
>>> for i in range( 5 ):
        p,x = ScoreParam(data,i)
        print '%.2f %.2f'%(p,x)

0.80 0.59
0.60 0.46
0.50 0.60
0.80 0.49
0.85 0.52
```

Code 14-28

The data is now divided into two subgroups, and other parameters can be used to separate each group. The left child group is considered in Code 14-29. In this case the last parameter is not considered since it was used in a parent node. The list *d1* contains only those data from the left-child of the first node. The fourth parameter indicates that a perfect separation of only the *d1* data is possible.

```
>>> d1 = [data[1],data[2],data[4],data[8],data[9],data[10],data[12],\
data[16],data[18]]
>>> for i in range( 4 ):
        p,x = ScoreParam(d1,i)
        print '%.2f %.2f'%(p,x)

0.94 0.83
0.94 0.08
0.94 0.96
1.00 0.79
```

Code 14-29

Each node divides the data as best it can, but it considers only a subset of the data from its parents. Figure 14-19 presents the final tree for this data. The list of the patients falling on either side of the decision is listed, and for those lists that have a mix of *True* and *False* another node is added. The nodes in this case cannot consider the parameter of its ancestors, although this is not a strict rule.

From the decision tree, a set of *if* statements that divide the data can be generated. In this case if *prms[4]>0.52* and *prms[3]<0.79*, then the patient is healthy. These decisions are first order, though, and each node makes a decision based on only one parameter. This may not be the case as some of the behavior of a system may be based on coupled parameters.

Complicated problems can be solved by *random forests*. A forest consists of a set of trees each constructed from a randomly selected subset of data. Nodes in the trees can be constructed from higher order relationships. In a decision tree, a node's decision was based on a first order decision such as $prms[k] > \gamma$. A node in a tree of a random forest will have higher order decisions such as $prms[m]*prms[n] > \gamma$.

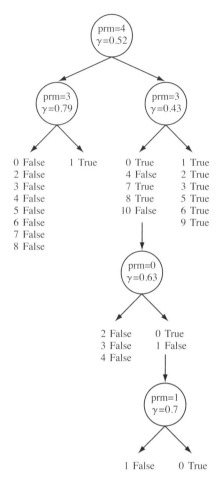

FIGURE 14-19 The decision tree for the generated data.

Of course the number of possible higher order combinations can be quite large. So each tree is constructed from a random selection of data and a random selection of decisions. Training of each tree proceeds similar to that of the decision trees in order to determine the threshold value for each node.

Some trees in the forest will perform better than others. These can be used as possible solutions or combined to create new solutions perhaps by employing a genetic algorithm.

14.6 Summary

Trees offer a method of efficiently storing and sorting large quantities of data according to a specified protocol. Some of the simplest trees are reviewed in this chapter, reserving the discussion of much more powerful trees to other texts.

Problems

1. Build a tree program that allows any number of branches per node.

2. Build a tree program that keeps track of the parents of each node.

3. Consider Figure 14-4 and modify the tree program so that when a node is added at the bottom of the tree it moves upward if its T value is less than that of its parent. The node keeps moving until it reaches a parent with a smaller T value or the node reaches the top of the tree.

4. Create a set of random DNA vectors. Sort these using UPGMA where the distance between two strings is the number of characters that are different.

5. Create a dictionary tree for a random set of DNA vectors.

6. Create a percolation tree, automating the process in Code 14-18.

15 Text Mining

Biological information tends to be more qualitative than quantitative. The result is that a lot of the information is presented as textual descriptions rather than equations and numbers. As a consequence, the field of mining biological texts for information is emerging. Like many topics in this book, this field is large in scope and evolving. Thus only a few introductory topics are presented here, and readers desiring more information should consider resources dedicated solely to this topic.

15.1 An Introduction to Text Mining

The goal of text mining in this chapter is to extract information from written documents. While that sounds fairly straightforward, it is a difficult task. A scientific document presents information in many different forms: from text and equations to tables, figures, and images. Each of these requires a separate method of extracting and understanding complex information in the text. For the purposes of this chapter, the concern will be limited to only the text.

Even if the text is extracted and statistically analyzed, the process does not represent a direct path to understanding the material contained within the document. Understanding the ideas in a text is an extremely difficult task that has kept researchers busy for several decades and will continue to do so. This chapter will consider simple methods of comparing documents and thereby associating documents. This represents only the basics of a burgeoning field.

15.2 Collecting Bioinformatic Textual Data

Written texts are now abundantly available from web resources such as CiteSeer. These are commonly provided as PDF documents that need to be converted to text files so they can be loaded into Python routines. Some PDF files allow users to save the file as a text and some will allow them to copy and paste the text into a simple text editor. There are also programs available that will convert PDF files into text files. Programs such as PyPDF can be employed to read PDF files directly into a Python program.

The text file will contain more than just the text. Symbols will appear where the original text had equations or images. Furthermore, the text contains punctuation, capitalization, and nonalphabetic characters. Since one purpose of text mining is to associate text between documents, it is prudent to remove unwanted characters and punctuation.

Code 15-1 shows the **Hoover** function, which cleans up the text string. It first converts all letters to lowercase and converts all newline characters to spaces. Each letter has an ASCII integer equivalent. The space character is 32 and '*a-z*' is 97-122. The **chr** function converts the integer into a character. This function replaces all characters that do have the correct ASCII code with an empty string, effectively removing these characters. In this example 11% of the characters were removed from the original text.

```
# miner.py
def Hoover( txt ):
    # clean it up
    work = txt.lower()
    work = work.replace( '\n',' ' )
    valid = [32] + range( 97,123)
    for i in range( 256 ):
        if i not in valid:
            work = work.replace( chr(i), " )
    return work

>>> fp = file('data/pdf.txt')
>>> txt = fp.read()
>>> fp.close()
>>> clean = Hoover( txt )
```

Code 15-1

15.3 Creating Dictionaries

Among the many tools that Python offers to manipulate long strings of data, the fastest is the dictionary. For example, it may be necessary to know the location of every word in a text. Each word is designated as a key, and the data for each key is a list of the locations of that word. The function **AllWordDict** in Code 15-2 creates a dictionary *dct* that considers each word in the list *work*. If the word is not in the dictionary, then an entry is created using the word as the key and a list containing the variable i as the data. If the word is already in the dictionary, then the list is appended with the value i.

```
# miner.py
def AllWordDict( txt ):
    dct = { }
    work = txt.split()
    for i in range( len( work )):
        wd = work[i]
        if wd in dct:
            dct[wd].append( i )
        else:
            dct[wd] = [i]
    return dct

>>> dct = AllWordDict( clean )
>>> len( dct )
745
```

Code 15-2

It should be noted that the variable *i* is the location in the list *work*, not a location in the string. In the example text, the word *'primitives'* appeared in three locations. The first 100 characters of the text are shown in Code 15-3, and the entry from the dictionary for the word *'primitives'* is also shown. As can be seen, the first returned value is 1, which corresponds to the second word in the text and not a position in the string. In many text mining procedures the distance between two words *a* and *b* is measured by the number of words instead of the number of characters between them.

```
>>> clean[:100]
'image primitives jason m kinser bioinformatics and computational biology
george mason university man'
>>> dct['primitives']
[1, 2098, 2509]
```

Code 15-3

In this example there are 745 individual words. However, many of them are simple words such as *'and'*, *'of'*, *'the'*, etc., which are not useful. Another concern is that some words are similar except for their ending: *'computations'*, *'computational'*, etc. Dealing with these issues is rather involved, but for the current discussion a simple approach is used that can be replaced later. In this simple approach only the first five letters of the words are used. Words that are shorter than five letters are discarded and words with the same first five letters are considered to be the same word.

Code 15-4 shows the function **FiveLetterDict**, which modifies **AllWordDict** to include only words of five letters or more and to consider only the first five letters. The number of entries in the dictionary used in this example are nearly half that of the previous dictionary.

```
# miner.py
def FiveLetterDict( txt ):
    dct = { }
    work = txt.split()
    for i in range( len( work )):
        wd = work[i]
        if len(wd)>=5:
            wd5 = wd[:5]
            if wd5 in dct:
                dct[wd5].append( i )
            else:
                dct[wd5] = [i]
    return dct

>>> dct = FiveLetterDict( clean )
>>> len( dct )
425
```

Code 15-4

15.4 Methods of Finding Root Words

The use of the first five letters is a very simple (and poorly) performing solution to a complicated problem. This section presents a few other approaches that could be used.

15.4.1 Porter Stemming

Porter Stemming is a method that attempts to remove or replace common suffixes such as *-ing*, *-ed*, *-ize*, and *-ance* from English words (Porter, 2007) This is not an easy task as the rules do not remain constant. For example, the word *petting* should be reduced to *pet* whereas *billing* reduces to *bill*. In one case one of the double consonants is removed and in the other case it is not. Still more confounding are words that have one of the target suffixes but it is part of the root word, such as *string* which ends with *ing*.

Computer code for almost any language is found at Porter (2007) and the Python code in particular is found at Porter Code (2007). While this program works well for many different words, it is not perfect. Code 15-5 shows some of the more disappointing results. These are not shown here to belittle Porter stemming but rather to demonstrate that algorithms do not perform perfectly and that readers should be aware of performance issues related to the programs that they use. Although many of the words attempted were properly rooted, this example shows that it is a very difficult task.

```
# porter.py
>>> ps = PorterStemmer()

>>> w = 'running'
>>> ps.stem( w, 0, len(w)-1)
'run'
>>> w = 'gassing'
>>> ps.stem( w, 0, len(w)-1)
'gass'

>>> ps.stem( 'conditioning',0,11)
'condit'
>>> w = 'conditioner'
>>> ps.stem( w, 0, len(w)-1)
'condition'

>>> w = 'runnable'
>>> ps.stem( w, 0, len(w)-1)
'runnabl'
>>> w = 'doable'
>>> ps.stem( w, 0, len(w)-1)
'doabl'
>>> w = 'excitable'
>>> ps.stem( w, 0, len(w)-1)
'excit'

>>> w= 'atomizer'
>>> ps.stem( w, 0, len(w)-1)
'atom'
```

Code 15-5

15.4.2 Suffix Trees

As we saw in Section 14.4.4 a suffix tree is a common tool for organizing string information that comes in several varieties. The example presented here is designed to identify suffixes that can be removed. Given a string of letters, a suffix tree builds branches at locations in which words begin to differ. An example is (*battle*, *batter*, *bats*) in Figure 15-1. In this case the first node in the tree is *bat* because all words in

15.4 Methods of Finding Root Words

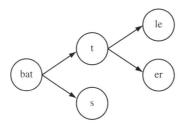

FIGURE 15-1 A suffix tree for three words.

the list begin with *bat*. A split occurs in the tree as two words have a *t* for the fourth letter and one word has an *s*. Along the *t* branch there is another split at the next position. The goal is to identify groups of nodes that commonly follow a stem. In this case, three of the four subsequent nodes are common suffixes and the other node (*t*) is a common addition before some stems.

The Porter Stemming method attempts to identify a suffix by examining a single word. In this suffix tree other types of suffixes also associated with this root are considered.

15.4.3 Combining Simplified Porter Stemming with Slicing

The goal of extracting roots of words is a complicated task and may be too hard to perform perfectly within the scope of this book. A simpler goal is thus to find unique roots. In this case the idea of finding the exact set of letters that form the root is relaxed. Consider the two words *computer* and *compute*, where the root word is *compute*. If the function used here extracts only *comput*, it has still found a unique string that can be used to find the various words with the word *compute*.

This is accomplished by combining the ideas of stemming and the earlier version of slicing. The function **StemSlice** is a very easy function that searches the last letters of a word for endings that are in the list *suff*. If a word is found, then the ending and one extra letter is removed and the root is stored in *temp*. If the length of *temp* is sufficient ($>=3$), then the root becomes *temp*. The results are shown at the bottom of Code 15-6. Again the attempt is not to find the perfect root but to find roots that are unique. Basically, if *atomi* is found in the text, then words such as *atomizing* or *atomizer* are found.

```
def StemSlice( wrd ):
    suff = ['ing','ings','ed','ly','ance','ence','s','ize', 'le', 'er', 'ers']
    nwrd = wrd # new word
    for i in suff:
        if i == nwrd[-len(i):]:
            # a hit
            temp = nwrd[:-len(i)-1]
            if len(temp)>=3:
                nwrd = temp
    return nwrd

>>> StemSlice( 'lasing')
'lasing'
>>> StemSlice( 'lasting')
'las'
>>> StemSlice( 'running')
'run'
```

```
>>> StemSlice( 'gassing')
'gas'
>>> StemSlice( 'conditioning')
'conditio'
>>> StemSlice( 'conditioner')
'conditio'
>>> StemSlice( 'runnable')
'runna'
>>> StemSlice( 'doable')
'doa'
>>> StemSlice( 'atomizer')
'atomi'
```

Code 15-6

15.5 Document Analysis

This section focuses on the simple task of comparing documents according to word frequencies. It does not explore the far more complicated topic of document analysis in depth, and interested readers are encouraged to examine research beyond the scope of this book.

The tasks to be accomplished here are to extract the frequencies of words, to find words that are seen more (or less) frequently than normal, and to isolate words that are indicative of the topic.

15.5.1 Text Mining Ten Documents

To demonstrate the examples presented here, a small data set of ten documents were selected from CiteSeer (2008). Five of the documents cover research in a target field of gene finding, and these are categorized as positive documents. The other five documents are not in the target field and are categorized as negative documents. In this experiment the positive documents are assigned numbers [0, 1, 4, 5, and 6]. Each document is cleansed using the **Hoover** function and converted to a dictionary using the **FiveLetterDict** program, as shown in Code 15-7.

```
>>> nms = os.listdir('data/10docs')
>>> dcts = []
>>> for i in nms:
        if '.txt' in i:
            txt = file( 'data/10docs/' + i ).read()
            clean = Hoover( txt )
            dcts.append( FiveLetterDict( clean ))
```

Code 15-7

15.5.2 Word Frequency

The word frequency matrix *wfm* will contain the frequency of every word found in a document. The element *wfm[i, j]* represents the jth word in the ith document. The construction of *wfm* begins with the word count matrix *wcm*, which collects the number of times the jth word is seen in the *i*th document. Because each document has a different set of words, it is prudent to collect the list of words from all documents before allocating space for *wcm*.

15.5 Document Analysis

The word list is created from **GoodWords**, as shown in Code 15-8. This program loops through the individual dictionaries and collects all of the words into *gw*. Since words can appear in more than one document, the **set** function is used to pare the list down to one copy of each individual word. The **list** function is used to convert the set back to a list for processing in subsequent functions. The total size of all dictionaries is 6,350 entries but there are only 2,840 unique words in the ten documents.

```
#miner.py
def GoodWords( dcts ):
    ND = len( dcts )
    gw = []
    for i in range( ND ):
        gw = gw + dcts[i].keys()
    gw = set( gw )
    gw = list( gw )
    return gw

>>> s = 0
>>> for i in range( 10 ):
    s+= len( dcts[i] )

>>> s
6350

>>> gw = GoodWords( dcts )
>>> len( gw )
2840
```

Code 15-8

The dimensions of *wcm* is $ND \times NW$, where ND is the number of documents and NW is the number of unique words. Code 15-9 shows the function **WordCountMat**, which determines the values of ND and NW and then flows into a nested loop. The *for i* loop considers each document and the *for j* loop considers each word. Recall that the entry for a dictionary is a list containing the locations of the word in the text. So the number of times that a word appears in the text is simply in the length of the list for the dictionary entry. The example shows the first 12 words in the dictionary and the number of times that each appears in the individual documents. Recall that the dictionary will rearrange its contents, and so the words are not in alphabetical order.

```
# miner.py
def WordCountMat(dcts ):
    # dcts is a list of dct
    ND = len( dcts )   # number of dicts
    # find all words in all dicts
    goodwords = GoodWords( dcts )
    LW = len( goodwords )
    wcmat = zeros( (ND,LW), int )
    for i in range( ND ):
        for j in range( LW ):
            if goodwords[j] in dcts[i]:
                wcmat[i,j] = len( dcts[i][goodwords[j]] )
    return wcmat

>>> wcm = WordCountMat( dcts )
>>> wcm.shape
(10, 2840)
```

CHAPTER 15   TEXT MINING

```
>>> wcm[:,:12]
array([[ 0,  0,  1,  0,  0,  2,  0,  9,  9,  0,  1,  0],
       [ 0,  2,  0,  0,  4, 10,  0, 66, 11,  0,  0,  0],
       [ 0,  0,  0,  0,  0,  0,  0,  0,  0,  0,  0,  0],
       [ 1,  0,  0,  0,  0, 15,  0,  0,  8,  5,  5,  0],
       [ 0,  0,  0,  0,  1,  6,  1,  1,  8,  0,  1,  0],
       [ 0,  0,  0,  0,  0,  0,  0,  2,  0,  0,  0,  0],
       [ 0,  2,  0,  0,  2,  5,  0, 25,  0,  0,  1,  0],
       [ 0,  0,  0,  1,  0, 24,  0, 11,  8,  0,  0,  0],
       [ 0,  0,  0,  0,  0,  1,  0,  5,  4,  0,  1,  0],
       [ 0,  0,  0,  0,  0,  0,  0,  0,  0,  0,  0,  1]])
>>> gw[:12]
['inexp', 'elega', 'asymp', 'resor', 'surve', 'selec', 'votes', 'accur',
 'const', 'consu', 'regar', 'cutan']
```

Code 15-9

In the example the word *inexp* is seen just once in doc[3]. A word such as *accur* is more interesting in that it has a high count in some documents (66 in doc[1]) and a low count in others (9 in doc[0]). This does not mean that *accur* is more frequent in doc[1] than it is in doc[0]. If doc[1] were ten times larger than doc[0], this word would be seen more frequently in doc[0] even though its entry in *wcm* is smaller. The frequency of a word is the number of times a word is seen divided by the total number of words. This is defined as the probability

$$P(W_{i,j}) = \frac{wcm_{i,j}}{\sum_j wcm_{i,j}}. \tag{15-1}$$

Code 15-10 shows the **WordFreqMatrix** function, which converts the *wcm* to the word frequency matrix *wfm* by performing the first-order normalization from Equation (15-1) on each row of the matrix. The example shows the probabilities of words 6, 7, and 8 appearing in each of the ten documents. The *wfm* shows that the word *accur* occurs with a frequency of 1.07% in doc[1].

```
#miner.py
def WordFreqMatrix( wcmat ):
    V = len( wcmat )
    pmat = zeros( wcmat.shape, float )
    for i in range( V ):
        pmat[i] = wcmat[i]/float( wcmat[i].sum() )
    return pmat

>>> from numpy import set_printoptions
>>> set_printoptions( precision=4 )
>>> wfm = WordFreqMatrix ( wcm )
>>> wfm[:,6:9]
array([[ 0.     ,  0.0030,  0.0030 ],
       [ 0.     ,  0.0107,  0.0018 ],
       [ 0.     ,  0.    ,  0.      ],
       [ 0.     ,  0.    ,  0.0016 ],
       [ 0.0006,  0.0006,  0.0046 ],
       [ 0.     ,  0.0066,  0.      ],
       [ 0.     ,  0.0050,  0.      ],
       [ 0.     ,  0.0024,  0.0017 ],
       [ 0.     ,  0.0038,  0.0031 ],
       [ 0.     ,  0.    ,  0.      ]])
```

Code 15-10

15.5 Document Analysis

The next step determines whether this frequency is above or below the normal, which requires that the average of each column be computed. The probability of a word occurring in any document is computed by

$$P(W_j) = \frac{\sum_i wfm_{i,j}}{\sum_{i,j} wfm_{i,j}}. \tag{15-2}$$

The **WordProb** function in Code 15-11 normalizes each column and computes the probability of each word occurring in any document. The results for three of the words are printed, and the probability of the word *accur* appearing in any document is 0.0042.

```
def WordProb( wcmat ):
    vsum = wcmat.sum( 0 )  # number of times each word is seen
    tot = vsum.sum()  # total number of words
    pvec = vsum/float(tot)
    return pvec

>>> wpr = WordProb( wcm )
>>> wpr[6:9]
array([  3.52883055e-05,   4.19930835e-03,   1.69383866e-03])
```

Code 15-11

It is easy to determine if a word is appearing more frequently than normal by comparing values in *wfm* to values in *wpr*. The probability of *accur* appearing in any document is 0.0042 and the probability of it appearing in doc [1] is 0.0107, which is more than double the first value. Formally, the voracity of a word is computed by

$$\Omega_{i,j} = \frac{p(W_{i,j})}{p(R_j)}. \tag{15-3}$$

and the example is shown in Code 15-12. In doc [1] the word *accur* (middle column) has a voracity score of 2.53, which means that it is seen 2.53 times more than usual.

```
>>> voracity = pmat/wpr
>>> voracity[:,6:9]
array([[  0.    ,   0.7141,   1.7705],
       [  0.    ,   2.5386,   1.0489],
       [  0.    ,   0.    ,   0.    ],
       [  0.    ,   0.    ,   0.9326],
       [ 16.2117,   0.1362,   2.7019],
       [  0.    ,   1.5770,   0.    ],
       [  0.    ,   1.1852,   0.    ],
       [  0.    ,   0.5638,   1.0165],
       [  0.    ,   0.9137,   1.8123],
       [  0.    ,   0.    ,   0.    ]])
```

Code 15-12

15.5.3 Indicative Words

The search then is to find words that occur frequently in the positive documents and less so in the negative documents. There is an additional caveat: Words must be seen a minimum number of times before they should be considered. The function **IndicWords** (Code 15-13) computes the ratio of voracity of positive documents to negative documents with the included *mask*. It receives the *wcm* and *voracity*. It also receives two lists—*pdoc* and *ndoc*—that indicate which documents are positive and negative. The *mask* is 1 for all documents that have *mincount* occurrences of the word. The *paccum* will accumulate the *voracity* values for the positive documents. Likewise the *naccum* accumulates the values from the negative documents. Both are normalized by the number of documents of each. The score of each word is the ratio of the positive and negative values. The 1 is added since some values of *naccum* can be 0.

```
def IndicWords( wcm, voracity, pdoc, ndoc, mincount=5 ):
    # mask for words with minimum number of occurrences
    vsum = wcm.sum(0)
    mask = greater( vsum, mincount ).astype(int)
    # allocate
    ND, NW = wcm.shape
    paccum = zeros( NW, float )
    for i in pdoc:
        m = greater( voracity[i], 1 ).astype(int)
        paccum += m * voracity[i]
    naccum = zeros( NW, float )
    for i in ndoc:
        naccum += voracity[i]
    paccum /= len( pdoc )
    naccum /= len( ndoc )
    scores = mask * paccum/(naccum+1)
    ag = argsort( scores )[::-1]
    return ag, scores
```

Code 15-13

The example is run in Code 15-14, and the results for the best ten words are shown. The best word is *gw[1670]* with a score of 6.39. This word is "storm", which may seem to be an unusual word for gene finding. Recall that the words are only five letters long by choice. In four of the five positive documents, the author G. D. Stormo is referenced and his name is shortened to *storm* for this project. Had the negative documents discussed the topic of weather, then the statistics for this word would have changed drastically. Several of the best words in this example stem from author's names, which in this small example are indicative of gene finding documents.

```
>>> ag, scores = IndicWords( wcm, voracity, pdoc, ndoc )
>>> ag[:10]
array([1670, 2251, 2309, 864, 2645, 1538, 836, 2791, 1944, 233])
>>> scores[ag[:10]]
array([ 6.3869, 5.4581, 4.5708, 4.5344, 4.4820,
        4.4249, 4.3868, 4.3599, 4.2609, 4.2192])
```

```
>>> take( gw, ag[:10] )
array(['storm', 'salam', 'solov', 'dynam', 'gelfa', 'untra', 'knuds',
       'hauss', 'lawre', 'snyde'],
      dtype='|S5')
```

Code 15-14

15.5.4 Document Classification

The indicative words can be used to classify documents. The theory is that a new document can be classified as a gene-finding document if it has a sufficient number of indicative words appearing in abundance. The complication is that this new file will contain words not in *gw* so it is not possible simply to add another row to *wcm*. Code 15-15 loads the new file, cleanses it, and creates its dictionary, which has 194 unique words.

```
>>> zt = file( 'data/zhangstat.txt' ).read()
>>> zt = Hoover( zt )
>>> ztdict = FiveLetterDict( zt )
>>> zlen = len( ztdict )
>>> zlen
194
```

Code 15-15

The indicative words from Code 15-14 are then considered in Code 15-16. If these are in *ztdict*, then the scores are accumulated. The lists *hits* and *nohits* accumulate the words that were considered as indicative. The *hits* collects those words that are also found in the new document. The probabilities of these words are accumulated in *sc*. The example shows that 40 indicative words were also found in this document, and 140 indicative words were not found in the document. Ten of the words that were found are printed out. Certainly words such as promoter, yeast, splice, region, coding, and biology are words associated with gene splicing. (The underscoring marks the first five characters, which is all that this program uses.) This indicates that the document certainly has something to do with biology. The document written by Zhang (1998) considered statistical features of human exons for the purposes of gene finding.

```
>>> hits, nohits = [], []
>>> i = 0
>>> sc = 0
>>> while scores[ag[i]]>1:
        if gw[ag[i]] in ztdict:
            hits.append( gw[ag[i]] )
            sc += len( ztdict[gw[ag[i]]] )/float(zlen)
        else:
            nohits.append( gw[ag[i]] )
        i+=1

>>> len( hits )
40
>>> len( nohits )
140
```

```
>>> hits[:10]
['untra', 'promo', 'yeast', 'splic', 'regio', 'findi', 'codin', 'termi',
'biolo', 'accor']
>>> sc/len(hits)
0.0135
```

Code 15-16

This was just one example and the training set was small. A more thorough study would start with many more documents and a wide range of topics in the negative documents. It is expected that words such as authors' names would eventually become less important since authors with similar names would appear in the negative documents and that a large range of author's names would be encountered in positive documents. Once a large system was trained then unclassified, documents would then be scored similar to Code 15-16 and those with higher scores would be considered as the best candidates for discussing the target topic.

A more involved study would consider negative documents from a sister topic. For example, if the target topic were gene finding, negative documents would cover other topics in bioinformatics. The indicative words would then become those unique to the specialty of gene finding. This type of application has significant benefits to researchers who wish to automatically scan several documents in from bioinformatics resources to find documents covering their target field.

15.6 Summary

Biological information has always been more qualitative than quantitative and a plethora of such information is contained in written texts rather than in numerical arrays. Text mining has therefore emerged as a powerful tool used to extract meaning from written texts in the field of bioinformatics. This is a difficult task and this field is nowhere near the completion of this goal.

Basic text mining algorithms rely on word frequencies and relative positions of words. This process is complicated by the fact that words can have different endings or that meanings are sometimes extracted from combinations of words that are not necessarily contiguous.

Bibliography

CiteSeer. (2008). *CiteSeer*. Retrieved from http://citeseer.ist.psu.edu/.

Porter. (2007). *Porter*. Retrieved from http://www.tartarus.org/~martin/PorterStemmer.

Porter. (2007). *Porter Code*. Retrieved from
http://www.tartarus.org/martin/PorterStemmer/python.txt.

PyPDF. (2007). *PyPDF*. Retrieved from http://pybrary.net/pyPDF.

Zhang, M. Q. (1998). Statistical features of human exons and their flanking regions. *Human Molecular Genetics, 7* (5), 919–932.

Problems

1. Find the most common words in several text documents covering the same topic.

2. Get multiple papers concerning two different topics. Perform word counts in each. Are there words that are common in one type of document but not the other?

3. Do the frequencies of simple words (*a*, *the*, *and*, etc.) change for documents concerning different topics?

4. Collect ten papers on the same topic. Find the most frequent words. Find the words that are common in one document but not the others.

5. In the example presented in Code 15-14, the words important to the positive documents were recovered. The negative documents in this case covered a topic very different from gene finding. Run the programs again but use documents 2, 3, 7, 8, and 9 as the positive documents. Recover the important words.

6. Build a suffix tree for text from a document and manually identify nodes that are suffixes. It would be best to use a small document as the number of nodes can become quite large and manual analysis would be unfeasible.

7. Collect documents on the topics "gene sequencing" and "gene functionality." Determine which words are indicative for each of these topics.

16 Measuring Complexity

The genome consists of many regions that have varying complexity. Some are full of repeating elements, some are seemingly random, and some code for functions. Measures of complexity of strings can lead to insights into functionality of portions of the genome. This chapter will explore several simple methods of measuring complexity.

16.1 Linguistic Complexity

Linguistic complexity is a loosely defined term that measures variations in a string. For a text string, this measure can be the ratio of the number of unique words to the total number of words. Obviously, if no words are repeated, then this ratio is 1. If a large text contains very few unique words, then this measure becomes quite small. This is an easy program to create as long as the text does not contain punctuation. In the previous chapter we saw how Code 15-1 used the **Hoover** program to clean up text by removing nonalpha characters and converting all characters to lowercase. In Code 16-1 the function **EngComplexity** measures the complexity of this text. The variable *uniq* counts the number of unique words and *ratio* compares the number of unique words with the total number of words.

```
# complexity.py
import miner
def EngComplexity( fname ):
    txt = file( fname ).read()
    txt = miner.Hoover( txt )
    stxt = txt.split()
    uniq = set( stxt )
    ratio = len( uniq)/float(len(stxt ))
    return ratio

>>> EngComplexity( 'data/pdf.txt' )
0.244
```

Code 16-1

This measure is performed over the entire file. By measuring this over a sliding window, it is possible to measure local complexity throughout a text and the complexity of different parts of the string. Code 16-2 shows the function **SlideEng**, which measures this type of complexity over a sliding window. The window size is defined by *wind* and the amount of slide is determined by the *for* loop. Figure 16-1 shows that the complexity of text changes.

Chapter 16 Measuring Complexity

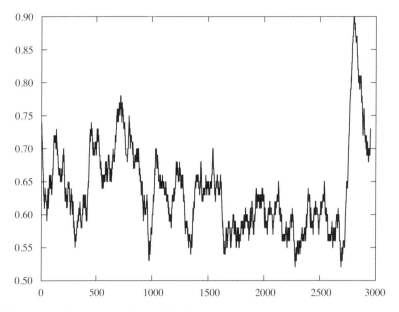

FIGURE 16-1 Complexity measures of *'data/pdf.txt.'*

```
# complexity.py
def SlideEng( fname, wind ):
    # wind = window size
    txt = file( fname ).read()
    txt = textmine.Hoover( txt )
    stxt = txt.split()
    L = len( stxt )-wind
    vec = zeros( L, float )
    for i in range( L ):
        uniq = set( stxt[i:i+wind] )
        vec[i] = len( uniq)/float(len(stxt[i:i+wind] ))
    return vec

>>> seng = SlideEng( 'data/pdf.txt',100 )
>>> akando.PlotSave('fig.txt', seng )
```

Code 16-2

Written texts have natural word breaks. There are, however, long streams of data that have no natural breaks. One example is a high precision value of pi. In Code 16-3 the function **SlidePiChuting** reads a file that contains more than 1,500 digits of pi. In this case the stream is separated into individual characters using the **list** function. The *wind* controls the number of characters to be viewed in a single window. Figure 16-2 shows the result and indicates that pi maintains a similar complexity through the first 1,520 digits.

```
# complexity.py
def SlidePiChuting( fname='data/pi.txt', wind=10 ):
    txt = file( fname ).read().split()[1]
```

```
        L = len( txt )-wind
        vec = zeros( L, float )
        for i in range( L ):
            a = list( str( txt[i:i+wind]))
            uniq = set( a )
            vec[i] = len( uniq)/float(len(a ))
        return vec
```

Code 16-3

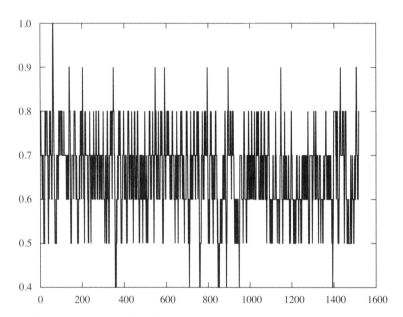

FIGURE 16-2 Complexity measures of the first 1,520 digits of pi.

In the final step, DNA streams are converted into a set of words of a specific length *wordsize*. For example, for a word size of 2 the string *ATGT* becomes [*'AT'*,*'TG'*,*'GT'*]. The number of possible words is the length of the alphabet raised to the power of *wordsize*. DNA consisting of four letters will create $4^3 = 64$ unique words of length 3. It is now necessary to have a minimum *wind* size. If the window size is less than 64 in this example, it will not be possible to achieve a complexity of 1. Code 16-4 shows the function **DNAcomplexity**, which calculates the complexity over a sliding window of size 256. Code 16-5 drives this code for a case of slightly more than 10,000 DNA characters. At location 5000 a repeat of *'at'* is inserted, which results in a dip in complexity (Figure 16-3).

```
# complexity.py
def DNAcomplexity( seq, abet, wordsize, wind ):
    labet = len( abet )
    L = len( seq ) - wind
    vec = zeros( L, float )
    mx = float(labet**wordsize)
```

```
    for i in range( L ):
        # gather words
        words = []
        for j in range( i, i+wind+1 ):
            words.append( seq[j:j+wordsize] )
        ct = len( set( words ))
        vec[i] = ct/mx
    return vec
```

Code 16-4

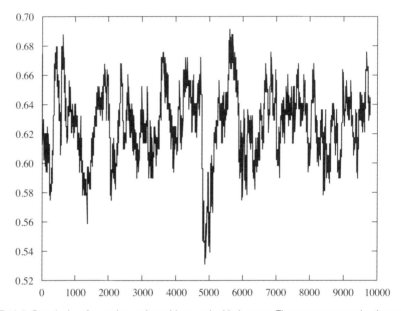

FIGURE 16-3 Complexity of a random string with an embedded repeat. The repeat causes the decrease in the complexity near $x = 5000$.

```
>>> from numpy import take, random
>>> r = (random.ranf( 10000 )*4).astype(int)
>>> seq = take( list('acgt'),r )
>>> seq = ''.join( seq )
>>> seq = seq[:5000] + 'at'*25 + seq[5000:]
>>> vec = DNAcomplexity( seq,'atgc', 4, 256 )
```

Code 16-5

16.2 Suffix Trees

Suffix trees (see Chapter 14) are useful in measuring string complexity. Because the number of nodes in a suffix tree is related to the complexity of the string, an easy implementation is to count the number of nodes needed for each N base in the string. Code 16-6 loads a rather large Genbank file and computes the suffix tree for every 100 bases. Figure 16-4 shows the results for *Heliobacter pylori HPAG1*. There are obvious regions of unusual complexity measures, as shown by the spikes. These, of course,

16.2 Suffix Trees 245

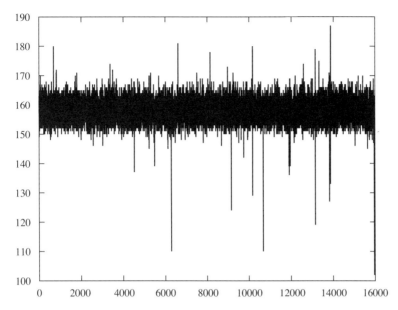

FIGURE 16-4 Complexity as measured by the size of suffix trees for *Heliobacter pylori HPAG1*.

should be examined to determine the reasons that complexity changed (see Problem 4 at the end of this chapter).

```
>>> import genbank, suffixtree
>>> data = genbank.ReadGenbank( 'genbank/cp000241.gb.txt')
>>> seq = genbank.ParseDNA( data )
>>> klocs = genbank.FindKeywordLocs( data )
>>> glocs = genbank.GeneLocs( data, klocs )
>>> a = []
>>> for i in range( 0, len(seq), 100 ):
        tree = suffixtree.SuffixTree( seq[i:i+100] )
        a.append( len( tree ))
>>> akando.PlotSave('fig4.txt', array( a ) )
```
Code 16-6

Code 16-7 performs a final test computing the complexity around the beginning of genes. The **ComplexityAtStarts** function receives *glocs* and *dna*, which come from reading the Genbank file. This routine looks only at genes that do not have the complement flag set. For each of these regions, 1,000 bases are extracted about the start region and the complexity for each is computed for several shifts. In the example, the complexities for the first 20 genes are gathered in the list *cas*. This is converted into a matrix to easily compute the average of all of the vectors, which is plotted in Figure 16-5. In this plot the start of the genes is at $x=500$.

```
# complexity.py
def ComplexityAtStarts( glocs, dna, wind ):
    # glocs from genbank.GeneLocs
    hits = []
```

```
        for i in glocs:
            print i, i[1]
            if i[1] == False:
                # considering only non-complements
                start = i[0][0][0]
                seq = dna[start-500:start+500]
                vec = []
                for j in range( len(seq)-wind):
                    if j%100==0: print j,
                    t = suffixtree.SuffixTree( seq[j:j+wind])
                    vec.append(len(t))
                vec = array( vec )
                hits.append( vec )
        return hits

>>> data = genbank.ReadGenbank( 'genbank/ae005174.gb.txt')
>>> seq = genbank.ParseDNA( data )
>>> klocs = genbank.FindKeywordLocs( data )
>>> glocs = genbank.GeneLocs( data, klocs )
>>> cas = ComplexityAtStarts( glocs[:20], seq, 100 )
>>> a = array( cas )
>>> akando.PlotSave('fig5.txt', a.mean(0))
```

Code 16-7

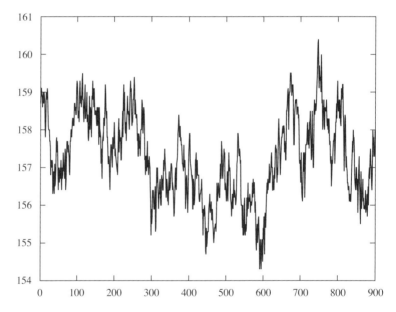

FIGURE 16-5 Average complexity near the start of genes.

16.3 Superstrings

Given a set of strings, the *superstring* is the smallest string that contains all of the initial strings (Teng and Yao, 1993). For example, given the strings *'abc'*, *'def'*, and *'cde'*, the superstring *'abcdef'* contains all of the original three strings in the shortest

16.3 Superstrings

possible representation. The compression ratio of a superstring is the length of the superstring divided by the sum of the lengths of the substrings. Formally, a set of strings $S = \{s_1, \ldots, s_n\}$ is used to create a superstring a. The compression ratio is

$$Q = \frac{|a|}{\sum_i |s_i|} \qquad (16\text{-}1)$$

The complexity measure of a single sequence begins by creating several substrings. These substrings must overlap, and every character in the original sequence must be in at least one of the substrings. As an example, the previous two sentences are used as an original string (with punctuation removed). Preparation of this original string is shown in Code 16-8.

```
>>> orig = 'To measure the complexity of a single sequence it is first used to
create several substrings. These substrings must overlap and every character in
the original sequence must be in at least one of the substrings'
>>> orig = miner.Hoover( orig )
>>> orig = orig.replace( ' ',"  " )
```

Code 16-8

The function **CutItUp** in Code 16-9 receives the original sequence and the number of substrings desired. Two optional arguments are the minimum and maximum lengths of the substrings. The number of substrings is actually a minimum number. The function has two loops, and the first one extracts random substrings. The array *unseen* keeps track of the elements used in the substrings. The second loop creates more substrings that will capture those elements not initially used. The first loop creates *ncuts* number of substrings. However, the number of substrings can increase through the second loop.

It is up to the user to ensure that there is a sufficient overlap in the substrings. In this case the length of the original string is 174 characters. If the goal is to have five-fold sampling (on average an element is used in five substrings), the total number of characters in the substrings should be about 870. In this example the average substring has a length of 30, and so the number of strings needed is about 29. The array *folds* keeps track of the number of times each element is used.

```
# superstring.py
def CutItUp( seq, ncuts, mincut=20, maxcut=40 ):
    answ = []
    lseq = len( seq )
    unseen = ones( lseq ) # tracks elements that have been used
    folds = zeros( lseq )
    for i in range( ncuts ):
```

```
            # get a list of elements not yet used
            nz = nonzero( unseen )[0]
            # randomly select a starting point
            beg = int( random.rand() * lseq-mincut )
            # randomly select an ending point
            lg = int( random.rand()*(maxcut-mincut)) +mincut
            end = beg + lg
            if end >= lseq:
                end = lseq
            # make this cut
            answ.append( seq[beg: end] )
            # change unseen
            unseen[beg:end] = zeros( end-beg)
            folds[beg:end] += ones( end-beg )
        # it must be complete
        nz = nonzero( unseen )[0]
        while len(nz)>0:
            beg, end = nz[0], nz[0]+mincut
            if end>=lseq: end = lseq
            answ.append( seq[beg:end] )
            unseen[beg:end] = zeros( end-beg)
            nz = nonzero( unseen)[0]
            folds[beg:end] += ones( end-beg )
        return answ, folds

>>> cuts, folds = CutItUp( orig, 29 )
```

Code 16-9

The plot in Figure 16-6 graphs the *folds* and shows the number of times that each element was sampled. The list *cuts* contains the substrings. In this example there are 30 substrings, so one additional substring was added to capture elements not previously used. As seen in Figure 16-6, the only elements with just one sample is at the beginning of the original string.

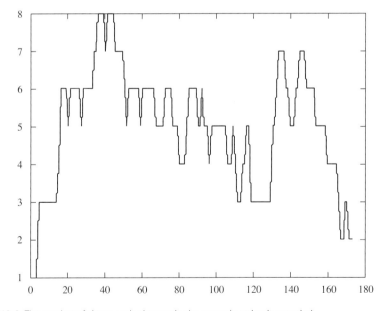

FIGURE 16-6 The number of times each element in the example string is sampled.

16.3 Superstrings

The construction of the superstring can follow several philosophies. A greedy or nongreedy approach would use the assembly routines in Chapter 12. The greedy assembly approach would require a change though. In this case when sequences are added to a contig, it is absolutely required that all elements in the contig align perfectly. Partial matches would not be allowed.

In this case, it is easy to create a program for the nongreedy approach. This approach uses a GA (see again Section 11.3). The gene in a GA consists of a set of index numbers. In this case there are 30 substrings, and so the gene would consist of the numbers 0 to 29 in some order. The gene provides a recipe for creating a superstring. A substring is compared with the superstring and if there is a perfect overlapping region then the string is located at that position. Otherwise it is added to the end of the string.

Code 16-10 shows the function **SuperFromGene**, which converts a gene into a superstring. The process builds the superstring by considering each substring sequentially according to the order stored in *gene*. Thus, *gene* is a list of integers, and *cuts* is the list of substrings. The superstring *sst* is initialized by copying the first substring. The program then considers each substring and searches for a match of the first three characters in the substring with the *sst*. The *snip* is the first three characters and the *cnt* is the number of times that the *snip* exists in *sst*. If the *snip* does not occur, the new substring is appended to the *sst*.

If the *snip* does exist in *sst*, it is possible that there will be a match. Each location where the three-character *snip* exists in *sst* is denoted by the value of *me* in the *j*-loop. The end of the substring may or may not extend beyond the end of *sst*, and so the value of L is used to limit the length of the exact match. The two cases are shown in Figure 16-7. The overlapping region of the substring and the *sst* are compared for a perfect match. If a perfect match exists, then any portion of the substring that extends beyond the end of *sst* is appended to *sst*.

There may be several locations in which *snip* matched a segment of *sst*. If none of them produce a perfect match, then the substring is appended to the *sst*. The length of *snip* is 3, and therefore the strings need to overlap by a minimum of three elements.

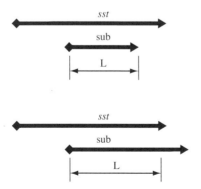

FIGURE 16-7 Two cases illustrating the alignment of a substring with *sst*.

```
# superstring.py
def SuperFromGene( gene, cuts ):
    sst = cuts[gene[0]]
    for i in range( 1, len(gene)):
        # look for cuts[i][:3] in sst
        snip = cuts[gene[i]][:3]
        cnt = sst.count( snip )
        # for all possible matches see if it is perfect
        if cnt == 0:
            # append
            sst += cuts[gene[i]]
        else:
            hit = 0 # set to 1 if a perfect match is found
            k = 0
            for j in range( cnt ):
                me = sst.find( snip, k )
                k = me + 1
                # is it a perfect match?
                L1 = len( sst )-me # length of remainder of sst
                L2 = len( cuts[i] )
                L = min((L1,L2) )
                if sst[me:me+L] == cuts[i][:L]:
                    # a perfect match
                    hit = 1
                    # attach
                    if L2>L1: sst += cuts[i][L1:]
                    break
            if hit==0:
                # all locations were viewed and none matched
                sst += cuts[gene[i]]
    return sst
```

Code 16-10

Code 16-11 creates a gene from a set of integers and builds the superstring from this gene. The elements of *gene* are shuffled, and a different superstring is created. The lengths of the superstrings are dependent upon *gene*. It will soon be the task of the GA to find the *gene* that creates the shortest superstring.

```
>>> gene = range( len( cuts ))
>>> sst = SuperFromGene( gene, cuts )
>>> len( sst )
493
>>> random.shuffle( gene )
>>> sst = SuperFromGene( gene, cuts )
>>> len( sst )
718
```

Code 16-11

The first superstring is better than the second because it is shorter. These values, however, do not indicate if it is a good superstring. The maximum length of a superstring is the combined length of all of the substrings. The function **LenS** in Code 16-12 computes this length and shows that the maximum length is 718 for this case. Results will vary since the **CutItUp** function randomly cuts the string. The minimum length of a superstring is not known. In this case the length of the original string was 174 but a superstring could certainly be shorter if the original had repeating regions.

16.3 Superstrings

```
# superstring.py
def LenS( S ):
    ls = 0
    for i in S:
        ls += len( i )
    return ls

>>> LenS( cuts )
718
>>> len( orig )
174
```
Code 16-12

As in all GA applications, the cost function needs to be created. In this case the cost function is simply the length of the superstring. The function **ScoreAll** in Code 16-13 receives a list of genes, obtains the superstring generated from the gene, and then archives the length of the superstring in *sc*.

```
# superstring.py
def ScoreAll( genes, cut):
    LG = len( genes )
    sc = zeros( LG, int )
    for i in range( LG ):
        sst = SuperFromGene( genes[i], cut)
        sc[i] = len( sst )
    return sc
```
Code 16-13

The entire process is driven by **DriveGA** (Code 16-14), which is similar to Code 11-23. It receives the substrings in the list *cuts* and the number of GA genes that the user wants, *ngenes*. This follows the typical GA protocol in that the genes are scored and the best are more likely to create the next generation of genes. The changes to this system include the assurance that every kid is legal by sending each one through **gasort.Legalize** and that the best vector is retained by using *oldbest* and *oldgene* to keep track of the best score and gene from the previous generation. If this gene is better than every new gene, it is inserted back into the list of genes.

```
# superstring.py
def DriveGA( cuts, Ngenes ):
    # create genes
    a = range( len( cuts ))
    folks = []
    for i in range( Ngenes ):
        random.shuffle( a )
        folks.append( copy.copy(a))
    # drive
    oldbest = 999999
    fcost = ScoreAll( folks, cuts )
    valid = range( len (cuts))
    for i in range( 2000 ):
        kids = ga.CrossOver( folks, fcost )
        for j in range( Ngenes ):
            kids[j] = list( kids[j] )
            gasort.Legalize( valid, kids[j] )
```

```
            kcost = ScoreAll( kids, cuts )
            ga.Feud( folks, kids, fcost, kcost )
            ga.ShuffleMutate( folks[1:], 0.05 )
            fcost = ScoreAll( folks, cuts )
            best = fcost.min()
            if fcost.argmin() != Ngenes-1:
                random.shuffle( folks[-1] )
            if oldbest < best:
                folks[0] = copy.copy( oldgene )
                fcost[0] = oldbest
            else:
                oldbest = best
                oldgene = copy.copy( folks[ fcost.argmin()] )
            #if i%50==0: print best
        besti = fcost.argmin()
        return folks[besti]
```

Code 16-14

This routine has many options for the user. For example, the number of iterations, the mutation rate, and the number of genes are commonly manipulated to find an optimal result. In this example, 2,000 iterations produced the gene g, which returned a score of 363 (Code 16-15). Furthermore, optimization can perhaps be achieved through more iterations.

```
>>> best = DriveGA( cuts, 30 )
>>> sst = SuperFromGene( g, cuts )
>>> len( sst )
363
```

Code 16-15

The length of a superstring is a measure of the complexity of a string. Multiple runs of this algorithm will produce a set of measurements for the same initial string. Strings with repeating regions will have a greater probability of creating a shorter superstring. Strings that have no repeating regions will create superstrings that are the same value as those returned by **LenS**. Thus a measure of complexity is the compression ratio Q.

In the next test ten genes were considered. For each gene, 500 bases before and after the start location were extracted. Thus there is a 1,000-base string for every string, and the start of the gene is in the middle. Code 16-16 shows the function **SSTComplexAtStarts** that measures the complexity of these strings using superstrings. The *glocs* is a list of ten genes from Code 16-16. The *i*-loop considers each gene, and the *seq* is the 1,000 bases. The *j*-loop considers 100 base substrings and computes the superstring and its length. The *vec* stores the lengths of the superstrings for the 900 sliding windows. These are stored in the list *hits*.

```
# superstring.py
def SSTComplexAtStarts( glocs, dna, wind ):
    # glocs from genbank.GeneLocs
```

16.3 Superstrings

```
    hits = []
    for i in glocs:
        print 'I',i, i[1], i[0][0][0]
        if i[1] == False and i[0][0][0]>500:
            # considering only non-complements
            start = i[0][0][0]
            seq = dna[start-500:start+500]
            vec = []
            print i, start, len(dna), len(seq)
            for j in range( len(seq)-wind):
                print j,
                cuts, folds = CutItUp( seq[j:j+wind], 20,10,20 )
                gene = DriveGA( cuts, 16 )
                sst = SuperFromGene( gene, cuts )
                vec.append( len(sst)/float(LenS(cuts)) )
            print ''
            vec = array( vec )
            hits.append( vec )
    return hits
>>> hits = SSTComplexAtStarts( glocs[:10], dna, 100 )
>>> mat = array( hits )
>>> akando.PlotSave('fig26_8.txt', mat.mean(0)/mat.std(0))
```
Code 16-16

The graph in Figure 16-8 shows the ratio of mean to standard deviations for the ten genes. Large spikes in this graph indicate regions in which the superstrings had similar lengths or large lengths. One extreme example occurs a little more than 300 bases in front of the start location ($x=500$). When measuring complexity, large spikes indicate a lower complexity, and so this region should be investigated further.

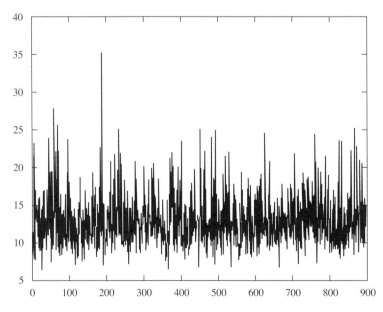

FIGURE 16-8 The ratio of mean to standard deviation of the lengths of the superstrings for ten genes.

16.4 Summary

Complexity of strings is another field of analysis used in bioinformatics. Applications are based on the ideals that different regions in the genome will have different string complexities. Regions with structure are less complex, and therefore it is believed that the coding regions, for example, are less complex than random regions. This chapter reviewed a few of the many methods to measure a string's complexity. The performances of these methods were expressed through examples of coding versus noncoding regions in a string.

Bibliography

Teng, S. H., and F. Yao. (1993). Approximating shortest superstrings. *Procedings of the IEEE*, 158–164.

Problems

1. Compare the complexity of a string of Shakespearean text versus text taken from a newspaper article.

2. Using **SlideEng**, compute the complexity of a large DNA string for different window sizes. Explain the effect of window size on the measurement of complexity.

3. Repeat Problem 2 for a long string in which there are multiple repeats of length N embedded in the string.

4. Look again at Section 16.2. Repeat the model shown in Figure 16-4 for a Genbank file that contains many genes. Match the locations of the large spikes (both positive and negative) to events in the file. Do the spikes align with coding regions, repeats, or other behavior?

5. Run the superstring program for different values of *Ngenes*. Plot superstring length versus *Ngenes*.

6. Run the superstring program for different values of the mutation rate. Plot superstring length versus the mutation rate.

7. A superstring was constructed using a GA and a nongreedy method. Create a program that builds a superstring using a greedy method.

8. Use the nongreedy approach to building superstrings. Consider several different values for the number of genes. For each value, run five trials with each trial returning a *best* gene. Plot the average superstring lengths versus the number of genes. What is the optimal value for the number of genes?

9. Optimize the nongreedy approach for building superstrings by finding the best mutation rate.

17 Clustering

Measurements extracted from biological systems may be dependent on a large number of variables in manners that are not yet understood. One method of analyzing such data sets is to group data vectors that are similar. Once a group is collected, it can be further analyzed to find the reasons for the similarity. The process of *clustering* is often used to create these groups, and the most common of these methods is the k-means clustering algorithm. This chapter will focus on the development and use of the k-means method and some useful extensions.

17.1 The Purpose of Clustering

Given a set of data vectors $\{\mathbf{X} : \vec{x}_1, \vec{x}_2, \ldots, \vec{x}_N\}$, the object is to group the vectors so that each group contains only those vectors that are similar to each other. The measure of similarity is defined by the user for each particular application. The number of clusters can either be fixed or dynamic depending on the algorithm chosen. The result of the algorithm will be a set of groups, and the *constituents* of each group is the set of self-similar vectors.

Code 17-1 creates a simple function that generates random data for clustering. Because purely random data would be inappropriate for clustering, this algorithm generates a small number of random seeds and then generates data vectors that are random deviations from these seeds. In this fashion some of the vectors should be related to each other through a common seed. These vectors should thus find a reason to cluster. The variable N is the number of vectors to be generated, and L represents the length of the vectors.

```
# clustering.py
# generate data to be used in clustering
def CData( N, L, scale = 1, K=-1 ):
    # N = number of data vectors
    # L = length of data vectors
    # create a random seeds
    if K==-1:
        K = int( random.rand()*N/20 + N/20)
    seeds = random.ranf( (K,L) )
    # create random data based on deviations from seeds
    data = zeros( (N,L), float )
    for i in range( N ):
        pick = int( random.ranf() * K )
        data[i] = seeds[pick] +scale*(0.5*random.ranf(L)-0.25)
```

```
            return data

# use random.seed only to replicate the results in this text
>>> random.seed( 3996 )
>>> data = CData( 100, 10 )
```

Code 17-1

Code 17-2 presents a simple algorithm for comparing one vector with a set of vectors. The comparison is performed by the absolute subtraction

$$score = \sum_i |\vec{t} - \vec{d}_i|, \tag{17-1}$$

where the vector **t** is the target and **d**$_i$ is the i-th data vector. In Code 17-2 *diffs* is a matrix that contains the subtraction of the target vector from all of the vectors in *vecs*. This command looks a bit odd in that the two arguments of the subtraction do not have the same dimensions. Python understands this predicament and performs the subtraction of the target vector with each row of *vecs*. The result is *diffs*, which is the same dimension as *target*. The **sum** command only sums along axis #1, which is the second dimension in *diffs*.

```
# clustering.py
# compare all of the vectors to a target: abs-subtraction
def CompareVecs( target, vecs ):
    N = len( vecs ) # number of vectors
    diffs = abs( target - vecs )
    scores = diffs.sum( 1 )
    return scores
>>> scores = CompareVecs( data[0], data )
```

Code 17-2

The executed command in Code 17-2 computes the comparison of the first vector with the entire data set. A perfect match of a vector with the target will produce a score of 0. Code 17-3 sorts the scores and creates a plotting file that is shown in Figure 17-1. The **argsort** function returns an array of indices for the data sorted from lowest to highest. Thus *scores[ag[0]]* is the lowest score, and *scores[ag[-1]]* is the highest score. The *ag* is an array, and it is used as an index in *scores[ag]*. This will extract the values of *scores* according to the indices of *ag*.

```
>>> import akando
>>> ag = argsort( scores )
>>> akando.PlotSave( 'plot.txt', scores[ag] )
gnuplot> plot 'plot.txt'
```

Code 17-3

This plot is typical for data such as this that uses a different vector as a target. There are a few vectors that are similar to the target and many that are dissimilar. There

17.1 The Purpose of Clustering

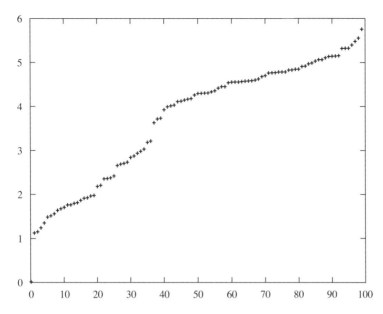

FIGURE 17-1 The comparison scores of vectors to a target vector, which are sorted from lowest to highest.

seems to be a sharp differentiation between $2<y<3$. Thus the threshold 2.5 is chosen, which means that any score less than the threshold is considered to be a good match.

As a control experiment, a simple greedy algorithm is created. One vector is chosen as the target, and all of the vectors that are close to it (scoring below the threshold value) are collected as a single group. Vectors that belong to a group are not considered for further grouping. This program has obvious problems related to the fact that a vector may not belong to the best group. Consider a case in which vector **C** is similar to vector **A** and very similar to vector **B**. Vector **A** is chosen as the first target, and vector **C** would thus be chosen to belong to that group, preventing vector **C** from joining the **B** group for which it was better suited. This is merely a control algorithm to which better algorithms can be compared.

Code 17-4 displays the simple function for clustering data by this greedy method. The data is converted to the list *work* to take advantage of some of the properties of lists. The **pop** function removes a vector from the list, and thus *target* becomes this vector and it no longer exists in *work*. The **nonzero** function will return a tuple containing indices of those scores that are less than the threshold, and the [::-1] reverses the indices so that the largest is first. A *group* is created out of the vectors deemed similar to the target. Once *group* has been collected, it is appended to the list *clusters*.

The ordering of *nz* from highest to lowest is necessary for the *for* loop. Consider a case in which the ordering is from lowest to highest and the vectors 2 and 4 in this example are deemed close to the target. Inside the *for* loop, the *pop* function will add vector 2 to the group and remove it from the *work* list. In doing so vector 4 now becomes vector 3 and in the next loop the *pop* of vector 4 will be incorrect. By considering the vectors from highest to lowest, this problem is averted.

```
# clustering.py
# greedy clustering algorithm
def CheapClustering( vecs, gamma ):
    # vecs: array of data vectors
    # gamma: threshold. Below this is a match
    clusters = [ ]  # collect the clusters here.
    ok = 1
    work = list(vecs) # copy of data that can be destroyed
    while ok:
        target = work.pop( 0 )
        # score for all remaining vecs in work
        scores = CompareVecs(target, work )
        # threshold vectors
        nz = nonzero( less( scores, gamma ) )[0][::-1]
        group = []
        group.append( target )
        # add vectors to group
        for i in nz:
            group.append( work.pop( i ) )
        clusters.append( group )
        if len( work )==1:
            clusters.append( [work.pop(0)])
        if len(work)==0:
            ok = 0
    return clusters

>>> clusts = CheapClustering( data, 2.5 )
>>> map( len, clusts ) # print length of each cluster
[26, 21, 19, 11, 21, 2]
```

Code 17-4

In this particular experiment six clusters were created, and they contained the following number of members: 26, 21, 19, 11, 21, and 2. These clusters will be compared to the k-means clusters generated in the next section. A good cluster would collect vectors that are similar, and thus a single cluster should have a small cluster variance as measured by

$$\omega_k = \frac{1}{N_k} \sum_i \sigma_{k,i}^2, \qquad (17\text{-}2)$$

where $\sigma_{k,i}^2$ is the variance of the ith element of the kth cluster, and N_k is the number of vectors in the kth cluster. For each cluster, the variance of the vector elements are computed and summed. This scalar measures the variance of the vectors in a cluster. For the example case, the variances of the six clusters are shown in Code 17-5.

```
# kmeans.py
# Measure the variance of a cluster
def ClusterVar( vecs ):
    a = vecs.std( 0 )
    a = (a**2).sum()/len(vecs[0])
    return a

>>> for i in range( 6 ):
        print i, "%.3f" %ClusterVar( array( clusts[i] ))
```

```
0 0.027
1 0.020
2 0.019
3 0.016
4 0.020
5 0.012
```
Code 17-5

17.2 *k*-Means Clustering

k-means clustering is an extremely popular and easy algorithm in which the user defines the number of clusters, k, and a method by which these clusters are seeded. The algorithm will then perform several iterations until the clusters do not change. Each iteration consists of two steps. The first is to assign each vector to a cluster, thus creating the cluster's constituents. The second is to compute the average of each cluster. If a vector is determined to belong to a different cluster, it changes the constituency of the clusters, and the averages will thus be different in the next iteration. If the averages are different, other vectors may shift to new clusters. The process iterates until vectors do not change from one cluster to another.

The steps for this process are as follows:

1. Initialize k clusters.
2. Iterate until there is no change.
   2.1 Assign vectors to clusters.
   2.2 Compute the average of each cluster.
   2.3 If none of the clusters have changed, create a stop condition.

Each cluster is constructed from an initial seed vector, which can be a random vector, one of the data vectors, or some other method as defined by the user. Usually, the measure of similarity between a vector and a cluster average is a simple distance measure, but again the user has the opportunity to alter this if an application needs a different measure.

Code 17-6 displays two possible initiation functions. The function **Init1** receives the number of clusters and the length of vectors, and it generates only random vectors. The problem with this approach is that there is no guarantee that a cluster will collect any constituents. The function **Init2** randomly selects one of the data vectors as a seed for each cluster. It generates a list of indices and shuffles them in a random order. The first k indices of this shuffled order are used as the seed vectors. In this function the **take** function contains two arguments. The first is a list of indices to be taken. The second is the axis argument, and this forces the **take** function to extract row vectors from *data* instead of scalars.

```
# kmeans.py
# initialize with random vectors
def Init1( K, L ):
    # K is the # of clusters
    # L is the length of the vectors
    """K=Number of clusters, L=Length of vectors"""
    clusts = random.ranf( (K,L) )
    return clusts

# initialize with random data vectors
def Init2( K, data ):
    """K=Number of clusters"""
    r = range( len(data) )
    random.shuffle( r )
    clusts = data.take( r[:K],0 )
    return clusts
```

Code 17-6

Once an initial set of clusters is generated, the next step is to assign each vector to a cluster. This assignment is based on the closest Euclidean distance from the vector to each cluster. Code 17-7 displays a function that computes these assignments. In this function, *mmb* is a list that collects the constituents for each cluster, and it contains *k* lists. Thus *mmb[0]* is a list of the members of the first cluster that contains the vector identities. If *mmb[0] = [0,4,7]*, *data[0]*, *data[4]*, and *data[7]* are then members of the first cluster. There are two *for* loops in this function. The first initializes *mmb* and the second performs the comparisons and assigns each vector to a cluster. In the second loop the score for each cluster is contained in the vector *sc*, and *mn* indicates which cluster has the best score.

```
# kmeans.py
# Decide which cluster each vector belongs to
def AssignMembership( clusts, data ):
    NC = len( clusts )
    mmb = []
    for i in range( NC ):    for i in range( len( data )):
        sc = zeros( NC )
        for j in range( NC ):
            sc[j] = sqrt( ((clusts[j]-data[i])**2 ).sum() )
        mn = sc.argmin()
        mmb[mn].append( i )
    return mmb
```

Code 17-7

The next major step is to recompute each cluster as the average of all of its constituents. Thus, if *mmb[0] = [0,4,7]*, then *clust[0]* will become the average of the three vectors *data[0]*, *data[4]*, and *data[7]*. Code 17-8 displays this function as **ClusterAverage**. Within the *for* loop, *vecs* is the set of vectors for the *i*th cluster. Recall that *vecs* is actually a matrix where the rows are the data vectors. Thus the *k*th element of the average vector is the average of the *k*th column of *vecs*. The *vecs.mean(0)* uses the 0 to indicate that it is only the first axis (vertical) that is being averaged. This is used instead of *vecs.mean( )*, which would compute the average of all of the elements in the matrix.

```
# kmeans.py
# compute the average of the clusters
def ClusterAverage( mmb, data ):
    K = len( mmb )
    N = len( data[0] )
    clusts = zeros( (K,N), float )
    for i in range( K ):
        vecs = data.take( mmb[i],0 )
        clusts[i] = vecs.mean(0)
    return clusts
```

Code 17-8

These are the major functions necessary for k-means clustering. The next step is to create the iterations. Code 17-9 demonstrates the entire k-means algorithm. It creates *clust1*, which is the initial clustering. The *ok* flag is set to *True* and will keep the iterations ongoing until the algorithm detects that no changes were made. The data vectors are assigned to clusters, and a new set of clusters, *clust2*, is created. The variable *diff* computes the differences between these two clusters. If they are the same, no changes were made in this iteration and *ok* is set to *False*. At the end of each iteration, the *diff* is printed to the console. It should be monotonically decreasing, and quite often only a few iterations are required to create the clusters.

```
# kmeans.py
# typical driver
def KMeans( K, data ):
    clust1 = Init2( K, data )
    ok = True
    while ok:
        mmb = AssignMembership( clust1, data )
        clust2 = ClusterAverage( mmb, data )
        diff = ( abs( ravel(clust1)-ravel(clust2))).sum()
        if diff==0:
            ok = False
        print 'Difference', diff
        clust1 = clust2 + 0
    return clust1, mmb
```

Code 17-9

Code 17-10 displays an example using the same data and the same number of clusters from the previous section. The variances of these clusters are also printed. These variances are on the whole smaller than those from the greedy algorithm, which indicates that the members of these clusters are more closely related than in the previous case. The **take** command takes two arguments. Without the second argument, the **take** function will extract only the first element from each member of *mmb*. By including the second argument, the full vectors are extracted.

```
>>> random.seed( 8193 )
>>> clust1, mmb = KMeans( 6, data )
Difference 7.41782845541
Difference 2.70456785889
Difference 0.180388645499
Difference 0.0

>>> for i in range( 6 ):
      print i, "%.3f" %ClusterVar( data.take(mmb[i],0) )

0 0.014
1 0.020
2 0.018
3 0.021
4 0.019
5 0.0177
```

Code 17-10

One of the clusters does have a higher variance than the other clusters. In the k-means algorithm, every vector will be assigned to a cluster. Even a vector that is not similar to any other vector must be assigned to a cluster. Often this algorithm will end up with one cluster that collects outliers and has a higher variance. The solution to this is discussed in Section 17.4. However, it is important to first discuss how to solve more difficult problems in Section 17.3.

17.3 Solving More Difficult Problems

The Swiss roll problem is one in which data is organized in a spiral. One thousand data points are shown in Figure 17-2. The data is created by **MakeRoll** in Code 17-11, which then displays the generation of the data.

```
# swissroll.py
from numpy import random, pi, cos, sin
# create the data file
def MakeRoll( N=1000 ):
    # N = number of data points
    data = zeros( (N,2), float )
    for i in range( N ):
        r = random.rand( 2 )
        theta = 720*r[0] *pi/180
        radius = r[0] + (r[1]-0.5)*0.2
        # convert to x-y coords
        x = radius * cos( theta )
        y = radius * sin( theta )
        data[i] = x,y
    return data

# use the random.seed to replicate the results in the text
>>> random.seed( 284554 )
>>> data = MakeRoll()
>>> akando.PlotMultiple( 'swiss.txt', data )
gnuplot> plot 'swiss.txt'
```

Code 17-11

17.3 Solving More Difficult Problems

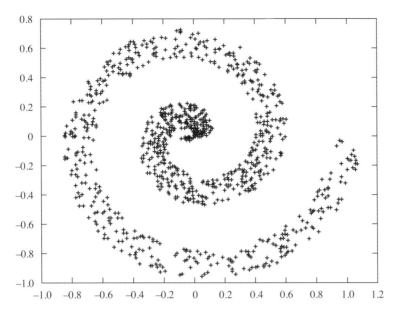

FIGURE 17-2 One thousand data points in a Swiss roll pattern.

Using ordinary k-means, it is possible to cluster the data. Code 17-12 shows the process while Figure 17-3 shows the results. The function **RunKMeans** follows the standard protocol for k-means. The first cluster is initialized and then the process iterates the two steps until a solution is reached. The function **GnuPlotFiles** will create a file for each cluster and store the data for each cluster in two columns in a text file. This is readable by GnuPlot. The results show that some clusters have several members from two separate parts of the spiral.

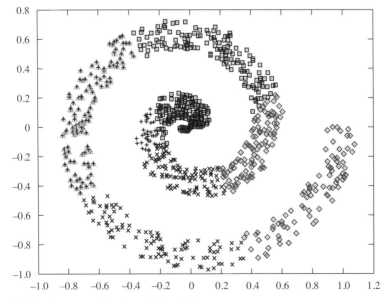

FIGURE 17-3 Four clusters resulting from Swiss roll data.

```
# swissroll.py
def RunKMeans( data, K=4 ):
    clust1 = kmeans.Init2( K, data )
    dff = 1
    while dff > 0:
        mmb = kmeans.AssignMembership( clust1, data )
        clust2 = kmeans.ClusterAverage( mmb, data )
        dff = ( abs( clust1.ravel()-clust2.ravel() )).sum()
        print dff
        clust1 = clust2 + 0
    return clust1, mmb

def GnuPlotFiles( mmb, data, fname ):
    # mmb is from RunKMeans
    # fname is a partial filename
    NC = len( mmb )   # number of clusters
    for i in range( NC ):
        filename = fname + str(i) + '.txt'
        fp = open( filename, 'w' )
        for j in mmb[i]:
            x,y = data[j]
            fp.write( str(x) + ' ' + str(y) + '\n')
        fp.close()

>>> clust, mmb = RunKMeans( data, 4 )
>>> GnuPlotFiles( mmb, data, 'mp' )
gnuplot> plot 'mp0.txt', 'mp1.txt', 'mp2.txt', 'mp3.txt'
```

Code 17-12

According to these results, the vectors representing the clusters are not on the bands. For example, the average of the first cluster is located at (0.58, −0.27), between the two sections of points denoted by the diamonds. The algorithm worked as it was told to do, and all of the points nearest this cluster vector are denoted as belonging to the cluster. However, the algorithm has selected two different arms of the spiral to belong to the same cluster — a result that is undesirable.

Before the k-means algorithm is applied to a problem, it is essential to understand the nature of the problem itself. Quite often the data needs to be processed before applying the algorithm. This is true for almost any algorithm. To solve this problem properly, there are two approaches. The first is to preprocess the data before sending it to the k-means engine. The second is to modify the k-means engine. Both approaches are valid.

17.3.1 Preprocessing Data

Knowing that the data is in some sort of spiral is evidence that a different representation of the data is warranted. In this case the preprocessing step is merely to convert the data to polar coordinates. In other applications the data may need to be transformed by more involved mathematics. Code 17-13 shows the function **GoPolar**, which uses simple mathematics to convert Cartesian coordinates to polar coordinates using

$$r = \sqrt{x^2 + y^2} \tag{17-3}$$

and

$$\theta = \tan^{-1}\left(\frac{y}{x}\right). \tag{17-4}$$

In this program the function **atan2** is used instead of **atan**. In most languages **atan2** represents the arc-tangent that is sensitive to quadrants. The answer has a range of 360 degrees, whereas the **atan** function has a range of 180 degrees. The result is that each *pdata[k]* is the polar coordinate of each *data[k]*. This function makes one small adjustment in that it multiplies the radius by a factor of 10, which puts the radial and angular values on the same scale.

```
# swissroll.py
from numpy import sqrt
from math import atan2
def GoPolar( data ):
    N = len( data )  # number of data points
    pdata = zeros( (N,2), float )
    for i in range( N ):
        x,y = data[i]
        r = sqrt( x*x + y*y )
        theta = atan2( y,x)
        pdata[i] = r, theta
    pdata[:,0] *=10      # scale the radius
    return pdata
>>> pdata = GoPolar( data )
```

Code 17-13

The converted data is now clustered by the same *k*-means algorithm, as shown in Code 17-14. Note that the data sent to **GnuPlotFiles** is the Cartesian data, not the polar data. This is necessary since the data is plotted in Cartesian coordinates. However, the clusters are defined from the polar data. The results are shown in Figure 17-4. By simply casting the data into a different coordinate space, the clustering is significantly different and in this case produces the desired result.

```
>>> clust, mmb = RunKMeans( pdata, 4 )
>>> GnuPlotFiles( mmb, data, 'mp' )
gnuplot> plot 'mp0.txt', 'mp1.txt', 'mp2.txt', 'mp3.txt'
```

Code 17-14

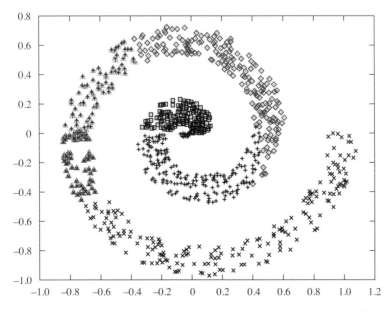

FIGURE 17-4 Multiple clusters generated from the polar version of the data shown in Code 17-14.

17.3.2 Modifications of *k*-Means

Another approach is to realize that in this case the Euclidean distance between data points is not the desired metric of similarity. The clusters should follow the trend of the data that is defined by the proximity of data points. Readers will see a spiral, but this is merely an illusion created by the density of data points.

Thus, for this case, a better metric is to measure the geodesic distances to data points. Two points that are neighbors have a distance measured by the Euclidean distance, but two points that are farther apart measure their distance as the shortest distance that connects through intermediate points. Thus, if there are three points (A, B, and C), the distance between A and C is the distance from A to B and then B to C. The geodesic distance is the shortest path that connects data points.

In order to accomplish this modification, it is necessary to compute the shortest distance between all possible pairs of points. The Floyd-Warshall algorithm performs this task in very few steps (Cormen, Leiserson, and Rivest, 2000). The algorithm contains three nested *for*-loops, which in Python would run very slowly. So, the Python algorithm uses an outer-addition algorithm that contains two of the *for*-loops. This function performs

$$M_{i,j} = a_i + b_j; \quad \forall i, j. \tag{17-5}$$

The **FastFloyd** function in Code 17-15 computes the shortest geodesic distance to all pairs of points. Even this more efficient version of the Floyd-Warshall algorithm can take a bit of time, and the **print** statement is used merely to show the user the progress of the algorithm.

```
# swissroll.py
from numpy import add, greater, less
def FastFloyd( w ):
    d = w + 0
    N = len( d )
    oldd = d + 0
    for k in range( N ):
        print k,
        newd = add.outer( oldd[:,k], oldd[k] )
        m = greater( newd, 700 )
        newd = (1-m)*newd + m * oldd
        mask = less( newd, oldd )
        mmask = 1-mask
        g = mask*newd + mmask * oldd
        oldd = g + 0
    return g
```

Code 17-15

The input to **FastFloyd** is a matrix of all the Euclidean distances for all pairs of points. The Floyd-Warshall algorithm will then search for shorter distances using combinations of intermediate data points. Code 17-16 shows the conversion of the *data* to Euclidean distances and then the call to **FastFloyd**. The result is a matrix that contains the geodesic distances for all possible pairs of points.

17.3 Solving More Difficult Problems

```
# swissroll.py
def Neighbors( data ):
    ND = len( data )
    d = zeros( (ND,ND), float )
    for i in range( ND ):
        for j in range( i ):
            a = data[i] - data[j]
            a = sqrt( ( a*a ).sum() )
            d[i,j] = d[j,i] = a
    return d

>>> dists = Neighbors( data )
>>> floyd = FastFloyd( dists )
>>> f = floyd**2
```

Code 17-16

Finally, the *k*-means algorithm is modified. In the original version the vectors were assigned to the cluster that was closest to the vector in a Euclidean sense. In this new version the vector is assigned to the cluster that is closest in a geodesic sense. So, the **AssignMembership** algorithm is modified. It first finds the data point that is closest to each cluster and then adds that distance to the geodesic distance of each data point to this closest point. This is the distance from the cluster to all of the data points. These distances are computed for all clusters. The last *for*-loop considers each data point and finds the cluster that is closest and assigns the data point to that cluster.

Code 17-17 displays the new **AssignMembership** function, followed by the Python commands to run the new *k*-means algorithm. Note that the **ClusterAverage** function comes from the *kmeans* module whereas the **AssignMembership** function uses the newly defined function. Displaying the results from this modification, Figure 17-5 shows that the clusters tend to capture points along the spiral arm, which is the desired result.

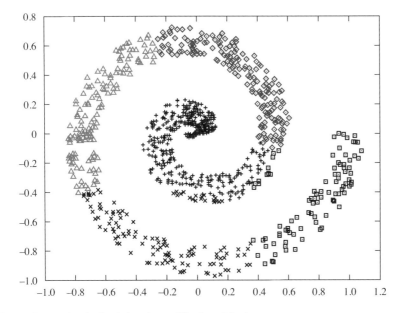

FIGURE 17-5 The results obtained after the modification of the *k*-means.

```
# swissroll.py
def AssignMembership( clusts, data, floyd ):
    mmb = []
    NC = len( clusts )
    ND = len( data )
    for i in range( NC ):
        mmb.append( [] )
    # for each cluster:get the distance to all data points
    dists = zeros( (NC,ND), float )
    for i in range( NC ):
        # find the data point closest to the cluster
        d = zeros( ND, float )
        for j in range( ND ):
            t = clusts[i] - data[j]
            d[j] = sqrt( (t*t).sum() )
        mn = d.argmin() # index of closest point
        mndist = d[mn]
        # use floyd distances
        dists[i] = mndist + floyd[mn]
    #for each data point:get the closest cluster and assign
    for i in range( ND ):
        # find the cluster with the min distance
        mn = dists[:,i].argmin()
        # assign
        mmb[mn].append( i )
    return mmb

>>> import kmeans
>>> diff = 1
>>> c1 = kmeans.Init2( 5, data )
>>> while diff > 0:
    mmb = AssignMembership( c1, data, f )
    c2 = kmeans.ClusterAverage( mmb, data )
    diff = sum( abs( ravel(c1)-ravel(c2)))
    print diff
    c1 = c2 + 0
>>> GnuPlotFiles( mmb, data, 'mp' )
```

Code 17-17

17.4 Dynamic *k*-Means

The number of clusters in the *k*-means algorithm is established by the user, and it is usually done with very little information. If too few clusters are created, the clusters tend to collect vectors that are not similar to the rest because all vectors have to belong to a cluster. If there are too many clusters, clusters can be built that are quite similar to each other. One method of approaching this problem is to dynamically change the number of clusters. The system needs to detect when there are too many or too few clusters and to make the appropriate adjustments.

The variance is measured by Equation (17-2) and remains small as long as the cluster contains similar constituents. Dissimilar vectors will raise the variance, but Equation (17-2) does not indicate which vector is dissimilar. This can in fact be determined. If there is more than one outlier, however, the isolation of the outliers does not necessarily indicate the required number of new clusters that are needed. Thus a simple approach is to detect that a cluster has a high variance and to random split its vectors into two new clusters and then allow the *k*-means iterations to sort it out.

To detect if two clusters are similar, the cluster average vectors are compared with one another. If they are similar, the constituents of the two clusters can be combined into a single cluster. This is also a simple but effective approach. Consider Code 17-18, which generates a set of data with five seeds. The data is shown in Figure 17-6.

```
>>> random.seed( 234)
>>> data = CData( 1000, 2, 0.3, 5 )
>>> akando.PlotMultiple( 'plot.txt', data )
gnuplot> plot 'plot.txt'
```

Code 17-18

Even knowing that there are five clusters does not guarantee that the k-means will cluster correctly. Code 17-19 presents an example run, and the results are shown in Figure 17-7. The first cluster is marked by crosses and shares a block of data points with two other clusters. The second cluster is marked with x's and includes two blocks of data. Even though the data is well separated, the clustering does not perform as expected.

```
>>> c1 = kmeans.Init2( 5, data )
>>> clust, mmb = RunKMeans( data, 5 )
>>> swederoll.GnuPlotFiles( mmb, data, 'mp' )
gnuplot> plot 'mp0.txt', 'mp1.txt', 'mp2.txt', 'mp3.txt'
```

Code 17-19

Code 17-20 shows the intercluster variances and then the intracluster differences. In the latter case the cluster numbers are printed before the difference between them. The intercluster variance is used to measure the similarities within a cluster. If the cluster gets too large, then it should be split. In the example case it is evident that the

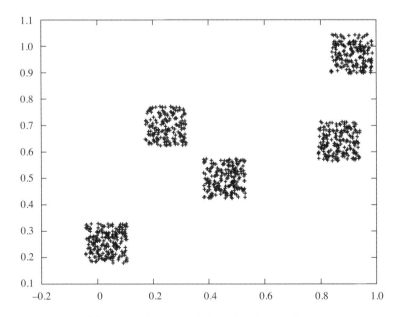

FIGURE 17-6 Five groups of data randomly generated about six different points.

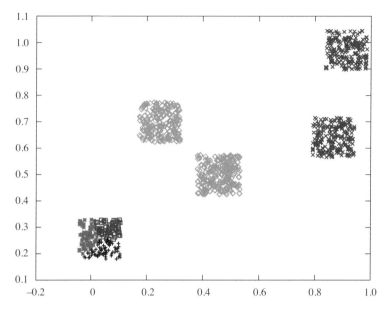

FIGURE 17-7 The clusters generated from the *k*-means algorithm.

first cluster should be split. Thus a threshold between 0.007 and 0.010 is needed to define the clusters that need to split.

The intracluster difference measures the similarity between cluster average vectors. If this is too small, the cluster vectors are close together and the clusters should be joined. In the example problem of Figure 17-7 there are three clusters that need to be joined into a single cluster. Code 17-20 indicates that the difference between these two cluster average vectors is 0.4 while all other vector pairs have a distance greater than 1.

```
>>> for i in range( 5 ):
        print "%.6f" %kmeans.ClusterVar( data[mmb[i]] )

0.000749
0.015767
0.000763
0.000518
0.012182

>>> for i in range( 5 ):
        for j in range( i ):
            a = clust[i] - clust[j]
            d = sqrt( (a*a).sum() )
            print i,j , "%.3f" % d

1 0 1.023
2 0 0.080
2 1 1.046
3 0 0.083
3 1 0.963
3 2 0.084
4 0 0.490
4 1 0.567
4 2 0.495
4 3 0.417
```

Code 17-20

17.4 Dynamic k-Means

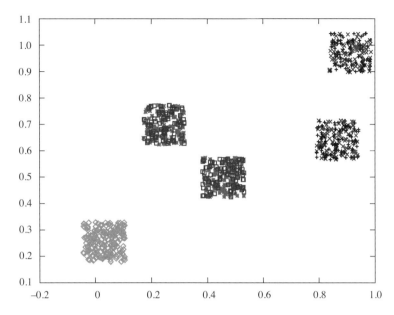

FIGURE 17-8 New clusters after splitting and combining the previous clusters based on inter- and intracluster variances.

Dynamic clustering will then separate clusters *mmb[1]* and *mmb[4]* into two clusters and combine clusters *mmb[0]*, *mmb[2]*, and *mmb[3]*. The splitting of a cluster is performed randomly. Recall that *mmb* is a list, and within that list is another list for each cluster. Randomly splitting a list involves creating two new lists and placing the constituents in either one. In Code 17-21 *m1* and *m2* are the split of *mmb[1]*. Likewise *m3* and *m4* are the split of *mmb[4]*. The *m5* is the combination of the other three clusters. Figure 17-8 shows the results. The combination works well but the splitting was done randomly, and so vectors from both groups are in both clusters.

```
# kmeans.py
def Split( mmbi ):
    # mmbi is a single mmb[i]
    m1, m2 = [], []
    N = len( mmbi )
    for i in range( N ):
        r = random.rand()
        if r < 0.5:
            m1.append( mmbi[i] )
        else:
            m2.append( mmbi[i] )
    return m1, m2

>>> m1, m2 = kmeans.Split(mmb[1] )
>>> m3, m4 = kmeans.Split( mmb[4] )
>>> m5 = mmb[0] + mmb[2] + mmb[3]
>>> mmb = [ m1, m2, m3, m4, m5 ]

>>> GnuPlotFiles( mmb, data, 'mp' )
```

Code 17-21

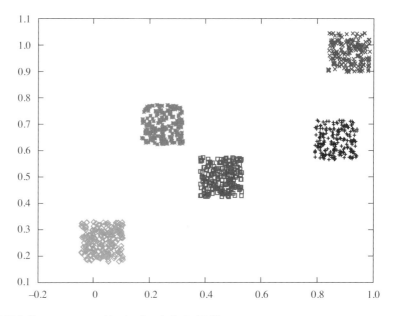

FIGURE 17-9 Clusters generated by the data in Code 17-22.

The final step is to run the k-means as shown in Code 17-22. Figure 17-9 shows the results, which are more in line with the expected results.

```
>>> c2 = kmeans.ClusterAverage( mmb, data )
>>> c1 = c2 + 0
>>> diff = 1
>>> while diff > 0:
        mmb = kmeans.AssignMembership( c1, data )
        c2 = kmeans.ClusterAverage( mmb, data )
        diff = ( abs( ravel(c1)-ravel(c2))).sum()
        print diff
        c1 = c2 + 0
>>> GnuPlotFiles( mmb, data, 'mp' )
```

Code 17-22

17.5 Comments on k-Means

As shown in the previous example, k-means may not solve the simplest cases without some aid. Or can they? The final solution shown in Figure 17-9 is interpreted as being correct by humans. It is entirely possible that the solution of Figure 17-7 is better suited for the application. In reality, the interpretation of the final results is completely up to the user. The potential danger of using k-means (or any clustering algorithm) occurs when the results are trusted without testing. Sometimes a different initialization will produce very different clusters. So when designing a problem that will be solved by k-means, it is also necessary to design a test to see if the clusters are as desired. It may be necessary to compute new clusters, change the data, change the algorithms, or split and combine clusters.

Finally, large problems may consume too much computer time, and so a process of *hierarchical clustering* can be employed. Basically, the data is clustered into a small number of clusters (thus keeping computations to a minimum). Once those clusters are computed, the data in each can be clustered again into smaller subclusters.

17.6 Summary

Clustering is a method in which a generic class of algorithms attempts to organize data in terms of self-similarity. This is a difficult task as similarity measures may be inadequate. One of the most popular methods of clustering involves the k-means algorithm, which requires the user to define the number of desired clusters and the similarity metric. The algorithm iterates between defining clusters and moving data vectors between clusters. It is an easy method to implement and can often provide sufficient results.

However, more complicated problems will require modifications to the algorithm. This will require that users understand the nature of the data and are able to define data conversions to improve performance.

Users should be aware that there is no magic clustering algorithm. It is necessary to understand the problem and the source and nature of the data, and to have expectations of results. Clustering results should be tested to determine if the clusters have the desired properties as well.

Bibliography

Cormen, T. H., C. E. Leiserson, and R. L. Rivest. (2000). *Introduction to Algorithms*. Cambridge, MA: MIT Press.

Problems

1. Create a set of vectors of the form $cos(0.1*x + r)$. Each vector should be N in length. The x is the index (0, 1, 2, ... N–1) and r is a random number. Cluster these vectors using k-means. Plot all of the vectors in a single cluster and in a single graph. Repeat for all clusters. Using these plots, show that the k-means clustered.

2. Repeat Problem 1 using $cos(0.1*x) + r$. Compute the clusters using k-means and plot. Explain what the clustering algorithm did.

3. Modify k-means so that the measure of similarity between two vectors is not the distance but the inner product.

4. Modify *k*-means so that it will cluster strings instead of vectors. Create many random DNA strings of length N. Cluster these strings. Each cluster should have a set of strings in which some of the elements are common. In other words, the first cluster contains a set of strings, and all of the mth elements are T. For each cluster find the positions in the strings that have the same letter.

5. Repeat Problem 4 but for each cluster find the positions in the strings that have common letters. For example, 75% of the mth element in the strings in the nth cluster were A.

6. Create a hierarchical clustering system. Generate data similar to Figure 17-6. Run *k*-means for $K=2$. Treat each of the clusters as a new data set. Run *k*-means on each of the new data sets. Plot the results in a fashion similar to Figure 17-9.

18 Self-Organizing Maps

The *k*-means algorithm does a fairly good job of organizing data, but as seen in the previous chapter it does not always provide the best results. A more powerful solution is the self-organizing map (SOM), a data organization algorithm that creates a mapping space that can be used as either a clustering tool or an associative memory (Kohonen 1982, 1990). This chapter will review the construction of the SOM with a simple application.

18.1 SOM Theory

The SOM is a set of vectors arranged in a lattice. The elements of the vectors are altered through training so that the boundaries are created within the lattice. These boundaries define the cluster regions.

The process begins with a set of data vectors $\{\mathbf{X} : \vec{x}_1...\vec{x}_N\}$ that all have the length D. The SOM lattice consists of a two-dimensional array of vectors $\{\mathbf{Y} : \vec{y}_{1,1}...\vec{y}_{1,P},\vec{y}_{2,1},...\vec{y}_{P,P}\}$ that are also of length D. The user decides on the value of P, which is a trade-off between the variations in the data vectors and the need to keep the computation time short. The value of P should be greater than 20 to provide sufficient resolution in the SOM. More complicated problems will require a larger value of P, which will result in a larger computational cost.

Initially, the vectors in $\mathbf{Y}$ have random or user-defined values. The SOM training considers each data vector in $\mathbf{X}$ and changes the values of some of the vectors in $\mathbf{Y}$. Each data vector is considered several times with differing SOM parameters, and eventually the vectors in $\mathbf{Y}$ become organized. The following steps occur during the SOM process:

1. Assume that a set of data vectors $\mathbf{X}$ has been given.
2. Create the SOM space $\mathbf{Y}$ and the value of the learning radius *hrad*.
3. Iterate until *hrad* is 1.
   a. Iterate over each vector in X.
      i. Find the vector in Y that best matches $\vec{x}$. (This is the BMU.)
      ii. Update the vectors in $\mathbf{Y}$ surrounding the BMU.
   b. Reduce the value of *hrad*.

Within the double loop (Steps 3.a.i and 3.a.ii), the program is considering a single data vector $\vec{x}$ and comparing it with all of the vectors in $\mathbf{Y}$. One of the $\mathbf{Y}$ vectors will be found to best match $\vec{x}$ and will be designated as the best matching unit (BMU). The method by which $\vec{x}$ is compared with each of the **Y** vectors is up to the user, but quite commonly the Euclidean distance is used. Once the BMU is located, all of the vectors in the SOM that are within a radius of *hrad* from the location of the BMU are updated by

$$\vec{y}[t+1] = \vec{y}[t] + h[t](\vec{x}_j - \vec{y}[t]), \tag{18-1}$$

where *t* is the iteration index and *h[t]* is the scalar that controls the training speed. Usually this value is much closer to 0 than it is to 1. After all of the vectors in **X** have been considered, the value of *hrad* is reduced and the process repeats. This shrinks the radius about the BMU. The SOM training ends when *hrad=1*.

During the training, the **Y** vectors are changed to become more like the **X** vectors. Initially, *hrad* is large (slightly larger than *P*), and so each **Y** vector is affected by many vectors in **X**. As *hrad* shrinks, the number of vectors that alter a vector in **Y** also shrinks. Eventually, each **Y** vector is affected by only one **X** vector (or a group of very similar **X** vectors). All **Y** vectors within a radius of *hrad* are encouraged to be similar to each other, but as *hrad* shrinks, so does this influence. Eventually, vectors in **Y** lose influence on each other, and boundaries that define the clustering regions are formed.

After training is complete, each **X** vector is located in the SOM by computing its final BMU. The **X** vectors that have BMUs in the same region of the SOM are clustered together.

18.2 An SOM Example

A good visual example of the SOM is to cluster vectors extracted from a red-green-blue (RGB) image. Each pixel in the image contains three values (red, green, and blue), which creates a vector of length 3. The number of training vectors is $V \times H$. (*V* is the vertical dimension of the image and *H* is the horizontal dimension.) This creates the set of training vectors **X**. The SOM will consist of a set of **Y** vectors that are of length 3. After training is complete, the SOM is expected to have regions that reflect the colors inherent in the image.

18.2.1 Reading an Image

The image used for this example is shown in Figure 18-1. Each pixel in this image is a three-element vector that can be used to train the SOM. The function **ReadImage** in Code 18-1 will convert the image into a single data cube. The output is a three-dimensional cube that is $149 \times 188 \times 3$ (where the first two numbers reflect the dimension of the image).

FIGURE 18-1 An input image that illustrates the use of a self-organizing map. (Used with permission by Ted Borodofsky, MD.)

```
# som.py
from PIL import Image
def ReadImage( fname ):
    mg = Image.open( fname )
    r,g,b = mg.split()
    h,v = r.size
    data = zeros( (v,h,3), float )
    data[:,:,0] = akando.i2a( r )/255.
    data[:,:,1] = akando.i2a( g )/255.
    data[:,:,2] = akando.i2a( b )/255.
    return data

>>> data = ReadImage( '/sojazz.png')
```

Code 18-1

18.2.2 Initializing the SOM

The SOM is a two-dimensional array of **Y** vectors in which the length of a vector is the same as the length of the **X** vectors. The number of **Y** vectors depends on the variance in the **X** vectors. Because there are very few different colors in this case, the size of the SOM can be kept small. The initial values of the vectors can be random or similar to some of the **X** vectors. The function **SOMinit** in Code 18-2 creates a random array that is $30 \times 30 \times 3$, which means that there are 900 SOM vectors of length 3. The random seed is used so that the results can be replicated by the reader, but users would generally not use this command.

```
# som.py
from numpy import random
def SOMinit( N, dm ):
    som = random.ranf( (N,N,dm) )
    return som

>>> random.seed( 4244 )
>>> som = SOMinit(30,3)
```

Code 18-2

Since the SOM vectors are only three elements long, it is easy to view the SOM visually. Each vector is encoded into RGB coordinates, and the image is created from the set of vectors. Code 18-3 shows the function **SOMmg**, which converts the SOM to a viewable image. For applications in which the SOM vector is longer than 3, a modification to this program will be required. Figure 18-2 shows the SOM map that is currently a set of random vectors.

```
# som.py
import akando
def SOMmg( som ):
    r = akando.a2i( som[:,:,0] )
    g = akando.a2i( som[:,:,1] )
    b = akando.a2i( som[:,:,2] )
    mg = Image.merge( 'RGB', (r,g,b) )
    return mg

>>> mg = SOMmg( som )
>>> mg.save('/fig15_2.png')
```

Code 18-3

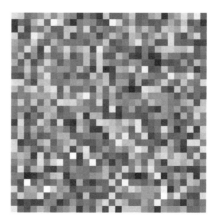

FIGURE 18-2 The SOM map after random initialization.

The image in Figure 18-1 provides the training data for the SOM. In some cases it is prudent to consider each training vector before adjusting $h(t)$. However, in this case there are 28,012 possible training vectors and many of them are identical. To reduce the training time, vectors are selected at random from the image, and *hrad* is lowered after each vector but at an extremely slower rate. The function **RandomVec** in Code 18-4 displays a simple method of extracting a random vector from the data cube.

```
# som.py
def RandomVec( data ):
    V,H,n = data.shape
    y = int( H*random.rand() )
    x = int( V*random.rand() )
    vec = data[x,y]
    return vec
```

Code 18-4

18.2.3 The Best Matching Unit (BMU)

The BMU is the vector in the SOM that is most similar to an input vector. The function **GetBMU** in Code 18-5 uses a Euclidean measure to find this vector and returns the v, h location in the SOM of this vector. The calls to the function demonstrate that the vector returned as the BMU is similar to the randomly chosen training vector.

```
# som.py
from numpy import sqrt
def GetBMU( som, vec ):
    # measure the distance to all vectors
    t = som - vec
    dist = sqrt( ( t*t).sum(2) )
    # find the smallest distance
    H = som.shape[1]
    v,h = divmod( dist.argmin(), H )
    return v,h

>>> vec = RandomVec( data )
>>> bmu = GetBMU( som, vec )
>>> bmu
```

```
(24, 19)
>>> som[bmu]
array([ 0.17310117,  0.00995321,  0.03318832])
>>> vec
array([ 0.09411765,  0.03137255,  0.        ])
```

Code 18-5

This program uses some nice features of Python. Consider the task of subtracting a vector from a two-dimensional array of vectors, as shown in Code 18-6. This is a triple-nested loop that can take a bit of time in Python. Although it is accurate, it is not fast. The first line of **GetBMU** performs the same operation. The *som* is a data cube and *vec* is a vector. Since the last dimension of *som* matches the length of *vec*, Python will subtract *vec* from the two-dimensional array of vectors contained in *som*.

```
for i in range( V ):
    for j in range( H ):
        for k in range( N ):
            t[i,j,k] = som[i,j,k] - vec[k]
```

Code 18-6

The second line of **GetBMU** computes the Euclidean distance from *vec* to all of the vectors in the *som*. This distance can be computed by Code 18-7, but it is much faster to use the single line in the function. The comments in Code 18-7 relate back to portions of the single line in **GetBMU**. Thus, in these two lines the Euclidean distances from *vec* to all 900 vectors in the *som* are computed as the array *dist*.

```
# t1 = t*t
for i in range( V ):
    for j in range( H ):
        for k in range( 3 ):
            t[i,j,k] = t[i,j,k]*t[i,j,k]
# temp = t1.sum(2)
V,H,N = som.shape
temp = zeros( (V,H) )
for i in range( V ):
    for j in range( H ):
        for k in range( 3 ):
            temp[i,j] += t[i,j,k]
# dist = sqrt( temp )
dist = zeros( (V,H) )
for i in range( V ):
    for j in range( H ):
        dist[i,j] = sqrt( temp[i,j] )
```

Code 18-7

The goal is to find the vector in the *som* that is closest to *vec*, and this would be the one with the minimum distance. The variables *v, h* are computed to be the location where the minimum occurs in *dist*, which is also the location in the *som* that contains the vector that is most similar to *vec*. Finally, it should be noted that in the calls to the function, *bmu* is a tuple that can be used as indices to an array. So, *som[bmu]* returns a vector located at *som[16,17]*.

18.2.4 Updating the SOM

Once the BMU is located, the next step is to adjust all of the vectors in the SOM that are close to the BMU—that is, all the vectors that are within a spatial distance of *hrad*. A simple approach would use a double loop. For example, the BMU is determined to be the **Y** vector located at (v_b, h_b). A vector $\vec{y}[v, h]$ is within a distance *hrad* if $\sqrt{(v - v_b)^2 + (h - h_b)^2} < hrad$. This can be determined by Code 18-8. However, this is a double loop and is slow. A much faster approach relies on the function **indices**.

```
for i in range( V ):
    for j in range( H ):
        if sqrt( (vb-i)**2 + (hb-j)**2 ) < hrad:
            # Update the vector som[i,j]
```

Code 18-8

The **indices** function is shown in Code 18-9, and it receives a tuple that is two dimensions. It returns a three-dimensional array. Inside of it are two two-dimensional arrays in which one increments values along the rows and the other increments along the columns. Code 18-10 computes the linear distance from each pixel in an array to the pixel in the upper-left corner by using the **indices**. In this small example the distance from *dist[v, h]* to *dist[0, 0]* is given by the value *dist[v, h]*.

```
>>> indices( (4,5) )
array([[[0, 0, 0, 0, 0],
        [1, 1, 1, 1, 1],
        [2, 2, 2, 2, 2],
        [3, 3, 3, 3, 3]],

       [[0, 1, 2, 3, 4],
        [0, 1, 2, 3, 4],
        [0, 1, 2, 3, 4],
        [0, 1, 2, 3, 4]]])
```

Code 18-9

```
>>> from numpy import set_printoptions
>>> set_printoptions( precision=3)
>>> ndx = indices( (4,5))
>>> dist = sqrt( ndx[0]**2 + ndx[1]**2 )
>>> dist
array([[ 0.   ,  1.   ,  2.   ,  3.   ,  4.   ],
       [ 1.   ,  1.414,  2.236,  3.162,  4.123],
       [ 2.   ,  2.236,  2.828,  3.606,  4.472],
       [ 3.   ,  3.162,  3.606,  4.243,  5.   ]])
```

Code 18-10

The upper-left corner in this case is the centroid of the distance measures since it is the location in which both arrays of **indices** are 0. By shifting the location of the zeros, the spatial distance from all locations in an array to a specified location can be computed. In Code 18-11 the centroid is moved to (2, 3), and the distance from this location to all other squares in the matrix are easily computed.

```
>>> ndx = indices( (5,5))
>>> ndx[0] -= 2
>>> ndx[1] -= 3
>>> sqrt( ndx[0]**2 + ndx[1]**2 )
array([[ 3.606,  2.828,  2.236,  2.  ,  2.236],
       [ 3.162,  2.236,  1.414,  1.  ,  1.414],
       [ 3.   ,  2.   ,  1.   ,  0.  ,  1.   ],
       [ 3.162,  2.236,  1.414,  1.  ,  1.414],
       [ 3.606,  2.828,  2.236,  2.  ,  2.236]])
```

Code 18-11

Consider this approach applied to an array that is the size of the SOM. The distance array is computed so that the centroid is located at the BMU. It is clear which locations are within a radius of *hrad* by simply computing *less(dist, hrad)*. In this fashion the Python loops as shown in Code 18-8 are not used.

Code 18-12 contains the function **Update**, and the spatial distance from all SOM vectors to the BMU is computed. The vectors that will be updated are those whose distance to the BMU is less than *hrad*, and the variable *mask* delineates these vectors from the others. The *mask* is multiplied by 0.05, which is the chosen value for *h[t]*. The final step is a loop that performs the update for all of the SOM vectors. The last two lines in the function perform the actual update. As a final comment, the **multiply.outer** command is used to speed up the computation shown in Code 18-13.

```
# som.py
def Update( som, vec, hrad, bmu ):
    # compute the vectors inside of a radius
    V,H, n = som.shape
    bmuv, bmuh = bmu
    ndx = indices( (V,H) )
    ndx[0] -= bmuv
    ndx[1] -= bmuh
    dist = sqrt( ndx[0]**2 + ndx[1]**2 )
    mask = less( dist, hrad ).astype(float)
    # weight these vectors
    mask *= 0.05
    # update
    mask = multiply.outer( mask,ones(n,int) )
    som += mask*(vec-som)
```

Code 18-12

```
    for i in range( n ):
        som[:,:,i] = som[:,:,i] + mask*(vec-som)[:,:,i]
```

Code 18-13

18.2.5 SOM Iterations

Traditionally, a SOM will consider all of the training vectors, lower the value of *h[t]*, and then move on to the next iteration. In this case the training vectors are randomly

FIGURE 18-3 A map of the SOM.

selected from a larger set. The *h[t]* is thus lowered after each vector is considered but at a much smaller rate than the traditional method.

Code 18-14 contains the function **SOMiterate**, which performs iterations and lowers *hrad* until it becomes a value of 1. The call to the function creates the SOM shown in Figure 18-3. Since there is randomization in initial construction of this SOM, another run would produce a different map.

```
# som.py
def SOMiterate( data, som, hradinit, hrate ):
    hrad = hradinit
    while hrad > 1:
        vec = RandomVec( data )
        bmu = GetBMU( som, vec )
        Update( som, vec, hrad, bmu )
        hrad *= hrate
    return som
>>> random.seed( 1138 )
>>> som = SOMinit( 30, 3 )
>>> som = SOMiterate( data, som, 30, 0.999 )
>>> SOMmg( som ).save('fig3.png')
```

Code 18-14

18.2.6 Interpreting the SOM

This particular SOM contains 900 vectors, and it has changed the vectors so that they are organized into groups of similar vectors that reflect the training vectors. Groups are isolated by comparing them to a training vector. Code 18-15 picks one of the training vectors and compares it to all of the SOM. A value $a[i, j]$ is the absolute subtraction of the SOM vector at i, j to *vec*, the input vector. A perfect match would be indicated by a value of 0.

18.2 An SOM Example

```
>>> vec = RandomVec( data )
>>> vec
array([ 1.        ,  1.        ,  0.09411765])
>>> a = (abs( som-vec )).sum(2)
>>> akando.a2i( a ).save('fig4.gif')
```

Code 18-15

The results are shown in Figure 18-4, where the dark regions are associated with the lower scores, which in this case are the better matches to the vector. As shown in Code 18-15, the selected vector contained red and green but very little blue, which creates yellow. In Figure 18-3 the yellow region is in the upper-left corner, and in Figure 18-4 this is the darker region. The boundary of a region can be defined by setting a threshold and finding all elements in matrix *a* that are below this threshold.

Code 18-16 extracts all of the yellow vectors from the image. The BMU of each of these is computed and stored in *hits*, which is an array the same size as the SOM lattice. Figure 18-5 marks the locations of these BMUs, and it is seen that they are grouped together matching the same region that is yellow in Figure 18-3. The same procedure can be used to find the red vectors, as shown in Figure 18-6. Clearly there are distinct groups for each.

```
>>> yellow = []
>>> for i in range( 149 ):
        for j in range( 188 ):
            if abs(data[i,j]-array([1,1,0])).sum() < 0.3:
                yellow.append( data[i,j] )
>>> hits = zeros( (30,30), int )
>>> for i in yellow:
        bmu = GetBMU( som, i )
        hits[bmu] =1
>>> akando.a2i( hits ).show()
```

Code 18-16

The next step is to define the boundaries in the SOM. The function **FindBoundaries** in Code 18-17 creates the two arrays *used* and *groups*. The *used* array

FIGURE 18-4 A comparison of an input vector and the SOM.

284   CHAPTER 18   SELF-ORGANIZING MAPS

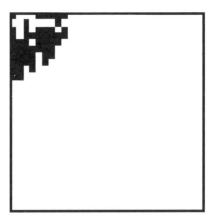

FIGURE 18-5 Black pixels indicating a BMU location for a yellow vector.

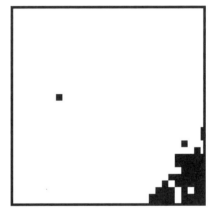

FIGURE 18-6 Black pixels indicating locations of BMUs for red vectors.

is full of zeros, but when a pixel v, h is determined to be in a group, the *used[v, h]* is set to 1. The *groups* array behaves the same way except that *groups[v, h]* is set to the group identification *ctrs*. The *used* array prevents pixels from being used twice. The *groups* array marks each group with a unique identification, as shown in Figure 18-7. Increasing the value of *gamma* will decrease the number of groups.

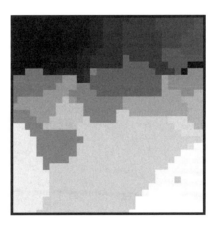

FIGURE 18-7 Boundaries in the SOM.

```
# som.py
def FindBoundaries( som, gamma=0.3 ):
    V,H,N = som.shape
    used = zeros( (V,H), int )
    groups = zeros( (V,H), int )
    ctr = 0
    while used.sum() < V*H:
        # pick a point
        for i in range( V ):
            for j in range( H ):
                if used[i,j]==0:
                    pi, pj = i,j
                    break
            if used[i,j]==0:
                break
        # find similar vectors
        probe = som[pi,pj]
        hits = abs(som-probe).sum(2)
        hits = less( hits, gamma ).astype(int)
        # define a group, update mask
        hits = logical_and( hits, (1-used) )
        used += hits
        if hits.sum() > 10:
            groups += ctr*hits
        ctr +=1
    return groups

>>> groups = FindBoundaries( som )
>>> akando.a2i( groups ).show()
```

Code 18-17

The final test clusters the data vectors. In the function **Cluster** in Code 18-18, the BMU of each data vector is again computed, but this time it is compared with the groups as defined by **FindBoundaries**. In addition, *mmbs* contains lists for each group ID. When vector *data[i, j]* is identified as belonging to group *k*, then it is appended to *mmbs[k]*. Thus, each list in *mmbs* becomes a list of the data vectors that belong to each group.

```
def Cluster( data, som, groups ):
    M = groups.max()
    mmbs = []
    for i in range( M+1 ):
        mmbs.append( [] )
    V,H,N = data.shape
    for i in range( V ):
        for j in range( H ):
            bmu = GetBMU( som, data[i,j] )
            groupid = groups[bmu]
            mmbs[groupid].append( data[i,j] )
    return mmbs

>>> mmbs = Cluster( data, som, groups )
```

Code 18-18

Did the SOM cluster well? Each list in *mmbs* is supposed to be a list of vectors that belong to the same group. In Code 18-19 the standard deviation of the data in each cluster is computed. Each vector has a length of 3, which means that there will be three standard deviation measures for each cluster. If the clusters are good, then the standard deviations should be low. The final line in Code 18-19 computes the standard deviation for all of the data vectors. The cluster standard deviations are seen to be much less than the overall standard deviation, and therefore it is concluded that clustering was achieved.

```
def StdDevs( mmbs ):
    for i in range(len(mmbs)):
        mat = array( mmbs[i] )
        print i, mat.std(0)

>>> StdDevs( mmbs )
0 [ 0.18444606   0.17936019   0.13097264]
1 [ 0.10980499   0.10272827   0.13415602]
2 [ 0.123451     0.10758153   0.08283947]
3 [ 0.12426433   0.10183875   0.08495605]
4 0.0
5 [ 0.10097848   0.08394196   0.05162767]
6 [ 0.14040437   0.0725663    0.10259251]
7 [ 0.03460168   0.08469498   0.13209471]
8 [ 0.06703896   0.07045247   0.08987883]
9 0.0
10 [ 0.02768175   0.02220421   0.02918107]

>>> data.reshape( (149*188,3) ).std(0)
array([ 0.30447813,   0.3159966 ,   0.16470184])
```

Code 18-19

The SOM can also be used as an associative memory, but most of the work is already accomplished. Basically, each cluster is assigned a description (perhaps as simple as "yellow cluster"). An unclassified vector $\vec{x}_U$ is presented to the SOM, and the BMU is calculated. The associative memory would return the group identification for $\vec{x}_U$, thus classifying it.

In the previous work the clusters were defined as those vectors that were close in a simple distance metric. Users have the option to create the SOM and groups using any other metric.

18.3 Summary

The self-organizing map (SOM) is a useful algorithm that incrementally modifies a set of vectors to become more like training vectors. The result is that the SOM vectors tend to organize themselves into groups in which each group has a similarity to one of the data vectors. Thus the SOM becomes a clustering engine.

By labeling such groups, the SOM can become an associative memory engine. Unclassified vectors can be presented to the SOM, which will locate the input vector into one of the labeled groups. Thus the previously unseen vector adopts a classification.

Bibliography

Biopython. (2007). *Biopython*. Retrieved from http://biopython.org.

Kohonen, T. (1982). Self-organized formation of topologically correct feature maps. *Biological Cybernetics*, 43, 59–69.

Kohonen, T. (1990). The self-organizing map. *Proceedings of the IEEE*, 78, 1464–1480.

Southern Jazz. (2007) *Southern Jazz*. Retrieved from http://www.southernjazz.com/.

Problems

1. Select a photographic image (with many more colors than Figure 18-1). Run the SOM to produce the map for this photo (as done in Figure 18-3).

2. Compute Figure 18-3 for Figure 18-1. Compute the histogram of the red data for Figure 18-1 (using **akando.RangeHistogram**). Compute the histogram of the red data in the SOM. Are the histograms similar?

3. Compute the SOM for Figure 18-1 (or an image of your choice). Repeat five times. Do the five maps match?

4. Modify the SOM so that the iterations stop too soon (*hrad > 1*). What does the SOM map look like? Describe the characteristics of a map when it stops too soon.

5. Create the SOM map for an image of your choice. After the SOM is built, compute the BMUs for all pixels of the image. Create a new array, *cnt*, that counts the BMUs for each *x, y* location in the SOM map. In other words, if a pixel produces a BMU of *v, h*, then *cnt[v, h] + =1*. Display *cnt* as an image.

6. Modify the problem so that pairs of pixels are combined to form a six-element vector. The first data vector is the RGB components of pixels [0, 0] and [0, 1], and the second data vector is the RGB component of pixels [0, 2], [0, 3], etc. Build a SOM with these six new element data vectors. Write a program that will convert this SOM with vectors of length 6 to an RGB display.

19 Principal Component Analysis

The data generated from experiments in bioinformatics may contain several dimensions and be quite complicated. However, the dimensionality of the data may far exceed its complexity. A reduction in dimensionality often allows simpler algorithms to analyze the data effectively. The most common method of data reduction in bioinformatics is *principal component analysis* (PCA).

19.1 The Purpose of PCA

Principal component analysis is an often-used tool that reduces the dimensionality of a problem. Consider a set of vectors that lie in R^N space. It is possible that the data is not scattered about but has an organization. When looked at in one view, the data looks scattered, but if viewed from a different orientation the data appears more organized. In this case the data does not need to be represented in the full R^N space but can be viewed in a reduced dimensional space as represented by this view. PCA is then used to find the coordinate axes that best orient this data, which thus possibly reduces the number of dimensions that are needed to describe the data.

In the following example there are three vectors of data, each with three dimensions. The data is three-dimensional, but the first dimension and the third are exactly the same. Thus one dimension is redundant, and in actuality the data is just two-dimensional.

$X_1 = \{\,2, 1, 2\,\}$
$X_2 = \{\,3, 4, 3\,\}$
$X_3 = \{\,5, 6, 5\,\}$

Let's consider a more complicated example: a data set that consists of P vectors, each with N elements and $N>>P$. In this situation a vector **v** can be written in terms of scalar coefficients: $\{a_1v_1, a_2v_2, \ldots, a_Nv_N\}$. It takes only N samples to compute all of the coefficients a_k. Yet, there are only P samples, and therefore it is not possible to determine all of the values of the coefficients. Basically, it is the same as the situation above in that some of the elements in the vectors are redundancies of the others. The dimensionality of the problem can be reduced.

The PCA approach is equivalent to rotating the coordinates of the space that could make some of the coordinates unimportant. Figure 19-1 shows a set of data points presented in two dimensions. Each data point requires an *(x, y)* coordinate to locate it in

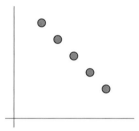

FIGURE 19-1 Data points in R^2.

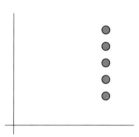

FIGURE 19-2 The same data points shown in Figure 19-1, after a rotation.

the space. Figure 19-2 shows that same data after rotation. In this case all of the x coordinates of the data are the same, and therefore this coordinate is no longer needed. The only necessary description of the data is the y coordinate. Thus through the process of rotation the dimensionality of the data has been reduced from R^2 to R^1.

Dimensionality can be reduced when one coordinate is very much like another or a linear combination of others. This type of redundancy becomes evident in the covariance matrix, which has the ability to indicate which dimensions are dependent on each other. The covariance matrix of a completely independent system will be diagonal; the covariance matrix of this new space will be as diagonal as possible for this data. If it were completely diagonal, then each data point would be found on an individual axis.

For cases in which there are some dependencies, methods such as eigenvectors or SVD (singular valued & decomposition) are used to find a new system that optimally diagonalizes the covariance matrix. Thus before PCA is demonstrated, it is necessary to discuss eigenvectors.

19.2 Eigenvectors

Equation (19-1) is the eigenvector equation for a given matrix $\mathbf{A}$. The matrix-vector multiplication of $\mathbf{A}$ with an eigenvector $\vec{v}$ is equal to the scalar-vector multiplication of an eigenvalue μ with the same vector,

$$\mathbf{A}\vec{v} = \mu\vec{v} \tag{19-1}$$

Generally, the matrix $\mathbf{A}$ is square, and for a matrix of $N \times N$ there are N different eigenvector or eigenvalue pairs.

The *numpy* package provides an eigenvector solution engine. Code 19-1 creates a matrix **A** that is square and symmetric (which emulates the type of matrices that will be used in the PCA analysis). The call to **linalg.eig** returns the eigenvalues and eigenvectors. Since **A** is 3 × 3, there are three values and vectors. The eigenvectors are returned as a matrix, and in the newer versions of *numpy* the eigenvectors are columns in the matrix. (Earlier versions of the predecessor *Numeric* returned eigenvectors as rows in the matrix.) The last two lines in Code 19-1 show that Equation (19-1) holds for the first eigenvalue–eigenvector pair, and similar tests would reveal that it also holds for the other two pairs.

```
>>> from numpy import linalg, random, dot, set_printoptions
>>> set_printoptions( precision =3)
>>> d = random.ranf( (3,3) )
>>> A = dot( d,d.transpose() )
>>> A
array([[ 0.796,  0.582,  0.622],
       [ 0.582,  0.456,  0.506],
       [ 0.622,  0.506,  0.588]])
>>> evl, evc = linalg.eig( A )
>>> evl
array([ 1.774,  0.062,  0.004])
>>> evc
array([[ 0.656,  0.698,  0.284],
       [ 0.505, -0.127, -0.853],
       [ 0.560, -0.704,  0.436]])
>>> dot( evc[:,0], A )
array([ 1.165,  0.896,  0.993])
>>> evl[0]*evc[:,0]
array([ 1.165,  0.896,  0.993])
```

Code 19-1

Another property of the eigenvectors of a symmetric matrix is that they are *orthonormal*, which means that they have a length of 1 and are perpendicular to each other. Code 19-2 demonstrates that the length of the first eigenvector is 1.0 (the self dot product is 1) and that the first two vectors are perpendicular (the dot product is 0).

```
>>> dot( evc[:,0], evc[:,0] )
1.0
>>> dot( evc[:,0], evc[:,1] )
0.0
```

Code 19-2

19.3 The PCA Process

Since the eigenvectors are orthonormal, they define a new coordinate space. The number of eigenvectors is the same as the number of dimensions in the original problem. In the PCA process a decision is made to get rid of some of the eigenvectors because they are deemed as not important, thus reducing the dimension. The matrix that is fed into the **linalg.eig** is the covariance matrix that measures the variance among the different elements in the input vectors. Basically, if element i and element j of the vectors are linked, the element $C[i,j]$ (the covariance matrix) is a larger value. The eigen calculation finds a new space in which the covariance matrix is most diagonal. This is also the space in which different elements of the data vectors are least linked.

The PCA reduction follows four steps:

1. Compute the covariance matrix of the data.
2. Compute the eigenvalues and eigenvectors of the covariance matrix.
3. Decide which dimensions to keep.
4. Convert the data points into the new space.

19.3.1 Case 1: More Dimensions Than Vectors

This demonstration begins with a simple problem of five vectors, each of length 10. Since the number of vectors is less than the dimension, a dimensionality reduction is guaranteed and at least five of the dimensions can be eliminated.

The first step is to compute the covariance matrix, which measures the variation among pairs of dimensions. The covariance between d_i and d_j is large if these two dimensions are coupled. Basically, if it is possible to partially predict the behavior of d_j based on the values of d_i, then the dimensions are coupled. In Figure 19-1 the behavior of the x and y dimensions were coupled, and they had a significant covariance. A covariance is 0 if the two dimensions are completely independent. The covariance of two variables is computed by

$$cov(d_i, d_j) = E[(d_i - \mu_i)(d_j - \mu_j)]. \tag{19-2}$$

The μ_i is the average of the ith variable. Code 19-3 demonstrates the creation of data vectors and the computation of the covariance matrix.

```
>>> random.seed( 1132 )
>>> data = random.ranf( (5,10) )
>>> data -= data.mean(0)
>>> cv = dot( data.transpose(), data )
```

Code 19-3

The covariance matrix is square and symmetric. Code 19-3 shows that the **transpose** function is used to orient the data vector as columns. A quick check to ensure that the correct computation is made is that the dimensions of the covariance matrix should be $N \times N$, where N is the dimension of the vectors. Thus the *covmat* is 10×10 in this case.

The second step is to compute the eigenvectors and eigenvalues of the covariance matrix, as shown in Code 19-4. This computation produced eigenvectors and eigenvalues with imaginary components. The imaginary parts of the numbers were on the order of *1e−18*, which is 0. Because the data type is still complex, only the *real* part of the eigenvalues and vectors are used. As shown here, there are four eigenvalues that are not 0. Values in the range of *e−16* are 0 except for computer bit error.

```
>>> evl, evc = linalg.eig( cv )
>>> evl.real
array([  1.950e+00,   8.318e-01,   1.029e+00,   1.247e-01,  -9.809e-17,
        -9.809e-17,   8.851e-17,  -2.967e-17,   3.600e-17,   2.551e-17])
```

Code 19-4

These four eigenvalues indicate which of the eigenvectors are important. As indicated in Step 3, the other dimensions can be eliminated from further consideration. Consequently, the new space after rotation is only R^4 instead of the original R^{10}.

Step 4 requires that the data points be converted into the new space. The location of the data points are computed by projecting them onto the new coordinates, which is the dot product of the data points with the new axes. The first example uses all ten dimensions in the new space to demonstrate that the data points can be converted to the new space and back to the old space. The coefficients *cffs* in Code 19-5 represent the data points in the new space. Each row in *cffs* is a vector that locates the data point with respect to the new axes as defined by the eigenvectors. The next two lines convert one of the data points from the new space back to the old space. In this case the conversion is performed for the first data vector (since there is a 0 as the first index for *cffs*). The vector *z* should be the same as the original data vector *data[0]*. The final print shows that this is the case and that the data points have been projected into the new space and then back to the original space without error.

```
>>> cffs = zeros( (5,10), float )
>>> for i in range( 5 ):
        for j in range( 10 ):
            cffs[i,j] = dot( data[i], evc[:,j].real )
>>> z = zeros( 10, float )
>>> for i in range( 10 ):
        z += cffs[0,i] * evc[:,i].real
>>> data[0]
array([  4.874e-01,  -4.799e-01,  -4.206e-01,  -4.718e-03,  -3.912e-05,
        -2.388e-01,   2.894e-01,   3.530e-01,  -3.275e-01,   2.460e-01])
>>> z
array([  4.874e-01,  -4.799e-01,  -4.206e-01,  -4.718e-03,  -3.912e-05,
        -2.388e-01,   2.894e-01,   3.530e-01,  -3.275e-01,   2.460e-01])
```

Code 19-5

The PCA eliminates those eigenvectors that have small eigenvalues, so the PCA would reduce the space to R^4. In this example, that is accomplished by keeping only the first four eigenvectors. Code 19-6 shows the conversion to the new PCA space and the example of converting the first data vector back to the original space. The replication occurs again without error. This means that the data points in R^{10} space were converted to the new R^4 space and then converted back again without error. Even though the data was originally in R^{10}, there is actually only R^4 information contained therein. The dimensionality reduction was accomplished.

```
>>> cffs = zeros( (5,4), float )
>>> for i in range( 5 ):
        for j in range( 4 ):
            cffs[i,j] = dot( data[i], evc[:,j].real )

>>> z = zeros( 10, float )
>>> for i in range( 4 ):
        z += cffs[0,i] * evc[:,i].real
>>> z
array([  4.874e-01,  -4.799e-01,  -4.206e-01,  -4.718e-03,  -3.912e-05,
        -2.388e-01,   2.894e-01,   3.530e-01,  -3.275e-01,   2.460e-01])
```

Code 19-6

19.3.2 Case 2: Linear Combinations in the Data

In this case the number of dimensions and number of vectors are the same, but one of the vectors is a linear combination of the others. When this occurs, the dimensions can be reduced. The data in Figure 19-1 illustrates such a case, where all data points are merely one of the data points multiplied by a scalar. Code 19-7 sets up the problem forcing *data[4]* to be a linear combination of three other vectors. The covariance matrix is computed and the eigenvectors extracted. When the eigenvalues are printed, note that *evl[3]* is 0 and therefore *evc[:, 3]* is removed from further consideration.

```
>>> data = random.ranf( (5,5) )
>>> data[4] = data[0] + 0.3*data[1] - 0.7*data[2]
>>> cv = dot( data.transpose(), data )
>>> evl, evc = linalg.eig( cv )
>>> evl
array([  5.752e+00,   6.292e-01,   4.367e-01,
        -3.649e-16,   3.871e-02])
```

Code 19-7

Code 19-8 converts the data points to the new space and then converts the first data point back to the original space. This conversion is performed without error.

```
>>> data = random.ranf( (5,5) )
>>> data[4] = data[0] + 0.3*data[1] - 0.7*data[2]
>>> cv = dot( data.transpose(), data )
>>> evl, evc = linalg.eig( cv )
>>> evl
array([  5.752e+00,   6.292e-01,   4.367e-01,
        -3.649e-16,   3.871e-02])
```

```
>>> pca = zeros( (5,4), float )
>>> pca[:,:3] = evc[:,:3] + 0
>>> pca[:,3] = evc[:,4] + 0
>>> cffs = zeros( (5,4), float )
>>> for i in range( 5 ):
        for j in range( 4 ):
            cffs[i,j] = dot( data[i], pca[:,j].real )
>>> z = zeros( 5, float )
>>> for i in range(4):
        z += cffs[0,i] * pca[:,i].real

>>> z
array([ 0.092, 0.565, 0.815, 0.237, 0.638])
>>> data[0]
array([ 0.092, 0.565, 0.815, 0.237, 0.638])
```

Code 19-8

19.3.3 Case 3: Imperfect Dimensionality Reductions

Because errorless reduction is rarely possible, a key question must be asked: Is this error devastating? Usually, the answer is that it is not. A majority of the data is contained in the principal coordinates (those with the larger eigenvalues). In many applications the data is converted to PCA space and is then organized in a manner that makes it easy to analyze. Unless the analysis requires fine precision, the small errors do not interfere with the analysis.

Very often a problem can be reduced in dimension but not perfectly so. The data contains linear combinations with deviations due to errors or noise. In Code 19-9 the data from Code 19-8 has a little noise added to it. The same process is run, and the differences are noticeable. To begin with, there are no eigenvalues that are 0, but there is one that is much smaller than the others and is thus removed. The reproduction of the data in this case is not exact.

```
>>> data += (2*random.ranf((5,5))-1)*0.05
>>> cv = dot( transpose(data), data )
>>> evl, evc = linalg.eig( cv )
>>> evl
array([ 5.906e+00,  6.149e-01,  4.807e-01,
        8.184e-05,  5.0316e-02])
>>> pca = zeros( (5,4), float )
>>> pca[:,:3] = evc[:,:3] + 0
>>> pca[:,3] = evc[:,4] + 0
>>> cffs = zeros( (5,4), float )
>>> for i in range( 5 ):
        for j in range( 4 ):
            cffs[i,j] = dot( data[i], pca[:,j].real )
>>> z = zeros( 5, float )
>>> for i in range(4):
        z += cffs[0,i] * pca[:,i].real

>>> z
array([ 0.143, 0.571 , 0.821, 0.227, 0.676])
```

Code 19-9

19.3.4 Coordinate Selection

The reduction of dimensions often incurs an error since the eigenvalues are not usually 0. In this case a data set is generated that does have some linear dependencies but not completely so and a decision must be made to select which vectors will define the coordinates of the new space. Code 19-10 creates this data set and then computes the eigenvalues. The eigenvalues are sorted and plotted from the largest to the smallest, as shown in Figure 19-3. This plot behaves in a traditional manner in which a few of the eigenvalues are large and the rest are small. The user must decide which eigenvalues are significant and which are trivial. In this case there is justification for keeping just three of the eigenvectors.

```
>>> from numpy import take
>>> ag = evl.argsort( )
>>> data = random.ranf( (10,10) )
>>> for i in range( 1,9):
        data[i] = (data[i]+data[i-1])/2
>>> cv = dot( data.transpose(), data )
>>> evl, evc = linalg.eig( cv )
>>> ag = argsort( evl )
>>> akando.PlotSave('fig3.txt', take( evl,ag[::-1]))
```

Code 19-10

The list *ag* represents the indices of the eigenvalues from lowest to highest. Thus, the last three entries in *ag* indicate which of the eigenvalues are the largest. In this case it is the last three eigenvectors, and these are copied into *pca*. The data points are moved to the new space and the first data point is converted back to the original space. The values of the original data range between 0 and 1. The last line of code in Code

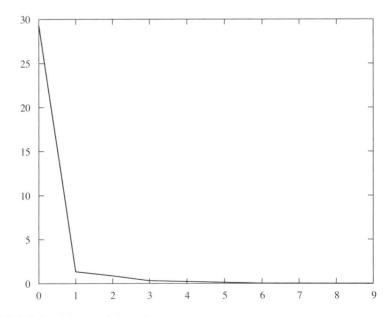

FIGURE 19-3 A plot of the sorted eigenvalues.

19-11 measures the average error between the original data vector and the reproduced data vector. Thus there is about a 4% error.

```
>>> ag[-3:]
array([2, 1, 0])
>>> pca = zeros( (10,3), float )
>>> pca[:,:3] = evc[:,:3] + 0
>>> cffs = zeros( (10,3), float )
>>> for i in range( 10 ):
        for j in range( 3 ):
            cffs[i,j] = dot( data[i], pca[:,j].real )
>>> z = zeros( 10, float )
>>> for i in range(3):
        z += cffs[0,i] * pca[:,i].real
>>> (abs(z-data[0]).mean() )
0.0415935634517
```

Code 19-11

19.4 Using SVD to Compute PCA

Singular valued decomposition (SVD) is another method of manipulating a matrix. Given a matrix **A**, the SVD is

$$\mathbf{A} = \mathbf{U}\mathbf{W}\mathbf{V}^T. \tag{19-3}$$

The matrix **W** is diagonal, and the matrices **U** and **V**ᵀ are Hermitian conjugate, which means that the inverse of the matrix is also the conjugate transpose. One of the uses of SVD is that it can easily invert a matrix. The inverse of matrix **A** may be difficult to compute directly, but when using the SVD method it is simply

$$\mathbf{A}^{-1} = \mathbf{V}\mathbf{W}^{-1}\mathbf{U}^T \tag{19-4}$$

This is verified in Code 19-12.

```
>>> a = random.ranf( (3,3) )
>>> u,w,vt = linalg.svd( a )
>>> ut = u.transpose()
>>> w1 = 1./w
>>> v = vt.transpose()
>>> ai = linalg.inv( a )
>>> zi = dot( v*w1,ut )
>>> (abs(zi-ai)).sum()
9.978e-015   # this is 0 error
```

Code 19-12

The SVD can also be used for PCA. Squaring the values in **W** replicates the eigenvalues as shown in Code 19-13. This example uses the *data* from Code 19-10. The rows in **V**ᵀ are the eigenvectors of the covariance matrix with the exception of a minus sign. This is verified in Code 19-14. The minus sign merely indicates that the vectors are collinear but pointing in opposite directions. This describes the same coordinate rotation with some mirrored dimensions.

```
>>> u,w,vt = linalg.svd( data )
>>> w**2
array([  2.958e+01,   1.455e+00,   9.595e-01,
         2.323e-01,   2.048e-01,   1.324e-01,
         6.668e-02,   4.118e-02,   7.450e-03,
         1.810e-04])
>>> evl
array([  2.958e+01,   1.455e+00,   9.595e-01,
         2.323e-01,   2.048e-01,   1.324e-01,
         6.668e-02,   4.118e-02,   7.450e-03,
         1.81037261e-04])
```

Code 19-13

```
>>> for i in range( 10 ):
        a = (abs(evc[:,i] + vt[i])).sum()
        b = (abs(evc[:,i] - vt[i])).sum()
        print min((a,b) )

2.942e-015
4.329e-015
3.726e-015
3.313e-014
9.230e-014
3.735e-014
3.880e-014
8.661e-014
1.366e-013
9.597e-014
```

Code 19-14

19.5 Describing Systems with Eigenvectors

Consider a system that contains a state vector that is altered in time through some sort of process. An example is the protein population within a cell. The state vector describes the population of proteins at a particular time. As time progresses the populations change, a process that is described as changes in the state vector.

Eigenvectors are a useful tool for describing the progression of a state vector in an easy-to-read format. In this case the state vector is $v[t]$ and the machine that changes the state vector is a simple matrix **M**. In reality the machine that changes the state vector can be quite complicated. The progress of the system is then expressed as

$$v[t+1] = v[t] + Mv[t]. \tag{19-5}$$

Code 19-15 runs the system for 20 iterations, storing each state vector as a row in *data*. The matrix **M** is forced to have a zero sum so that it does not induce energy into the system.

19.5 Describing Systems with Eigenvectors

```
>>> random.seed( 5239 )
>>> M = random.ranf( (5,5) )
>>> M = M - M.mean(0)
>>> data = zeros( (20,5), float )
>>> data[0] = random.rand(5)
>>> for i in range( 1, 20 ):
        data[i] = data[i-1] + dot( M, data[i-1] )
```

Code 19-15

This system contains 20 vectors, and it is not easy to display all of the information. The plot in Figure 19-4 shows just of few of the data vectors. The first element increases in value as time progresses. Some of the others increase and some decrease. Certainly, if the system contained hundreds of vectors and the relationships were complicated, it would be difficult to use such a plot to understand the system.

Code 19-16 computes the PCA of the covariance matrix. In this case the first two eigenvalues indicated that only the first two eigenvectors were to be kept. This data is plotted in Figure 19-5. The 20 data points represent the state of the system at the 20 time intervals. The first point *cffs[0]* is at 0.6,−0.1 and in this case the system is seen to create an outward spiral.

```
>>> cv = dot( transpose(data), data )
>>> evl, evc = linalg.eig( cv )
>>> cffs = zeros( (20,2), float )
>>> for i in range( 20 ):
        cffs[i,0] = dot( data[i], evc[:,0] )
        cffs[i,1] = dot( data[i], evc[:,1] )
>>> akando.PlotMultiple('fig5.txt', cffs )
>>> cffs[0]
array([ 0.604, -0.115])
```

Code 19-16

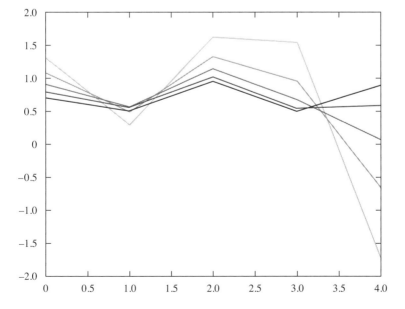

FIGURE 19-4 Time progression of a simple system.

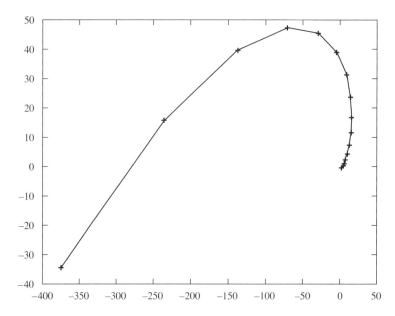

FIGURE 19-5 A plot of the data shown in Code 19-16.

The outward spiral indicates that the values in the system are increasing in magnitude. If this were to continue, the values of the system would approach infinity. This is an unstable system. An inward spiral would indicate that the system is tending toward a steady state in which the state vector stops changing.

The most interesting cases occur when the spiral does not expand outward or move inward. The system draws overlapping circles (or other types of enclosed geometries). This indicates that the system has obtained a stable oscillation. If the path exactly repeats its path, the oscillations are exactly repeated.

```
>>> from numpy import greater, sign
>>> data = zeros( (1000,5) )
>>> for i in range( 1, 1000 ):
        data[i] = data[i-1] + dot( M, data[i-1] )
        mask = greater( abs(data[i]), 1).astype(int)
        data[i] = (1-mask)*data[i] + mask*sign(data[i])
```
Code 19-17

Code 19-17 generates a system in which values are not allowed to exceed a magnitude of 1. As plotted in Figure 19-6, it starts in the middle and quickly begins looping to the left. Although this system was run for 1,000 time steps, only a few points are visible because it gets into an oscillation. Code 19-18 presents some of the values, showing that there is a regular cycle that repeats every 44 time steps. The hard corners appear because the system forces values to be no greater than 1, which is a nonlinear operation. The corners occur when one of the elements of the state vector drastically exceeds 1 and the nonlinear restriction is employed.

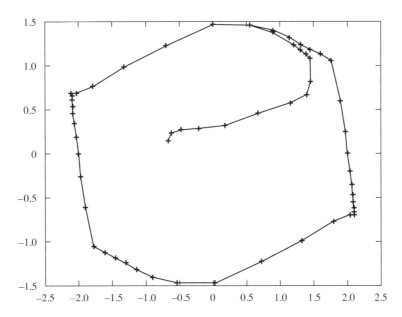

FIGURE 19-6 A closed loop system.

```
>>> cffs[0]
array([-0.042, -0.339])
>>> cffs[10]
array([ 0.893,  0.837])
>>> cffs[20]
array([ 0.983, -1.266])
>>> cffs[40]
array([ 1.359, -1.047])
>>> cffs[60]
array([ 1.526, -0.688])
>>> cffs[80]
array([ 1.681, -0.353])
>>> cffs[81]
array([ 1.484, -0.785])
```

Code 19-18

The systems shown in Figures 19-5 and 19-6 plot a data point for each time step, which also makes it possible to look at changes in the system. Code 19-19 creates a case in which two systems (that are slightly different) are applied to the same initial state vector. Both are analyzed, but this time the covariance matrix is computed on the transpose of the data forcing the variables to be the time steps. As shown in Code 19-19, the computation for *data* followed a similar path.

```
>>> random.seed( 5654 )
>>> data = zeros( (10,5), float )
>>> data[0] = random.rand(5)
>>> for i in range( 1, 10 ):
        data[i] = data[i-1] + dot( M, data[i-1] )
```

```
>>> cv = dot( data, data.transpose() )
>>> evl, evc = linalg.eig( cv )
>>> cffs = zeros( (5,2), float )
>>> for i in range( 5 ):
        cffs[i,0] = dot( data[:,i], evc[:,0] )
        cffs[i,1] = dot( data[:,i], evc[:,1] )
>>> data2 = zeros( (10,5), float )
>>> M2 = M + (2*random.ranf((5,5))-1)/33.
>>> data2[0] = data[0] + 0
>>> for i in range( 1, 10 ):
        data2[i] = data2[i-1] + dot( M2, data2[i-1] )
```

Code 19-19

In a *sensitivity analysis* the *cffs* are just five data points in two-dimensional space. The plot in Figure 19-7 shows a set of +'s which represent the five dimensions of the first system. The *'s represent the same five dimensions in the second system. The two data points that moved apart the most are located around $x = 12$, $y = -0.6$. According to the data in Code 19-19, this is the second variable of the five. The conclusion to draw is that the change in the system affected the second variable more than the others. Likewise, the change in the system barely affected the first and fourth variables.

In each case the system ran for ten time steps, and the plots in Figure 19-7 are sensitive to all of them. Thus the proximity of the data points infers that the first and fourth variables were similar throughout the ten time steps in both systems.

19.6 Eigenimages

Creating a space defined from a set of eigenimages adds a complication. An image that measures 256 × 256 will have 65,536 pixels. The covariance matrix computed from a set of images of this size would be 65,536 × 65,536. This is too large. An alterna-

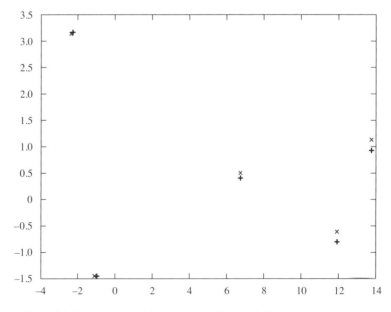

FIGURE 19-7 Plots of the five members of the state vector for two similar systems.

tive is to compute a related set of eigenvectors and convert back to the eigenimages (Turk and Pentland, 1991). The process begins with a set of N images, $\{\mathbf{M}_i; i = 1,...N\}$, in which the average is subtracted,

$$\mathbf{A} = \frac{1}{N}\sum_i \mathbf{M}_i, \tag{19-6}$$

$$\mathbf{M}'_i = \mathbf{M}_i - \mathbf{A} \quad \forall i. \tag{19-7}$$

A matrix $\mathbf{X}$ is constructed so that the ith column of the matrix is the set of pixels from the ith image. The covariance matrix can be computed by the inner product of pairs of columns,

$$c_{i,j} = \sum_k X_{k,i} X_{k,j} \tag{19-8}$$

This creates an extremely large matrix. The equation relating an eigenvector $\mathbf{v}_i$ with an eigenvalue μ_i is

$$C v_i = \mu_i \mathbf{v}_i. \tag{19-9}$$

This can be rewritten as

$$\mathbf{X}\mathbf{X}^T \mathbf{v}_i = \mu_i \mathbf{v}_i. \tag{19-10}$$

Now consider a slightly different eigenequation:

$$\mathbf{L}\mathbf{w}_i = \mathbf{X}^T\mathbf{X}\mathbf{w}_i = \mu_i \mathbf{w}_i. \tag{19-11}$$

The matrix $\mathbf{L}$ is $N \times N$, where N is the number of images. Usually, this is much smaller than the number of pixels, and so this equation is easier to solve. To connect 19.11 with 19.10, the 19.11 is left-multiplied by $\mathbf{X}$.

$$\mathbf{X}\mathbf{X}^T\mathbf{X}\mathbf{w}_i = \mu_i \mathbf{X}\mathbf{w}_i. \tag{19-12}$$

By noting that $\mathbf{C}=\mathbf{X}\mathbf{X}^T$ and that $\mathbf{X}\mathbf{w}$ creates a vector $\mathbf{v}$, the two equations are related. The eigenvalues of 18.11 and 18.9 are the same and the eigenimages are related by

$$\mathbf{v}_i = \sum_k w_{i,k} \mathbf{M}_i \tag{19-13}$$

To demonstrate this, a simple test will use three images, each image being 128×128, as shown in Figure 19-8. The first task is to load the images and subtract the average. Code 19-20 shows the **LoadImages** function, which receives a directory and a list of image file names. It returns a list of matrices.

```
# eigimage.py
from PIL import Image
def LoadImages( dr, mglist ):
    pics = []
    for i in mglist:
        mg = Image.open( dr +'/' + i )
```

```
            pics.append( akando.i2a( mg )/256. )
    return pics
>>> pics = LoadImages( 'data/', mglist )
```

Code 19-20

The average is subtracted in the **SubAvg** function in Code 19-21. The *avg* is needed later, and so it is returned. The subtraction is a centering process that moves the average of the data to the origin of the coordinate system. If this is not done, the first eigenvector becomes sensitive to the bias of the data.

```
# eigimage.py
def SubAvg( data ):
    N = len( data )
    avg = zeros( data[0].shape, float )
    for i in data:
        avg += i
    avg /= N
    #
    for i in range( N ):
        data[i] -= avg
    return avg, data

>>> avg, pics = SubAvg( pics )
```

Code 19-21

The covariance matrix for **L** is computed by the **Lcov** function shown in Code 19-22. This computes the inner product of the different pairs of images.

```
# eigimage.py
def Lcov( data ):
    N = len( data ) # number of images
    L = zeros( (N,N), float )
    for i in range( N ):
        L[i,i] = ( data[i]*data[i]).sum()
        for j in range( i ):
            L[i,j] = L[j,i] = ( data[i]*data[j]).sum()
    return L

>>> L = Lcov( pics )
```

Code 19-22

FIGURE 19-8 Three test images.

19.6 Eigenimages

FIGURE 19-9 The eigenimages created from the three test images in Figure 19-8.

The covariance matrix in this case is 3×3. It is this matrix that is used to determine the eigenvalues and eigenvectors. The eigenvectors are converted to eigenimages in **Eimage**, shown in Code 19-23, according to Equation (19-13). The resulting three eigenimages are shown in Figure 19-9.

```
# eigimage.py
def Eimage( L, data ):
    evls, evcs = linalg.eig( L )
    emg = []
    V,H = data[0].shape
    for j in range( N ):
        temp = zeros( (V,H), float )
        for i in range( N ):
            temp += evcs[i,j] * data[i]
        emg.append( temp )
    return emg, evls

>>> emg, evls = Eimage( L, pics )
```

Code 19-23

The eigenimages are used for multiple types of applications (see Section 23.4.4). One such example would be to use them as a classification resource. Coefficients of an image are created by multiplying the image with the eigenimages. In this case the original images can be reduced to just three scalars since there are three eigenimages. The eigenvalues are 34.8, 9.8, and 0, and the eigenimage associated with a very low eigenvalue can be removed from consideration. In this case, it is thus possible to reduce each of the three images to just two scalars. Code 19-24 shows this reduction.

```
>>> for i in range( 2 ):
        print ( emg[i]* pics[0] ).sum()
-17.1725333268
6.39438834731

>>> for i in range( 2 ):
        print ( emg[i]* pics[1] ).sum()
28.231001222
0.99506838928
```

```
>>> for i in range( 2 ):
        print ( emg[i]* pics[2] ).sum()
-11.058
-7.389
```

Code 19-24

In some applications these coefficients are used for recognition (Turk and Pentland, 1991). The theory is that similar images will be reduced to similar sets of coefficients, and it is much easier to compare the coefficients. The program **Coeffs** in Code 19-25 will compute the coefficients.

```
# eigimage.py
def Coeffs( pca, data ):
    D = len( pca )
    N = len( data )
    cffs = zeros( (N,D), float )
    for i in range( N ):
        for j in range( D ):
            cffs[i,j] = ( pca[j] * data[i]).sum()
    return cffs
```

Code 19-25

19.7 Summary

Principal component analysis (PCA) is an often-used tool that presents a new set of coordinates in which the data can be represented. These components are orthonormal vectors and are basically a rotation of the original coordinate system. However, the rotation minimizes the covariance matrix of the data, which means that some of the coordinates may become unimportant. Because these coordinates can be discarded, PCA space uses fewer dimensions to represent the data than the original coordinate system.

Principal components can be computed using an eigenvector engine or singular valued decomposition (SVD). The NumPy package offers both and the interface is quite easy. Eigenvectors are also used to explore the progression of a system. Limit cycle plots using eigenvectors indicate if the system is shrinking, expanding, or caught in some sort of oscillation.

Bibliography

Turk, M., Pentland, A. (1991). Eigenfocus for Recognition, *Journal of Cognitive Neuroscience*, 3(1), 71–86.

Problems

1. You have been given a set of N vectors in which the eigenvalues turn out to be $1, 0, 0, 0, \ldots$. What does this imply?

2. You have been given a set of N vectors, in which the eigenvalues turn out to be $1, 1, 1, 1, \ldots$. What does this imply?

3. Describe what you expect the eigenvalues to be in a set of purely random vectors. Confirm your prediction.

4. You have been given a set of N random vectors of length D. Compute the covariance matrix and the eigenvectors. Now compute the covariance matrix of the eigenvectors. Explain the results.

5. Repeat the work to obtain Figure 19-6, but add $\pm 5\%$ noise to each iteration. Explain the new system plot.

20 Species Identification

As we saw in earlier chapters, most amino acids can be created from several codons. This chapter will test the theory that different species can be identified by their codon distributions. By comparing the codon frequencies of genes from different species, systems can be developed to identify the species.

20.1 Data Collection

In a study conducted by Kanaya et al. (2001), 29 different bacterial species produced more than 59,000 genes, each with at least 100 codons. The study showed that species had signature codon frequencies and that it was possible to detect these patterns. The work in this chapter will use just five of the species as an instructional tool. The selected bacteria are shown in Table 20-1.

Each of the files contains several genes, and each file is read using the *genbank.py* routines. The DNA sequences for each gene are extracted and converted to codons. There are 64 different codons, and so each gene is converted to a vector that references the codons by a number between 0 and 63. Code 20-1 displays the function **FileReadConvert**, which reads in a Genbank file and extracts the coding DNA. Each of these is converted to a vector representing the codons. The i-loop considers each gene in the file and the j-loop extracts each codon in the ith gene. In the example, the first gene in the file is represented in *cods[0]*. It has a length of 700, so there are 2,100 bases in this gene. The indices of the first ten codons as well as the first two codons to these indices are shown.

Table 20-1 Accession Numbers for the Selected Bacteria.

Bacteria	Accession Number
Aquifex aeolicus	NC_00918
Pyrococcus abyssi	AL096836
Archaeoglobus fulgidus	AE000782
Helicobacter pylori	CP000241
Rickettsia prowazekii	AJ235269

```
import genbank

# codonfreq.py
def FileReadConvert( fname, codkeys ):
    # fname is the name of a Genbank file
    # codkeys is codons.keys( ) where codons is from genbank.Codons( )
    # read in the DNA sequences for the genes
    data = genbank.ReadGenbank( fname )
    dna = genbank.ParseDNA( data )
    klocs = genbank.FindKeywordLocs( data )
    glocs = genbank.GeneLocs( data, klocs )
    NG = len( glocs ) # number of genes
    codons = []
    for i in range( NG ):
        # extract DNA for this sequence
        cdna = genbank.GetCodingDNA( dna, glocs[i] )
        # convert to codons
        c = [] # codons for this gene
        for j in range( 0, len(cdna), 3 ):
            c.append( codkeys.index( cdna[j:j+3] ))
        codons.append( c )
    return codons

>>> fname = 'data/genbank/nc_00918.gb.txt'
>>> coddict = genbank.Codons()
>>> codkeys = coddict.keys()
>>> cods = FileReadConvert( fname, codkeys )
>>> cods[0][:10]
[42, 21, 51, 36, 22, 49, 47, 36, 1, 14]
>>> codkeys[42]
'atg'
>>> codkeys[21]
'gcg'
```

Code 20-1

The conversion of one gene to a codon frequency requires counting the number of times each codon appears and dividing those numbers by the total number of codons. This simple procedure is accomplished by **Freq64** in Code 20-2. The example shows the first codon being converted to a frequency vector. The output *f* is a vector of 64 numbers and the *f.sum( )* is 1.0. The first ten elements are printed. The first entry shows that the first codon appears in 0.57% of the gene.

```
from numpy import zeros

# codonfreq.py
def Freq64( inlist ):
    cnts = zeros( 64, float )
    for i in range( 64 ):
        cnts[i] = inlist.count( i )
    freq = cnts/cnts.sum()
    return freq

>>> f = Freq64( cods[0] )
>>> f[:10]
array([ 0.005,  0.021,  0.004,  0.042,  0.012,
        0.   ,  0.011,  0.002,  0.   ,  0.008])
```

Code 20-2

Of course, this needs to be repeated for all genes in the list *cods*. Code 20-3 displays **CodonFrequencies**, which repeatedly calls **Freq64** to convert each gene. The variable *lim* controls the minimum number of codons necessary to be used. Genes smaller than this length are discarded.

```
# codonfreq.py
def CodonFrequencies( cods, lim=100 ):
    # cods from FileReadConvert
    N = len( cods ) # number of genes
    codfreqs = [ ]
    for i in range( N ):
        if len( cods[i]) >= 3*lim:
            codfreqs.append( Freq64( cods[i] ))
    return codfreqs

>>> codfqs = CodonFrequencies( cods )
```

Code 20-3

This is repeated for all five files, as shown in Code 20-4. The result is a list that contains five items. Each item is a list of codon frequencies from each of the five files.

```
>>> fname = ['data/genbank/nc_00918.gb.txt', \
             'data/genbank/al096836.gb.txt',\
             'data/genbank/ae000782.gb.txt', \
             'data/genbank/cp000241.gb.txt',\
             'data/genbank/aj235269.gb.txt']
>>> cfs = []
>>> for i in range( 5 ):
        c = FileReadConvert( fname[i], codkeys )
        cfs.append( CodonFrequencies( c ))
```

Code 20-4

20.2 The First Clustering

The first test will determine if there is any validity to the theory that different species will have different codon distributions. Clustering is easily accomplished using the *k*-means algorithm, as demonstrated in Code 20-5. Since it is known that there are five species that generated this data, we can assume that $K = 5$.

```
>>> import kmeans
>>> K = 5
>>> data = cfs[0] + cfs[1] + cfs[2] + cfs[3] + cfs[4]
>>> clusts, mmb = kmeans.KMeans( K, array(data))
```

Code 20-5

The *k*-means algorithm will cluster any data fed to it, and so the next step is to determine if these clusters mean anything. Since the goal is to isolate species according to their codon distributions, a completely successful clustering would create five clusters, with each one containing vectors from a single species. The first file contains 681 vectors, which are also the first 681 vectors in the list *data*. If the clustering were perfect, then all 681 vectors would end up in the same cluster.

Code 20-6 counts the number of species that show up in each cluster. Since there are different numbers of genes for each Genbank file, the list *g* contains the boundaries for the different files. For example, the first species contains 681 genes, and so it is contained in *data[0:681]*; the second species contains the next 758 genes and therefore is in *data[681:1439]*. If the clustering were perfect, then each row and each column in *mat* would have only one nonzero entry. As can be seen, it is not perfect, but then again it was not expected to be. This is merely a preliminary study to determine if there is any justification to the idea that species have unique codon distributions. We can see that this study does contain evidence for that. For example, the last cluster (last row) contains mostly data from the first species. The first cluster (first row) is confused between two species, but it clearly rejects the other three species. With sufficient support in hand, it is prudent to further analyze this data.

```
>>> g = [0,681,1439,2339,3014,3042]
>>> mat = zeros( (5,5), int )
>>> for i in range( 5 ):
        for j in range( 5 ):
            g1 = greater( mmb[i], g[j] )
            g2 = less( mmb[i], g[j+1] )
            g1 = logical_and( g1, g2 ).astype(int)
            mat[i,j] = g1.sum()

>>> mat
array([[  0, 511, 728,   0,   0],
       [ 60, 132, 121,   0,   0],
       [ 23,  16,   9,  10,  27],
       [  0,   0,   5, 664,   0],
       [597,  98,  36,   0,   0]])
```
Code 20-6

20.3 Using Principal Component Analysis

This data set contains 3,402 vectors of length 64. Because this is a considerable amount of data for some algorithms to manipulate, a dimensionality reduction using principal component analysis (see Chapter 19) is pursued. Code 20-7 converts the *data* list into a matrix *dmat*. In this case the data vectors are rows in the matrix. The covariance matrix and its eigenvalues and eigenvectors are computed. Figure 20-1 plots the first ten eigenvalues, which indicates that the first four eigenvectors should be used in the PCA conversion. Code 20-8 finishes the conversion.

```
>>> from numpy import cov, linalg, transpose

>>> dmat = array( data )
>>> c = cov( transpose(dmat))
>>> evl, evc = linalg.eig( c )
```
Code 20-7

```
>>> cffs = zeros( (3402,4), float )
>>> for i in range( 3402 ):
        for j in range( 4 ):
            cffs[i,j] = dot( dmat[i], evc[:,j] )
```
Code 20-8

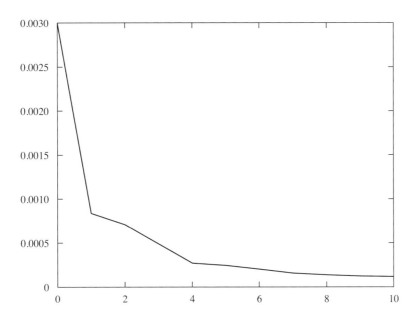

FIGURE 20-1 The first ten eigenvalues from Code 20-7.

20.4 The Second Clustering

The PCA did reduce the data from vectors of length 64 to vectors of length 4, which could possibly be detrimental to the goal and destroy valuable information. The data is thus clustered to ensure that important information has not been destroyed. Code 20-9 runs this clustering, and the results are similar to the previous case. The clusters show biases toward the different species. This indicates that the PCA process did not remove a significant amount of important data, but it does still confirm that the *k*-means did not perform a sufficient clustering.

```
>>> K = 5
>>> clusts, mmb = kmeans.KMeans( K, cffs )
>>> m = zeros( (5,5), int )
>>> for i in range( 5 ):
        for j in range( 5 ):
            g1 = greater( mmb[i], g[j] )
            g2 = less( mmb[i], g[j+1] )
            g1 = logical_and( g1, g2 ).astype(int)
            m[i,j] = g1.sum()
>>> m
array([[  0,  79, 677,   0,   0],
       [ 67, 116, 119,   0,   0],
       [ 27,  16,  24, 674,  27],
       [  8, 536,  72,   0,   0],
       [578,  10,   7,   0,   0]])
```

Code 20-9

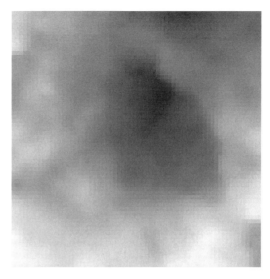

FIGURE 20-2 The SOM map that results from the data in Code 20-11.

20.5 Using a Self-Organizing Map

The self-organizing map (see Chapter 18) is next applied to the dimensionally reduced data. Code 20-10 runs the SOM training code for this data.

```
# codonfreq.py: MakeSOM
>>> net = som.SOMinit2( 64, 4 )
>>> hrad = 100
>>> for i in range( 100 ):
      print i,
      for j in range( 3402 ):
          bmu = som.GetBMU( net, cffs[j] )
          som.Update( net, cffs[j], hrad, bmu )
      hrad -=1
```

Code 20-10

The PCA reduced the data to four dimensions, but the computer can display only three color channels (such as red-green-blue). Thus, the four channels need to be reduced to three color channels. An easy method to do this is shown in Code 20-11. The red, green, and blue channels are sensitive to four channels in differing amounts. The resultant image is shown in Figure 20-2.

```
# codonfreq.py
from PIL import Image
def Color4( net ):
    # net from Make SOM
    # convert 4 channels to 3colors
    V,H,N = net.shape
    red = zeros( (V,H), float )
    green = zeros( (V,H), float )
    # first channel
    blue = net[:,:,0] + 0.67*net[:,:,1] + 0.33*net[:,:,2]
    green = 0.1*net[:,:,0] + 0.4*net[:,:,1] + 0.4*net[:,:,2]+0.1*net[:,:,3]
```

```
        red = net[:,:,3] + 0.67*net[:,:,2] + 0.33*net[:,:,1]
        #
        r = akando.a2i( red )
        g = akando.a2i( green )
        b = akando.a2i( blue )
        mg = Image.merge( 'RGB', (r,g,b) )
        return mg
>>> mg = Color4( net )
>>> mg.save( 'fig2.png' )
```

Code 20-11

Because it is difficult to determine whether the SOM has organized the data in this image, further analysis is needed. The function **LocateMany** computes the BMUs of a given data set and plots them in the same frame as the SOM. In the example shown in Code 20-12, the data from the first species is plotted, and Figure 20-3 shows the results. If the SOM organized the data, the marks should be in an isolated group. As seen in the image, almost all of the BMUs of the first species are located in a region on the SOM. This indicates that the codon frequencies of this species are not being confused with those from another species. In other words, this species is shown to have a unique set of codon frequencies. Species identification via codon frequency is possible.

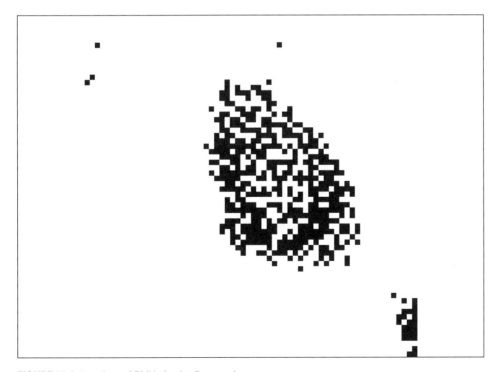

FIGURE 20-3 Locations of BMUs for the first species.

316   CHAPTER 20   SPECIES IDENTIFICATION

```
# locate points
# codonfreq.py
def LocateMany( net, data ):
    N = len( data ) # number of points to plot
    V,H,n = net.shape
    pix = zeros( (V,H) )
    for i in range( N ):
        bmu = som.GetBMU( net, data[i] )
        pix[bmu] = 1
    return pix

>>> pix = LocateMany( net, cffs[:681] )
```

Code 20-12

A concurrent plot of all five species is generated by running **LocateMany** for each species as shown in code 20-13. Figure 20-4 shows the BMU locations of all of the different species, with each species represented by a different gray level. The first four species define definite regions in the SOM space. The last species (shown with the whitest pixels), however, misbehaves and is scattered throughout the SOM. Except for the last species, the SOM was capable of sorting the species according to the codon frequencies.

```
>>> pix = zeros( (64,64) )
>>> for i in range( 5 ):
        pix += (i+1)*LocateMany( net, cffs[g[i]:g[i+1]] )

>>> akando.a2i( pix ).save('fig4.gif')
```

Code 20-13

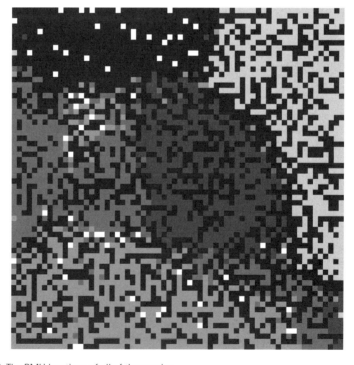

FIGURE 20-4 The BMU locations of all of the species.

20.6 Summary

This chapter considered the question of identifying species by their codon frequencies. Each species that was considered contained several genes, and the frequencies of each codon was computed. The results showed that species do tend to favor certain codons over others.

Three tools from previous chapters were used here: k-means, principal component analysis (PCA), and self-organizing maps. Having such tools made short work of this application.

Bibliography

Kanaya, S., et al. (2001). Analysis of codon usage diversity of bacterial genes with a self-organizing map (SOM): Characterization of horizontally transferred genes with emphasis on the *E. coli* O157 genome. *Gene*, 276, 89–99.

Problems

1. In the original research conducted by Kanaya et al. (2001), 29 different bacterial species were used. Repeat this experiment for as many of these 29 species as you can obtain.

2. Repeat the experiment discussed in Problem 1 for plant genomes.

3. Repeat the experiment discussed in Problem 1 for mammalian genomes.

4. Measure the average distance between two regions in an SOM. One way to do this would be to measure the distance from every point in one region to every point in the other region and to compute the average. Another way would be to find the "center of mass" of each region and measure the distance between them.

5. Write a program to compute the distances between all regions in an SOM (see Problem 4).

6. Bacterial genes have been organized into an evolutionary tree (see http://tolweb.org/Eubacteria/2 or other sources). Are neighboring regions in the SOM also neighboring on the evolutionary tree? Use the program from Problem 5 determine this.

21 Fourier Transforms

Any periodic signal with a wavelength of λ can be decomposed into a set of frequencies with wavelengths that are integer multiples of λ. In practice any digital signal can be converted to a set of frequencies without loss of information. A common way to compute these frequencies is through the use of the Fourier transform, but it is not the only method.

By converting information into Fourier space, it becomes possible to apply a large host of filters to keep the desired frequencies and dispose of the undesirable ones. Fourier filtering has been used to clean up noisy signals, to isolate targets, and to search for complicated structures in the signal. This chapter will focus on a few aspects of Fourier theory and some applications will be presented as instructional examples.

21.1 Fourier Theory

The Fourier transform of a function $f(x)$ converts the signal into a set of waves defined by *sine* and *cosine* functions:

$$f(x) = \frac{A_0}{2} + \sum_{m=1}^{\infty} A_m \cos mkx + \sum_{m=1}^{\infty} B_m \sin mkx. \tag{21-1}$$

The first term is the bias or DC (direct current) term, which is sensitive to the overall average of $f(x)$. The next two terms describe the set of sines and cosines. The wave number k is $2\pi/\lambda$ and m is an integer. The scalar weights A and B indicate the amount of each wave that is present in $f(x)$. The scalar weight A_m is dependent upon $f(x)$ and $\cos mkx$, and B_m is dependent upon $f(x)$ and $\sin mkx$. Modern usage begins with the relationship

$$e^{i\Theta} = \cos\Theta + \iota \sin\Theta, \tag{21-2}$$

and it represents the Fourier transform as

$$F(w) = \int_{-\infty}^{\infty} f(x) e^{-\iota wx} dx. \tag{21-3}$$

This states that each frequency $F(w)$ is dependent upon the integration of the function $f(x)$ masked with the wave of that frequency. In more basic terms, $F(w)$ indicates how much $\exp(-iwx)$ is in $f(x)$. The inverse Fourier transform is slightly different in that the functions swap places and the exponent has a different sign:

$$f(x) = \int_{-\infty}^{\infty} F(w)e^{-\iota wx}\,dw. \tag{21-4}$$

There is a considerable amount of theory that could accompany these equations, but since the signals that are used in bioinformatics are generally digital, there will be no further discussion of the analog Fourier transform here.

21.2 Digital Fourier Transform

Digital signals are sampled at discrete and evenly spaced increments. Because the data stored in computers is digitized, the digital Fourier transform (DFT) is particularly useful.

21.2.1 DFT Theory

The DFT is similar to the analog Fourier transform except that the integration becomes a summation and the exponent changes slightly. The forward DFT is

$$F[w] = \sum_{x=0}^{N-1} f[x] e^{-\iota 2\pi xw/N}. \tag{21-5}$$

The functions now use square brackets instead of parentheses to indicate that the signal is digital. The length of the vector $f[x]$ is N and the result of the computation produces $F[w]$, which has the same length. The first term $F[0]$ is the DC term and is sensitive only to the sum of $f[x]$. The term $F[1]$ uses the term $exp(-\iota 2\pi x/N)$, which has a wavelength of N. Thus $F[1]$ is sensitive to the amount of this wave present in $f[x]$.

The inverse transform is similar to the forward transform with the sign change in the exponent and a scaling factor:

$$f[x] = \frac{1}{N} \sum_{w=0}^{N-1} F[w] e^{-\iota 2\pi xw/N}. \tag{21-6}$$

The Fourier transform does not destroy information concerning the signal, which means that a signal transformed by Equation (21-5) and then Equation (21-6) should be returned to its original form.

21.2.2 Example with a Simple Sawtooth Signal

A sawtooth signal provides an easy demonstration of the Fourier transform. The signal is defined as

$$y = \begin{cases} 0 & x < \lambda/4 \\ 1 & \lambda/4 \le x < 3\lambda/4 \\ 0 & x \ge 3\lambda/4 \end{cases}. \tag{21-7}$$

The DFT of this signal creates a set of frequencies, $F[w]$. The first frequency has a wavelength that spans the length of the signal, and this is shown as the sine wave in Figure 21-1. According to Equation (21-6), the original signal can be reconstructed by the proper addition of the frequencies. In this case the second frequency is not present in the signal, and so $F[2]$ is 0. The $F[3]$ is present, and it creates a wave that is added

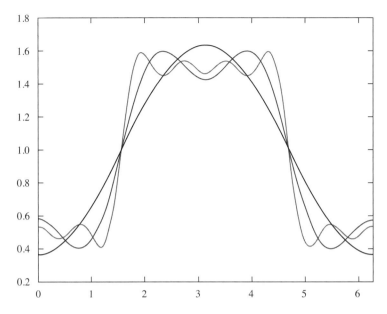

FIGURE 21-1 The summation of the first three nonzero frequencies.

to *F[1]*. This is shown in Figure 21-1 as the curve with two peaks. The third curve in this graph is the addition of *F[1]*, *F[3]*, and *F[5]*. As more of the frequencies are added, the resultant addition becomes more like the original sawtooth function.

The plot associated with *F[1]* can only be a wave of that frequency, and *F[1]* only describes the amplitude of the wave. This is the lowest frequency in the signal and is very broad. Low frequencies are present only if the original signal has broad regions. Had the sawtooth been much thinner, then this wide frequency would have been less present. As seen in this example, the higher frequencies tend to sharpen the edges.

21.2.3 Features of the DFT

There are three features of the DFT that are important to note. The first is that the DC term is the summation of the input signal

$$F[0] = \sum_x f[x]. \tag{21-8}$$

The second is that a forward transform followed by an inverse transform will recreate the original signal

$$f[x] = F^{-1}\{F\{f[x]\}\}. \tag{21-9}$$

The third is that, according to Parseval's theorem, energy in the Fourier representation is the same as it is in the original signal. For the digital domain, there is a scaling factor of N that is the length of $f[x]$. Parseval's theorem is

$$\sum |f[x]|^2 = \frac{1}{N} \sum |F[w]|^2. \tag{21-10}$$

21.2.4 Power Spectrum

In some applications it is necessary to know only the energy present in each frequency. This power spectrum is computed by

$$PS[w] = |F[w]|^2. \tag{21-11}$$

The power spectrum does destroy information, and it is not possible to reconstruct the original signal from the power spectrum. However, the power spectrum having only real valued terms can be plotted and analyzed visually.

21.3 Fast Fourier Transform

The DFT is a useful transformation but an inefficient computation. The fast Fourier transform (FFT) performs the same computations with significantly fewer computations. To understand the efficiency of the FFT, the inefficiencies of the DFT are first explored.

21.3.1 Duplicate Computations

The DFT contains many duplicate computations where the same elements are multiplied or added together numerous times. The efficiency of the DFT is that it eliminates these duplicate computations.

To facilitate this discussion, the following substitution is used to keep the equations a little cleaner:

$$W = \exp[\iota 2\pi/N]. \tag{21-12}$$

Consider a case in which $N=4$. Equation (21-5) becomes

$$\begin{aligned}
F[0] &= W^0 f[0] + W^0 f[1] + W^0 f[2] + W^0 f[3] \\
F[1] &= W^0 f[0] + W^1 f[1] + W^2 f[2] + W^3 f[3] \\
F[2] &= W^0 f[0] + W^2 f[1] + W^4 f[2] + W^6 f[3] \\
F[3] &= W^0 f[0] + W^3 f[1] + W^6 f[2] + W^9 f[3]
\end{aligned} \tag{21-13}$$

The W term is the wave, and it is cyclical with $W^k = W^{N+k}$. In the case where $N=4$, Equation (21-13) becomes

$$\begin{aligned}
F[0] &= W^0 f[0] + W^0 f[1] + W^0 f[2] + W^0 f[3] \\
F[1] &= W^0 f[0] + W^1 f[1] + W^2 f[2] + W^3 f[3] \\
F[2] &= W^0 f[0] + W^2 f[1] + W^0 f[2] + W^2 f[3] \\
F[3] &= W^0 f[0] + W^3 f[1] + W^2 f[2] + W^1 f[3]
\end{aligned} \tag{21-14}$$

The terms in these equations are rearranged and grouped to produce

$$\begin{aligned}
F[0] &= W^0(f[0] + f[2]) + W^0(f[1] + f[3]) \\
F[1] &= W^0(f[0] + W^2 f[2]) + W^1(f[1] + W^2 f[3]) \\
F[2] &= W^0(f[0] + f[2]) + W^2(f[1] + f[3]) \\
F[3] &= W^0(f[0] + W^2 f[2]) + W^3(f[1] + W^2 f[3])
\end{aligned} \tag{21-15}$$

In this form the duplications are seen. The equations for $F[0]$ and $F[2]$ both compute $(f[0] + f[2])$ and $(f[1] + f[3])$. There are duplications in $F[1]$ and $F[3]$ as well. The number of duplications rises rapidly for larger values of N. For a vector of length N, the DFT will require $N \times N$ multadds (multiply-additions). The FFT will require only $N \log_2 N$. These savings are quite significant for larger values of N. For example, if $N = 1024$, the DFT will require 1,048,576 multadds whereas the FFT will require only 10,240. That is a savings on two orders of magnitude for a modest size vector.

21.3.2 The FFT Method

The FFT method begins with a rearrangement of the original data. To demonstrate the FFT, the example of $N = 8$ is used, and the process is shown in Figure 21-2. The first column presents the eight elements of the vector. In the second column, the indices of these elements are converted to binary values. The binary values are reversed and then the data is resorted to these new binary values. These are then converted back to the decimal equivalents. In the end the data is resorted.

The FFT will then perform $\log_2 N$ iterations, so in the case of $N = 8$ there will be three iterations. In the first iteration the elements are grouped into pairs of two, and each pair is used to create a new pair, as shown in Figure 21-3. The first pair is $f[0]$ and $f[4]$. They create two new values of which the first one is $f[0] + f[4]$. The second value is the same except that each term is multiplied by a weighting factor. The weighting factors follow the same pattern as the rearranged indices. When this computation has ended, the new column contains eight complex valued numbers, and the old column can be disposed.

$f[0] >>> f[000] >>> f'[000] \longrightarrow f[000] >>> f[0]$
$f[1] >>> f[001] >>> f'[100] \longrightarrow f[001] >>> f[4]$
$f[2] >>> f[010] >>> f'[010] \longrightarrow f[010] >>> f[2]$
$f[3] >>> f[011] >>> f'[110] \longrightarrow f[011] >>> f[6]$
$f[4] >>> f[100] >>> f'[001] \longrightarrow f[100] >>> f[1]$
$f[5] >>> f[101] >>> f'[101] \longrightarrow f[101] >>> f[5]$
$f[6] >>> f[110] >>> f'[011] \longrightarrow f[110] >>> f[3]$
$f[7] >>> f[111] >>> f'[111] \longrightarrow f[111] >>> f[7]$

FIGURE 21-2 The reordering of the data as the first step in the FFT.

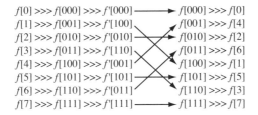

FIGURE 21-3 The first iteration of the FFT.

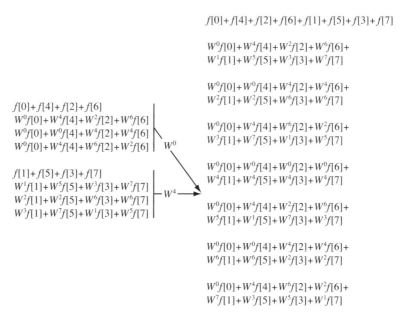

FIGURE 21-4 The second iteration of the FFT.

FIGURE 21-5 The third iteration of the FFT.

The second iteration is much like the first except that the elements are grouped by fours. This process is shown in Figure 21-4. The third iteration groups by eights and is shown in Figure 21-5.

The final result is a column of eight complex numbers, and the first number is $F[0]$, the second is $F[1]$, and so on. Each iteration requires N multadds, and the number of iterations is only $\log_2 N$. The FFT performed the same computation with a significant savings in work.

21.3.3 FFTs in SciPy

The SciPy package provides several Fourier routines, but only four will be reviewed here. These routines are in a module named *fftpack*, and the first line in Code 21-1 shows the import. This code creates a vector of eight random numbers and computes the FFT. The vector *af* is eight complex valued elements, and the rest of the code confirms the algorithm in Equations (21-8) through (21-10).

```
>>> from scipy import fftpack
>>> from numpy import random, conjugate
>>> a = random.rand( 8 )
>>> af = fftpack.fft( a )
# DC
>>> a.sum(), af[0]
(4.57069534941, (4.57069534941+0j))
# Forward and Reverse
>>> a2 = fftpack.ifft( af )
>>> (abs(a-a2)).sum()
4.05528839481e-016
# Parseval
>>> (a*a).sum(), (af*conjugate(af)).sum().real/8
(3.27026724041, 3.27026724041)
```

Code 21-1

As seen in the code, the inverse FFT is called with **fftpack.ifft**. The two-dimensional Fourier transforms are computed by **fftpack.fft2** and **fftpack.ifft2**.

21.3.4 The Swap Function

Code 4-17 presented the **akando.Swap** function, which is used to locate the lower frequencies into the center portion of the Fourier signal. In the case of a vector, the lowest frequencies are located at the ends of the vectors, and the highest frequencies are located in the center of the vector. In some filtering applications it is convenient to locate the lower frequencies in the middle of the vector. This is accomplished by simply swapping vector halves in which the first half of the vector is placed in the second half and vice versa.

For a matrix, the quadrants are swapped as shown in Figure 4-7. This places the lowest frequencies in the middle of the matrix and the higher frequencies toward the edge. This makes analyses such as those in Figures 22-7, 22-8, and 22-10 easier to manipulate.

21.4 Frequency Analysis

The Fourier transform converts the signal to frequency space but does not perform any analysis. However, it does make some types of analysis quite easy to accomplish. A few examples are presented in this section.

21.4.1 Simple Signals

The first example considers a simple signal that consists of a cosine function. The vector *y1* in Code 21-2 creates a simple cosine function that has eight peaks over the length of 256 elements. The Fourier transform is computed and plotted in Figure 21-6 (line without markers). The Fourier transform of this cosine function creates two peaks at $w = 8$ and $w = 248$, which are both eight units from the edge (excepting $F[0]$). The vector *y2* has 64 peaks, and therefore the two spikes (peaks with markers) are at $F[64]$ and $F[256-64]$. The peaks in the Fourier transform indicate the amount of each frequency present in the signal. It should also be noted that if *f[x]* is real-valued, the Fourier transform is symmetric (except for the $F[0]$ term).

```
>>> from numpy import arange, cos, pi
>>> from scipy import fftpack
>>> x = arange( 256 )
>>> y1 = cos( 2*pi/256*x*8 )
>>> Y = fftpack.fft( y1 )
>>> akando.PlotSave( 'fig16_7a.txt', Y.real )
>>> y2 = cos( 2*pi/256*x*64 )
>>> Y = fftpack.fft( y2 )
>>> akando.PlotSave( 'fig16_7b.txt', Y.real )
```

Code 21-2

The Fourier transform is also additive. In Code 21-3 a signal is generated from two cosines with different weighting factors. As seen in Figure 21-7, the Fourier transform is the addition of the two signals.

```
>>> y = cos( 2*pi/256*x*8 ) + 0.5*cos(2*pi/256*x*64)
>>> Y = fftpack.fft( y )
```

Code 21-3

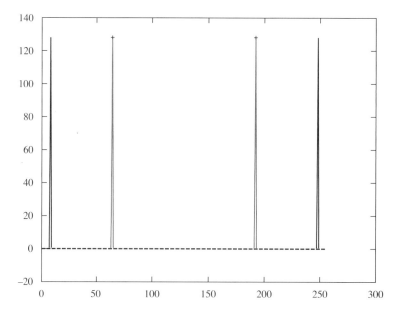

FIGURE 21-6 Two different Fourier transforms of signals with a single (but different) frequency.

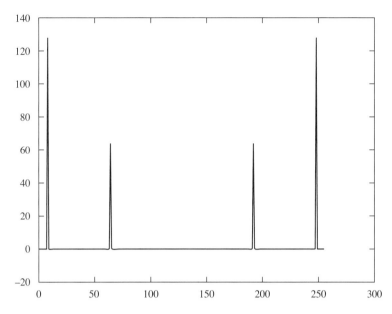

FIGURE 21-7 The Fourier transform of a signal composed of two different frequencies with different amplitudes.

21.4.2 DNA Coding Regions

The next example follows the work of Vanidyanathari and Yoon (2002). The idea is that coding regions in DNA contain a regular pattern of three codons and three bases. This regular pattern is not apparent in the noncoding regions. If proven to be correct, this method could be used to help identify coding regions in DNA streams.

Finding a regular pattern in a signal can be easily accomplished using Fourier transforms, but it is first necessary to collect a large amount of data. Such data can be obtained from Genbank files. Basically, the function **GetData** in Code 21-4 reads a Genbank file that contains a large DNA stream with 818 genes and a large amount of noncoding DNA.

The function **GetData** returns two items. The simpler item is the second item, which is the *dna* and it is merely a string of DNA characters that in this file is almost 1 million characters. The first item is *genes*, and it is a list of 818 items that indicate where the genes are in the *dna* string. Each item in this list is a tuple with two parts. The first is a list of locations for the beginning and ending of the gene; the second is a Boolean value that is true if the gene is actually a complement of the DNA substring. In the example shown, the first gene starts at *dna[98]* and ends at *dna[655]*, and it is not a complement. Some of the genes are spliced together from different portions of the DNA string, and so the list of locations may have several tuples.

```
# foura.py
import genbank
def GetData( fn ):
    gb = genbank.ReadGenbank( fn )
    dna = genbank.ParseDNA( gb )
```

```
        klocs = genbank.FindKeywordLocs( gb )
        genes = genbank.GeneLocs( gb, klocs )
        return genes, dna
>>> genes, dna = GetData('genbank/cp000049.gb.txt')
>>> len( genes )
818
>>> genes[0]
([(98, 655)], False)
```

Code 21-4

The function **Separate** in Code 21-5 creates two lists: *coding* and *noncoding*. Each item in the list is a string of DNA. The *coding* list contains the DNA coding strings (the genes). The function **genbank.GetCodingDNA** returns this string or its complement, depending on the flag. The *noncoding* list contains strings of DNA that are located between the genes (if they are at least 100 bases long). The result is two lists of coding and noncoding DNA. The goal of this work is to show that the coding DNA has a regular pattern not seen in the noncoding DNA.

```
# foura.py
def Separate( genes, dna ):
    Ngenes = len( genes )
    coding = []
    noncoding = []
    for i in range( Ngenes-1 ):
        cdna = genbank.GetCodingDNA( dna, genes[i] )
        coding.append( cdna )
        # noncoding
        start = genes[i][0][-1][-1]
        end = genes[i+1][0][0][0]
        nc = dna[start:end]
        if end-start>100: noncoding.append( nc )
    return coding, noncoding

>>> coding, nocoding = Separate( genes, dna )
```

Code 21-5

Currently, the data is in strings, but the Fourier transform requires numerical values. Chapter 23 considers methods of converting string information into numerical information, and one such method will be adopted here. The function **numseq.Chang** (Code 23.5) converts each of the four letters in DNA into one of four complex numbers. There is also a little smoothing (see Equation 23-1). The mechanics of this function are not important here as the goal is merely to convert a string to a numerical sequence. Code 21-6 demonstrates the conversion of a simple DNA string into a numerical sequence using this method.

```
>>> import numseq
>>> vec = numseq.Chang( 'ATGCTAGC' )
>>> vec
array([-0.25+0.75j, -1.25-0.75j,  0.25-1.25j,  1.25+0.25j, -0.25+1.25j,
       -1.25-0.25j])
```

Code 21-6

21.4 Frequency Analysis

Thus a single string can be converted into a numerical sequence. To convert each string in a list, a simple loop is required, as shown in **EncodeMany** in Code 21-7. Since there are two lists, the function **EncodeBoth** is created to manage both lists.

```
# foura.py
def EncodeMany( clist ):
    c = []
    for i in range( len( clist )):
        c.append( numseq.Chang( clist[i] ))
    return c

def EncodeBoth( coding, noncoding ):
    vcode = EncodeMany( coding )
    ncode = EncodeMany( noncoding )
    return vcode, ncode

>>> vcode, ncode = EncodeBoth( coding, nocoding )
```

Code 21-7

There are two more items that need to be considered before the results can be obtained. The first is that the Fourier transforms will produce complex valued results that are not easy to plot. Therefore the power spectrums shown in Equation (21-11) will be computed. The second is that the sequences have different lengths and the Fourier transforms of the vectors of different lengths are not compatible. The $F[1]$ terms of two vectors of different lengths represent two different wavelengths. The easy solution to this is to make all of the vectors the same length. In **PowerSpectrum** in Code 21-8, the vectors are padded to the size of 8,192 elements. This number is chosen because it is a power of 2 that is larger than the length of any of the genes.

The **PowerBoth** function computes the power spectrum for all vectors in both lists *vcode* and *ncode*. The result is two lists *vps* and *nps* that contain the power spectrums of all of the coding and noncoding sequences.

```
# foura.py
def PowerSpectrum( vecs ):
    N = len( vecs )
    ps = []
    for i in range( N ):
        pad = zeros( 8192, complex )
        T = len( vecs[i] )
        pad[:T] = vecs[i][:T]- vecs[i][:T].mean()+ 0
        a = fftpack.fft( pad )
        ps.append( (a*conjugate(a)).real )
    return ps

# foura.py
def PowerBoth( vcode, ncode ):
    vps = PowerSpectrum( vcode )
    nps = PowerSpectrum( ncode )
    return vps, nps

>>> vps, nps = PowerBoth( vcode, ncode )
```

Code 21-8

330   CHAPTER 21   FOURIER TRANSFORMS

The final step is to display the information. There are 818 genes, and it is not feasible to plot each one. Code 21-9 converts the list of equal length vectors to a matrix and then computes the average of each element. The plot in Figure 21-8 is this average, and the two spikes creating three equally spaced regions indicate that there is a repetitive nature to the vectors with a period of 3. Figure 21-9 is the plot for the noncoding regions, and the spikes are missing.

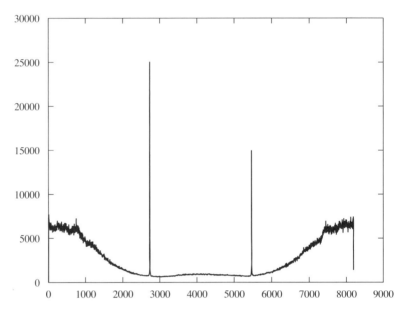

FIGURE 21-8 The power spectrum of the coding regions of several genes. The spikes indicate a repeating nature contained within the data.

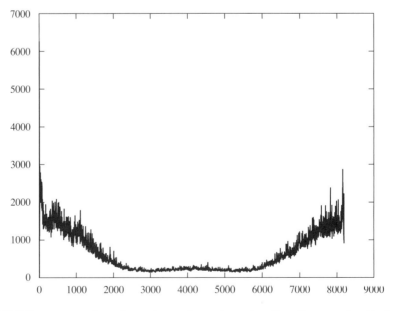

FIGURE 21-9 The power spectrum of several noncoding regions, where the spikes are noticeably absent.

These results support the claim from Vanidyanathari and Yoon (2002) that the coding regions have a repetitive nature. This was seen by converting the sequences to numerical vectors and examining the power spectra.

```
>>> mat = array( vps)
>>> ps = mat.mean(0)
>>> akando.PlotSave('fig16_8.txt', ps )
```

Code 21-9

21.5 Summary

Fourier transforms are powerful tools for extracting frequency-based information from signals. Filtering the frequencies allows for the extraction of important information as demonstrated in two examples. Like so much of this book, this chapter barely touched on the depth and power of these tools.

Bibliography

Vanidyanathari, P. P., and Byung-Jun Yoon. (2002). Digital filters for gene prediction applications. *Proceedings of the IEEE*, 306–310.

Problems

1. For the function shown in Figure 21-10, show that $A_0 = A_m = 0$ and that $B_m = \frac{2}{m\pi}(1 - \cos m\pi)$.

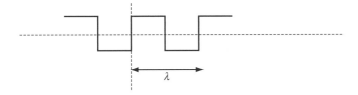

FIGURE 21-10 A periodic sawtooth function

2. What are the even-numbered frequencies for the Fourier transform in Equation (21-7)?

3. Given a vector v or real-valued numbers (not complex numbers), compute the Fourier transform. Show that $v[k] = v[-k]$ for all values of k except $k = 0$.

4. Compute the Fourier transform of a pure cosine function. Why are there two spikes?

5. Compute the Fourier transform of $vec1 + vec2$ where $vec1$ and $vec2$ are two cosine functions with different wavelengths. Plot the results. How many spikes should there be?

6. Review Problem 4. Create a vector f of length N that is all zeros except for two spikes at $f[k]$ and $f[-k]$ (where $0 < k < N/2$). Compute the inverse transform. Do you get a pure sine wave? Does the inverse transform create a pure sine wave?

7. Repeat Problem 6 with $f[k]=1$ and $f[-k]=0$. Compute the inverse transform. Explain the results.

8. Given a vector of length N of random numbers ranging from 0 to 1. Compute the Fourier transform. Set the DC term to 0. Compute the inverse transform. Compare the result with the original vector. Explain what the removal of the DC term accomplished.

22 Correlations

A correlation compares the relationship between two vectors and can detect similarities between them. It is a powerful technique that can be accelerated by the use of Fourier transforms. In this chapter two different applications are considered. The first correlates DNA strings, and the second filters an embryo image.

22.1 Correlation Theory

Convolution and correlation are two similar functions that are easily confused. Convolution is expressed by

$$c(p) = \int f(x)g(p-x)\,dx, \qquad (22\text{-}1)$$

and correlation by

$$c(p) = \int f(x)g^*(x-p)\,dx. \qquad (22\text{-}2)$$

The correlation considers the two functions $f(x)$ and $g(x)$ and uses the parameter p to create every possible alignment. For each alignment, the two functions are multiplied and the results summed. A small digital version of correlation is shown in Figure 22-1, which has two vectors, [5, 4, 1, 3, 2] and [1, 2, 4, 0]. In Part (a), the two have the particular alignment $p = 0$ and the numbers are multiplied (vertical lines) and summed to produce a result of 17. In Part (b), the second vector is shifted and a new result is computed. The final output is vector $c(p)$, which contains these correlation values.

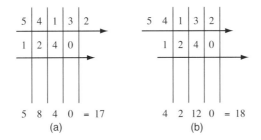

FIGURE 22-1 The diagram of a simple correlation.

Equation (22-2) is the slow method of computing correlations. The FFT can be used to accelerate the process. To begin this derivation, the Fourier transform

$$\mathcal{F}_p\{c(p)\} = \int c(p)\exp(-\iota p\alpha)\,d\alpha, \tag{22-3}$$

is applied to both sides of Equation (22-2) to produce

$$\mathcal{F}_p\{c(p)\} = \int_\alpha \int_x f(x)g^*(x-\alpha)dx\,\exp(-\iota p\alpha)\,d\alpha. \tag{22-4}$$

The substitution $z = x - a$ and $dx = -da$ is employed to produce

$$\mathcal{F}_p\{c(p)\} = \int_z \int_x f(x)g^*(z)dx\,\exp\!\left(-\iota p(x-z)\right)dz. \tag{22-5}$$

The integrals are swapped to get

$$\mathcal{F}_p\{c(p)\} = \int_x \int_z f(x)g^*(z)\exp\!\left(-\iota(x-z)\right)dz\,dx. \tag{22-6}$$

By removing the non-z terms from the z integral, the system becomes

$$\mathcal{F}_p\{c(p)\} = -\int_x f(x)\exp(-\iota px)\int_{-z} g^*(z)\exp(\iota pz))\,dz\,dx. \tag{22-7}$$

The terms in the z-integral are in fact the complex conjugate of the Fourier transform of $g(z)$. Furthermore, the terms in the x-integral are the Fourier transform of $f(x)$. By using capital letters to represent the Fourier transforms, the final form is

$$\mathcal{F}_p\{c(p)\} = F(p)G^*(p). \tag{22-8}$$

The result is that the Fourier transform of the correlation can be determined by computing the Fourier transform of the two functions by multiplying them together in an element by element fashion. The correlation is then the inverse transform of Equation (22-8). This provides a large savings in computation. For a vector of length N, the original correlation would require N^2 multadds whereas the Fourier method requires $3N \log_2 N + N$ multadds. For large N, this is a tremendous savings.

22.2 Random Signal Correlation

The simple example in Code 22-1 illustrates the usefulness of correlation. A signal of random numbers is created and correlated with itself (autocorrelation). The result is shown in Figure 22-2. In this case there is one alignment in which all of the elements are aligned with themselves, and the correlation value is quite large. For all other alignments, random numbers are aligning with random numbers and the correlation values are low. The spike indicates that an alignment has been detected.

22.2 Random Signal Correlation

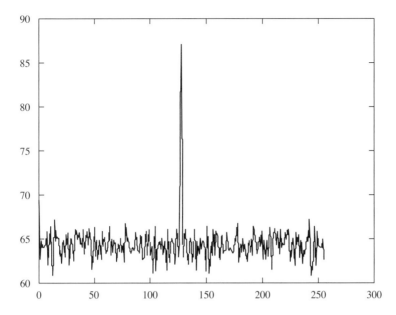

FIGURE 22-2 The autocorrelation of a random vector.

```
>>> from numpy import random
>>> y = random.rand( 256 )
>>> corr = akando.Correlate( y, y ).real
```
Code 22-1

Code 22-2 creates the vector a, which is the same as y except that the elements are shifted 100 places. The two vectors are correlated again and the same spike occurs (Figure 22-3) but in a different location. It is shifted 100 units from the center of the correlation, indicating that vector a and y aligned but there was a relative shift between them.

```
>>> from numpy import concatenate
>>> a = concatenate( (y[100:], y[:100]) )
>>> corr = akando.Correlate( y, a ).real
```
Code 22-2

22.3 Structured Signal Correlation

The random vector produced a sharp correlation spike, but that is not always the case. Code 22-3 shows a different situation in which the signal is a wide Gaussian function. Its correlation is shown in Figure 22-4. This time the correlation is a much wider peak. In this case the two functions align at $p = 0$ and produce the value at the top of the peak. At $p = 1$ there is only a small change in the values that align, and this relative

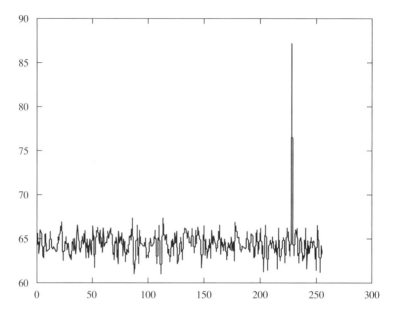

FIGURE 22-3 The correlation of two identical vectors with the exception that one vector is shifted with respect to the other.

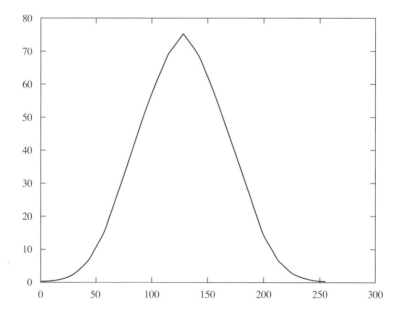

FIGURE 22-4 The correlation of a Gaussian function.

shift also produces a value that is large but not quite as large as the peak. The structure of the signal does affect the nature of the correlation output.

```
>>> from numpy import exp
>>> y = exp( -(x-96)**2/(2*25**2))
>>> corr = akando.Correlate( y, y ).real
```

Code 22-3

22.4 Correlation of DNA Strings

Correlations can be used to align DNA sequences but first the sequences must be converted to a numerical representation. Code 22-4 creates a matrix *mat* that has four rows relating to the four nucleotide bases. In column *k* of *mat*, one of the four elements is set to 1, depending on the value of the DNA string *g[k]*. The DNA is converted to a four-row matrix consisting of 1's and 0's. Each row is correlated with itself, and the four auto-correlations (correlations of a signal with itself) are summed to produce the result shown in Figure 22-5. This spike indicates the alignment of two similar DNA strings.

```
>>> from numpy import zeros
>>> import foura
>>> genes, dna = foura.GetData( 'genbank/cp000049.gb.txt')
>>> g = dna[98:98+256]
>>> abet = 'acgt'
>>> mat = zeros( (4,len(g)))
>>> mat
array([[ 0., 0., 0., ..., 0., 0., 1.],
       [ 0., 0., 0., ..., 1., 1., 0.],
       [ 0., 1., 0., ..., 0., 0., 0.],
       [ 1., 0., 1., ..., 0., 0., 0.]])
>>> for i in range( 256 ):
        ndx = abet.index( g[i] )
        mat[ndx,i] = 1
>>> corr = zeros( 256 )
>>> for i in range( 4 ):
        corr += abs(akando.Correlate( mat[i], mat[i] ))
```

Code 22-4

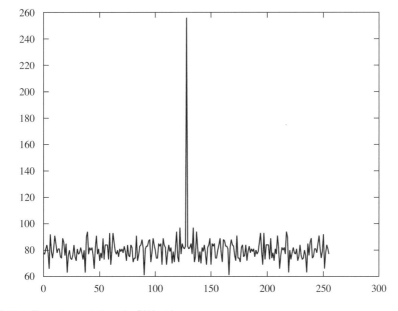

FIGURE 22-5 The autocorrelation of a DNA string.

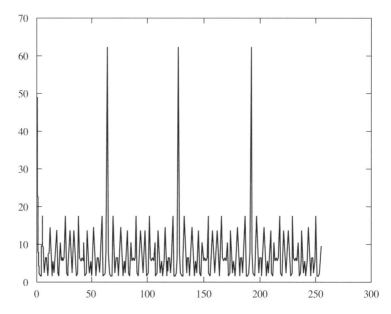

FIGURE 22-6 The correlation of a repeating sequence shows four spikes with the first at $x = 0$.

Another example considers the autocorrelation of a repeating sequence. Code 22-5 creates a string of four identical repeating segments and computes the correlation. Figure 22-6 shows four identical spikes representing the identification of the four repeating segments.

```
>>> d = dna[98:98+64]+dna[98:98+64]+dna[98:98+64]+dna[98:98+64]
>>> for i in range( 256 ):
        ndx = abet.index( d[i] )
        mat[ndx,i] = 1
>>> for i in range( 4 ):
        mat[i] -= mat[i].mean()
>>> corr = zeros( 256 )
>>> for i in range( 4 ):
        corr += abs(akando.Correlate( mat[i], mat[i] ))
```

Code 22-5

22.5 Higher Dimensions

Fourier transforms can easily be applied to matrices and higher-dimensional data cubes. The FFT will require that each dimension of the data structure be a power of 2, but the dimensions do not have to be the same length.

22.5.1 Two-Dimensional FFTs in SciPy

SciPy provides two functions that perform two-dimensional Fourier transforms in **fftpack.fft2** and **fftpack.ifft2**. It is possible to compute the two-dimensional Fourier transform by computing the transform of the rows and then the columns of a matrix, but the SciPy functions are faster.

For higher-dimensional structures, a one-dimensional FFT is applied along each axis. For example, a three-dimensional FFT is computed by performing the FFT on rows in the x direction, then all rows in the y direction, and then finally all rows in the z direction.

22.5.2 Image Frequencies

The two-dimensional Fourier transform of simple geometric shapes has unique properties. Code 22-6 creates a simple solid square in the array a. The two-dimensional transform is shown in the form of a weighted checkerboard in Figure 22-7. In this case $w = 10$, and so the square in the input was 20×20. Figure 22-8 presents another case in which $w = 4$ and the square was much smaller. The larger square produced the smaller features in Fourier space.

This code created the Fourier transform as *af* and then performed mathematical conversions to produce *showme*. The *af* has several properties that make it unpleasant for viewing: (1) the values are complex; (2) the DC term is much larger than the other terms; and (3) the lower frequencies are toward the corners of the frame. A computer display has only 256 different levels of intensity (in the gray scale). If the DC term is

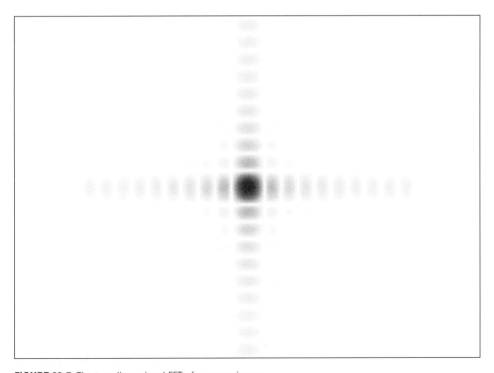

FIGURE 22-7 The two-dimensional FFT of a square image.

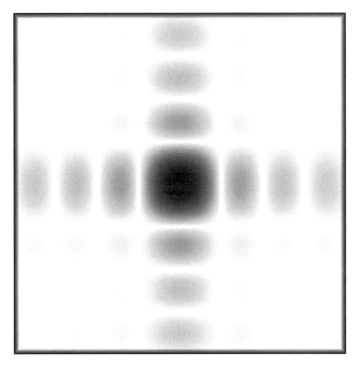

FIGURE 22-8 The two-dimensional FFT of a smaller square.

more than 256 times larger than the other values, then only the DC term will be visible. Thus the **log** is computed to squash the values into a more viewable range. The value of 1 is added to *abs(af)* so that the lowest value of the **log** is 0. The next-to-last line inverts the image so that black is the higher intensity. In some cases (such as paper print) it is easier to view the "negative" of the image.

```
>>> from numpy import ones, log
>>> a = zeros( (256,256) )
>>> w = 10
>>> a = zeros( (256,256) )
>>> a[128-w:128+w,128-w:128+w] = ones( (2*w,2*w) )/(4*w*w)
>>> af = fftpack.fft2( a )
>>> showme = log( abs(af)+1)
>>> showme = akando.Swap( showme )
>>> showme = showme.max() - showme
>>> akando.a2i( showme ).show()
```

Code 22-6

If the input had been a circle instead of a square, the Fourier transform would be a set of decaying rings. More complicated geometric shapes create a more complicated Fourier pattern.

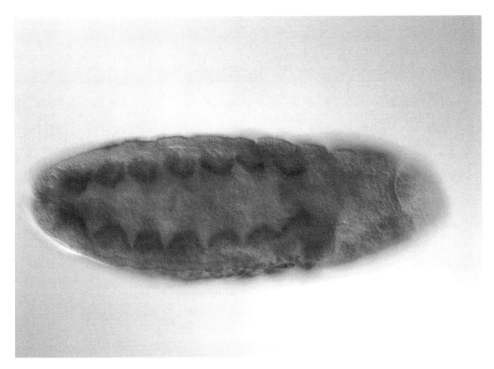

FIGURE 22-9 An image of a fly embryo (Courtesy of the Berkley Drosophila Genome Project).

22.6 The Onset of Image Processing

Figure 22-9 is the image of a fly embryo (Tomancak et al., 2007), which has a repetitive structure. The image is loaded in Code 22-7 and transformed to produce the image in Figure 22-10. The "crosshair" pattern is due to the rectangular frame that contains the image. The rest of the cloud is the Fourier transform of the image data. Although it is now difficult to see geometric patterns in the transform, all of the information is still there.

```
>>> from PIL import Image
>>> mg = Image.open( 'fig16_18.jpe')
>>> data = akando.i2a( mg.convert('L') )/255.
>>> data = 1-data
>>> af = fftpack.fft2( data )
>>> data = 1-data   # make background darker than the target
>>> af= fftpack.fft2( data )
>>> showme = log( abs(af)+1)
>>> showme = akando.Swap( showme )
>>> showme = showme.max() - showme
>>> akando.a2i( showme ).show()
```

Code 22-7

Since the spinal elements in the image are oriented in horizontal arrays, these frequencies can be isolated using a wedge filter, as shown in Code 22-8. This filter

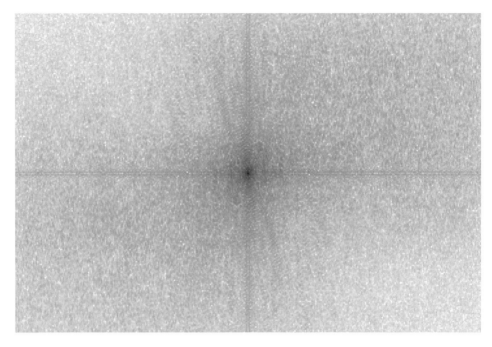

FIGURE 22-10 The FFT of a fly embryo.

creates two wedges from the center of the frame outward to the left and right. All of the frequencies within this wedge are kept in the second line, but all other frequencies are removed. The filtering is completed by converting the remaining frequencies back to image space and displaying them as shown in Figure 22-11. It is easy for the human eye to spot the spinal structure in Figure 22-9. However, creating a program to do so is harder. The Fourier image in Figure 22-11 has bright spots—which are far easier for a computer program to detect—at the location of most of the targets. Thus targets in the original image were isolated by keeping their frequencies and discarding others.

```
>>> wedj = akando.WedgeFilter((1080,1520), 0, 20
)+akando.WedgeFilter((1080,1520), -20, 0 )
>>> bf = akando.Swap(af ) * wedj
>>> bf = akando.Swap( bf )
>>> b = fftpack.ifft2( bf )
>>> akando.a2i( b.real ).save('fig20.gif')
```

Code 22-8

22.7 Two-Dimensional Correlations

Correlations can be performed in two dimensions as well. There is no difference in the theory except that two-dimensional FFTs are used. The **akando.Correlate** function identifies the dimensionality of the input and uses the correct Fourier transforms. This function can thus receive either vectors or matrices.

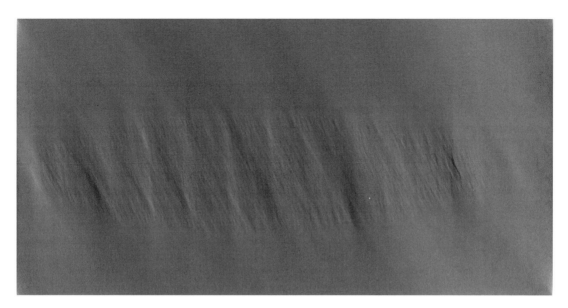

FIGURE 22-11 The filtered image, with bright spots marking the locations of the structures.

Summary

A correlation is a tool that easily compares all possible relative shifts between two signals. Many engineering and scientific applications have employed correlations for the purposes of filtering and detection. In the field of bioinformatics, correlations can also be a useful tool. In this chapter the simple application of comparing DNA sequences was demonstrated. However, correlations can also be useful in analyzing biological images and signals from a variety of sources.

Bibliography

Tomancak, P., et al. (2007). Retrieved from Tomanacak, P., et. al. (2007.) Systematic Determination of Patterns of Gene Expression during *Drosophila* Embryogenesis, Genome Biology 3, 1-14.

Problems

1. Compute the autocorrelation of the Gaussian function $y = Ae^{-\frac{(x-x_0)^2}{2\sigma^2}}$. What is the functional form of the result? What happens to the results when σ is increased?

2. Equation (22-1) defines convolution. What is the equivalent of Equation (22-8) for a convolution?

3. Create two images of circles with different radii (using **akando.Circle**). Compute the two-dimensional correlation (using **akando.Correlate**). Why is the two-dimensional correlation of these two images a ring?

4. Create a vector **v** of length 512. Set *v[i]=1* for *i=0, 32, 64, ….* Compute the Fourier transform. Explain why there are the number of spikes that are shown.

5. Repeat Problem 4 with *i=0, 30, 60, 90, ….* Why are the spikes of the Fourier transform different heights?

23 Numerical Sequence Alignment

The DNA and protein alignment algorithms that have been thus far discussed relied on representations of the biological information as strings with a specific alphabet, and string manipulation algorithms were required to align the sequences. An alternative approach is to represent the biological information as a vector of numerical values. In doing this a variety of numerical comparison algorithms become available for the purposes of comparing and analyzing data. This chapter will discuss some methods of representing the information with numerical vectors and basic comparison algorithms.

23.1 Alternative Encodings

The focus here will be on several numerical encodings of sequence data. Many of these contain advantages and disadvantages that should be taken into consideration when they are employed. This section will not use the encoding methods discussed here, but Section 23.2 will employ some of the better-performing methods.

23.1.1 Hydrophobicity

A protein molecule that is hydrophobic does not bind well to water molecules. Amino acids and their hydrophobicity are shown in Table 23-1. This data illustrates that it is easy to convert a protein into a numerical string by converting the amino acids into their hydrophobic values.

The **MakeChart** function in Code 23-1 builds a table for the hydrophobicity values. Of course, there are other measurements such as polarity, charge, and structure that are available (Wang and Atchley, 2006; Taylor et al., 2005). These can be used to create numerical representations of sequence data.

```
# hydro.py
def MakeChart( ):
    phob = []
    phob.append( ('Alanine','Ala','A',0.616) )
    phob.append( ('Arginine','Arg','R',
0))
    phob.append( ('Asparagine','Asn','N',0.236))
    phob.append( ('Aspartate','Asp','D',0.028))
    phob.append( ('Cysteine','Cys','C',
0.68))
```

```
        phob.append( ('Glutamate','Glu','E',0.043))
        phob.append( ('Glutamine','Gln','Q',0.251))
        phob.append( ('Glycine','Gly','G',
0.501))
        phob.append( ('Histidine','His','H',0.165))
        phob.append( ('Isoleucine','Ile','I',0.943))
        phob.append( ('Leucine','Leu','L',0.943))
        phob.append( ('Lysine','Lys','K',0.283))
        phob.append( ('Methionine','Met','M',0.738))
        phob.append( ('Phenylalanine','Phe','F',1))
        phob.append( ('Proline','Pro','P',
0.711))
        phob.append( ('serine','ser','s',0.359))
        phob.append( ('Threonine','Thr','T', 0.45))
        phob.append( ('Tryptophan','Trp','W',0.878))
        phob.append( ('Tyrosine','Tyr','Y',
0.88))
        phob.append( ('Valine','Val','V',0.825 ))
        abet = 'ARNDCEQGHILKMFPSTWYV'
        return phob, abet

>>> phob, abet = MakeChart()
```

Code 23-1

Converting a sequence to a numerical vector is quite easily accomplished by the **ConvertSequence** function shown in Code 23-2.

Table 23-1 The Hydrophobicity Values for Each Amino Acid

Acid			Hydrophobicity
Alanine	Ala	A	0.616
Arginine	Arg	R	0
Asparagine	Asn	N	0.236
Aspartate	Asp	D	0.028
Cysteine	Cys	C	0.68
Glutamate	Glu	E	0.043
Glutamine	Gln	Q	0.251
Glycine	Gly	G	0.501
Histidine	His	H	0.165
Isoleucine	Ile	I	0.943
Leucine	Leu	L	0.943
Lysine	Lys	K	0.283
Methionine	Met	M	0.738
Phenylalanine	Phe	F	1
Proline	Pro	P	0.711
Serine	Ser	S	0.359
Threonine	Thr	T	0.45
Tryptophan	Trp	W	0.878
Tyrosine	Tyr	Y	0.88
Valine	Val	V	0.825

23.1 Alternative Encodings

```
# hydro.py
def ConvertSequence( seq, phob, abet ):
    N = len( seq )
    vec = zeros( N, float )
    for i in range( N ):
        ndx = abet.find( seq[i] )
        vec[i] = phob[ndx][3]
    return vec

>>> vec = ConvertSequence( 'GQATDNWY', phob, abet )
>>> vec
array([ 0.501,  0.251,  0.616,  0.45 ,  0.028,  0.236,  0.878,  0.88 ])
```

Code 23-2

23.1.2 GC Content

The GC content is the percentage of elements that are either G or C within a certain window size. Promoter regions just in front of the start of a gene are believed to be rich in G's and C's (Down and Hubbard, 2002). Code 23-3 contains **GCContent**, which uses a direct method of converting a sequence to a GC percentage. The *win* is the window size. Thus, if *win=10*, then ten elements are used to compute the percentage.

```
# numseq.py
def GCContent( seq, win ):
    # win is the window size
    N = len( seq )
    D = N - win # number of windows
    vec = zeros( D, float )
    for i in range( D ):
        a = seq[i:i+win].count( 'A' )
        c = seq[i:i+win].count( 'C' )
        g = seq[i:i+win].count( 'G' )
        t = seq[i:i+win].count( 'T' )
        vec[i] = float(c+g)/(a+c+g+t)
    return vec

>>> vec = GCContent(
'AGATGCTGAGATCGGTAGCGTGTATGCTAGCTGACATCATTTTCACGCGCGCGCGCGCGCCGCGCGCGC
GCGGGCGATATATAGCT', 10 )
>>> vec
array([ 0.5,  0.5,  0.4,  0.5,  0.6,  0.6,  0.5,  0.5,  0.5,  0.6,  0.6,  0.6,
0.7,  0.6,  0.5,  0.4,  0.5,  0.6,  0.5,  0.4,  0.4,  0.5, 0.4,  0.5,  0.5,
0.6,  0.5,  0.4,  0.5,  0.5,  0.4,  0.3,  0.3,  0.2,  0.3,  0.2,  0.3,  0.4,
0.4,  0.5,  0.6,  0.7,  0.8,  0.9,  0.9,  1. ,  1. ,  1. ,  1. ,  1. ,  1. ,
1. ,  1. ,  1. ,  1. , 1. ,  1. ,  1. ,  1. ,  1. ,  1. ,  1. ,  1. ,  1.
,  1. ,  1. ,  0.9,  0.8,  0.7,  0.6,  0.5,  0.4,  0.3,  0.3,  0.3])
```

Code 23-3

This brute force method does work, but it repeats computations. It counts the elements for each window, but consecutive windows have all but one element in common. A faster method is to compute the first window and then just add or subtract values in the numerator as C's and G's are added and subtracted from the window. Code 23-4 shows the **FastGCContent**, with the results plotted in Figure 23-1. For *win=10* this new function is about four times faster. The savings increase as *win* increases.

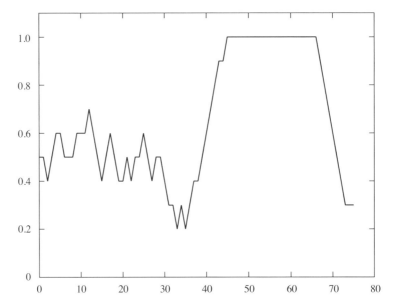

FIGURE 23-1 The GC content for the sequence in Code 23-3.

```
# numseq.py
def FastGCContent( seq, win ):
    # win is the window size
    N = len( seq )
    D = N - win # number of windows
    vec = zeros( D, float )
    # first window
    a = seq[:win].count( 'A' )
    c = seq[:win].count( 'C' )
    g = seq[:win].count( 'G' )
    t = seq[:win].count( 'T' )
    nmr = float( c+g )
    vec[0] = nmr/win
    w1 = win-1
    # subsequent windows
    for i in range( 1, D ):
        if seq[i-1]=='C' or seq[i-1]=='G':
            nmr -=1
        if seq[i+w1]=='C' or seq[i+w1]=='G':
            nmr += 1
        vec[i] = nmr/win
    return vec
```

Code 23-4

23.1.3 Numerical Methods

The numerical methods that follow directly convert DNA bases into numbers. These are pure mathematical conversions and they are not based on biological measurements.

A Simple Numerical Encoding An obvious conversion of a DNA sequence to a numerical vector is to simply set A = 1, C = 2, G = 3, and T = 4, but this is a poor choice. For example, if two sequences were to be compared by subtraction, this encoding would state that a mismatch of A to C would not be as bad as a mismatch of A to

T. In the first case, the score for the mismatch would be 1 and in the second case it would be 3.

A numerical conversion is a little more difficult than just assigning numbers to bases. The problem is that the bases are orthogonal to each other in a mathematical sense. The subtraction (or any other comparison method) must treat each pair of bases equally unless a bias is desired by the user.

A Binary Encoding An encoding that suits the orthogonality condition is to convert each base into a binary-element vector that contains three 0's and one 1. Figure 23-2 shows a sequence converted in this fashion. Each column represents one base. Although this method fits the orthogonality requirement, it is also inefficient.

A Three-Dimensional Equidistant Encoding It is possible to find four points in three-dimensional space that are equidistant from each other. These are points at the corners of a pyramid with a triangular base. The following values are four such points that are also equidistant to the center of the coordinate space:

$A = (-0.5, -0.2886, -0.204)$

$C = (0.5, -0.2886, -0.204)$

$G = (0, 0.5774, -0.204)$

$T = (0, 0, 0.6123)$

This encoding uses three numbers instead of four to encode each base. It is not much of a savings.

An Encoding Using Smoothed Complex Numbers A method shown in Chang (2006) converts the bases into four complex numbers that are the four corners of a box in Bode space where the *x*-axis represents the real part of an imaginary number and the *y*-axis represents the imaginary component. The numerical sequence considers three consecutive bases:

$A = 1 + j$

$T = 1 - j$

$C = -1 - j$

$G = -1 + j$

These are weighted according to the following equation:

$$y[n] = x[n] + \frac{x[n-1]}{2} + \frac{x[n-2]}{4} \qquad (23\text{-}1)$$

```
AGCTAGCTGCTT

100010000000
001000100100
010001001000
000100010011
```

FIGURE 23-2 A binary element conversion.

This function is shown in Code 23-5, where an example call is given.

```
# Chang's method
def Chang( seq ):
    a = [1+1j, 1-1j, -1-1j, -1+1j]
    abet = 'ATCG'
    N = len( seq )
    vec = zeros( N-2, complex )
    for i in range( N-2 ):
        n1, n2, n3 = abet.find(seq[i+2]), abet.find(seq[i+1]), abet.find(seq[i])
        vec[i] = a[n1] + 0.5*a[n2] + 0.25*a[n3]
    return vec

>>> vec = Chang( 'tacgtac' )
>>> vec
array([-0.25-0.75j, -1.25+0.75j,  0.25-0.75j,  1.25+0.75j, -0.25-0.75j])
```

Code 23-5

23.2 Numerical Alignments

Numerical alignments can employ a large host of tools not available for string processing. In this section a few powerful tools are examined, including Fourier analysis, which was discussed in Chapter 21. In a study conducted by Tempel et al. (2006), it was necessary to match sequences within a specific degree of agreement. This section will solve this problem by finding sequence matches within a specified threshold using Fourier transforms and correlations. In this problem there is a target sequence *a* of length 16. The goal is to find places in a larger string *b* where there is a substring that closely matches *a*. Any substring that contains 12 letters that exactly match *a* is considered a match. What makes this problem difficult is that *any* 12 of the 16 letters can match.

As an additional example, a smaller problem is considered. The target sequence is ACTTG. The data sequence is much longer, and the task is to find locations in *data* that matches four out of the five letters in the target. Again, this is a match of *any* of the four letters.

The solution is to convert both the target and that data into binary sequences, as shown in Figure 23-2. Each row is a set of 1's and 0's that indicate the locations of the four bases. One caveat to note: Because the lengths of the rows for both the target and the data must be the same, the target is padded with zeros. Code 23-6 shows the function **PartMatch**, which receives the target sequence, the data sequence, and the minimum number of matching letters required (*nhits*).

The array *query* contains the four rows of binary elements to encode the target sequence. The array *data* contains the binary elements for the data sequence. Initially, the array *corr* also has four rows—one for the correlation of each row of *query* and *data*. However, these are summed so that *corr* becomes a vector. This is the sum of the four correlations and peaks of height *nhits* or larger indicate regions that have a sufficient number of matches. The **nonzero** function is employed to find these locations. The function returns a vector in which each element is the location in *seq* where there is a good match.

```
# numseq.py
def PartMatch( targ, seq, nhits ):
    # targ = target sequence
    # seq = data sequence
    # nhits = minimum number of matches needed.
    D = len( seq )
    query = zeros( (4,D) )
    data = zeros( (4,D) )
    # convert targ
    a = array( list( targ ))
    lt = len( targ )
    query[0,:lt] = (a=='A').astype(int)
    query[1,:lt] = (a=='C').astype(int)
    query[2,:lt] = (a=='G').astype(int)
    query[3,:lt] = (a=='T').astype(int)
    # convert sequence
    a = array( list( seq ))
    data[0] = (a=='A').astype(int)
    data[1] = (a=='C').astype(int)
    data[2] = (a=='G').astype(int)
    data[3] = (a=='T').astype(int)
    # correlate
    corr = zeros( (4,D) )
    for i in range( 4 ):
        a = akando.Correlate( data[i], query[i] )
        corr[i] = akando.Swap(a.real)
    corr = corr.sum(0)
    # find the peaks
    hits = greater_equal( corr, nhits*0.9999 ).astype(int)
    nz = hits.nonzero( )[0]
    if D%2==0: return nz
    else: return nz+1
```

Code 23-6

The example shown in Code 23-7 indicates that there were two locations in which four out of five letters matched. Both *seq[8:8+5]* and *seq[21:21+5]* contain sequences that match four out of five characters in *targ*.

```
>>> targ = 'ACTTG'
>>> seq = 'TTAGCTAGACTTTTGACTGAGATTTG'
>>> pm = PartMatch( targ, seq, 4 )
>>> pm
array([ 8, 21])
>>> seq[8:8+5]
'ACTTT'
>>> seq[21:21+5]
'ATTTG'
```

Code 23-7

23.3 Measuring the Hurst Exponent

The Hurst exponent measures the complexity of a vector. In order to measure the complexity of a string, the string is first converted to a numerical vector. According to the method presented in Yu and Chen (2000), a DNA string is converted to numbers using the simple encoding (ACGT) = $(-2, -1, 1, 2)$, as presented in **TwoEncode** in Code 23-8.

```
# numseq.py
def TwoEncode( seq ):
    N = len( seq )
    vec = zeros( N, int )
    a = (-2, -1, 1, 2 )
    b = 'ACGT'
    for i in range( N ):
        ndx = b.find( seq[i] )
        vec[i] = a[ndx]
    return vec
```

Code 23-8

The length of the sequence is N and the computation of the Hurst exponent considers vectors of length n where $n=1,2,\ldots,N$. Thus, for each value of n, there is a set of three computations: (1) to compute the average of the sequence up to length n; (2) to subtract the average from the running data; and (3) to capture the range of the data and divide it by the standard deviation of the data.

Code 23-9 illustrates the **Hurst** function, which contains a loop over n. For each value of n, the average and the matrix X are computed. The $R[n]$ is the range of X for this value of n, and $S[n]$ is the standard deviation. The matrix M contains two columns used for plotting. The first column is the $log(n)$ values and the second column is the $log(R[n]/S[n])$ values. The Hurst exponent, H, is also computed but explained after a view of the plot.

```
# numseq.py
def Hurst( seq ):
    # n- averages
    N = len( seq )
    x = TwoEncode( seq )
    R = zeros( N, float )
    S = zeros( N, float )
    for n in range( 1,N ):
        avg = x[:n].mean()
        X = zeros( n, float )
        for t in range( 1, n ):
            X[t] = X[t-1] + (x[t-1] - avg )
        R[n] = X.max() - X.min()
        S[n] = sqrt( 1./n * ((x[:n]-avg)**2).sum())
    Z = R[3:]/S[3:]
    M = zeros( (N-3,2), float )
    M[:,0] = log(arange( 3,N))
    M[:,1] = log(Z)
    H = akando.linearRegression( M[:,0], M[:,1] )
    return M, H[0]
```

Code 23-9

Figure 23-3 shows the plot of $log(R[n]/S[n])$ versus $log(n)$, which is nearly linear. Thus, it is possible to estimate the data points using linear regression and to obtain the slope of this line. The final line in **Hurst** performs this task. The data used in this case is the same that was used in Yu and Chen (2000). DNA from a gene with accession number AF033620 from 1730 to 2560 was extracted and used as a test case. Code 23-10 shows the run. The sequence was converted to uppercase since the **Hurst** function

23.3 Measuring the Hurst Exponent

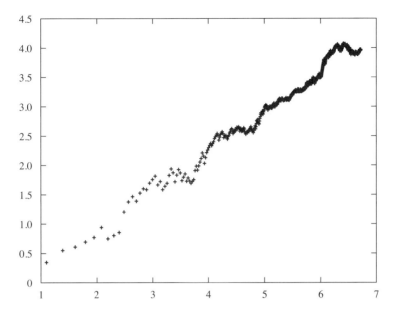

FIGURE 23-3 Plot of log(R[n]/S[n]) versus log(n).

expected it. The plot nearly matches that of Yu and Chen, and the calculated exponent is very close to their value of 0.673.

```
>>> data = genbank.ReadGenbank( 'data/genbank/AF033620.gb.txt')
>>> dna = genbank.ParseDNA( data )
>>> seq = dna[1729:2560]
>>> Seq = seq.upper()
>>> M, H = Hurst( Seq )
>>> H
0.69538
>>> akando.PlotMultiple('fig6.txt', M )
```
Code 23-10

A purely random sequence will produce a Hurst exponent of 0.5, and a more structured string will produce a higher value. If H > 0.5, the system is Brownian or persistent. If H < 0.5, the signal is antipersistent and there is a negative correlation among the elements. Code 23-11 shows a purely random case and an extremely structured case.

```
>>> rseq = RandomString( 830, 'ACGT' )
>>> M, H = Hurst( rseq )
>>> H
0.549660342251
>>> M, H = Hurst( 'ATGTCATG'*100 )
>>> H
0.0214231060783
```
Code 23-11

23.4 Chaos Representation

Large DNA strings (over 100,000 characters) can be represented as images. Two different approaches convert the long strings into similar images using chaos theory. Even though these strings cover several genes and large non-coding regions, they are still indicative of the species from which the string was taken. Thus, these types of images are useful for tasks such as species identification.

23.4.1 Representing the Data

The work of Deschavanne et al. (1999) represented the data using an image rather than a vector of numbers. The process conceptually begins with a 2 × 2 matrix, as shown in Figure 23-4(a). The first block contains the letter C and the value of the matrix for this cell is the number of Cs in the string. The other cells are similarly computed. Figure 23-4(b) divides each cell into 2 × 2 cells, and all of the two-word DNA combinations are represented. This part of the figure is created by duplicating the first matrix four times and then appending one of the four DNA letters to the respective duplicates. The process repeats, and Figure 23-4(c) shows the matrix for three-word combinations.

This block process can be continued for any word length. However, longer words will require much longer DNA strings in order to sufficiently populate the cells. In the case of eight-letter words, the matrix will be 256 × 256. There will be 65,536 cells, and the DNA string required to populate it must thus be much larger than that. In the following examples, a seven-letter word matrix was used, and the DNA strings had more than 1 million bases.

The first program is to generate the matrix as shown above. Because it may be desirable to build a matrix of any size, the program follows the procedure shown in Figure 23-4. Code 23-12 contains **AllPossStrings**, which receives an alphabet and the word size. The outer *for* loop will perform the iterations to increase the size of the matrix. Inside of this loop are two *for* loops. The first increases the size of the matrix by making four replicates; the second adds the appropriate letter to each of the new cells.

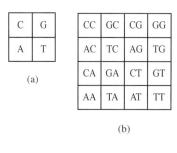

FIGURE 23-4 The first three sequence combinations for the chaos representation.

23.4 Chaos Representation

```
# chaos.py
def AllPossStrings( abet, SZ=8 ):
    # SZ is number of iterations. 8 will make a 256x256 array
    # is also the length of the strings
    a = array( [[abet[0],abet[1]],[abet[2],abet[3]]] )
    for n in range( SZ-1 ):
        N = len( a )
        grab = arange( 2*N )/2
        b = zeros( (2*N,2*N), 'S8' )
        # expand the matrix
        for i in range( N ):
            b[2*i] = take( a[i], grab )
            b[2*i+1] = take( a[i], grab )
        # left add the new character
        for i in range( 2*N ):
            for j in range( 2* N ):
                k, l = i%2, j%2
                m = k*2+l
                b[i,j] = abet[m] + b[i,j]
        a = copy.copy( b )
    return b
>>> aps = AllPossStrings( 'ACGT', 7 )
>>> aps.shape
(128, 128)
>>> aps[:5,:5]
array([['AAAAAAA', 'CAAAAAA', 'ACAAAAA', 'CCAAAAA', 'AACAAAA'],
       ['GAAAAAA', 'TAAAAAA', 'GCAAAAA', 'TCAAAAA', 'GACAAAA'],
       ['AGAAAAA', 'CGAAAAA', 'ATAAAAA', 'CTAAAAA', 'AGCAAAA'],
       ['GGAAAAA', 'TGAAAAA', 'GTAAAAA', 'TTAAAAA', 'GGCAAAA'],
       ['AAGAAAA', 'CAGAAAA', 'ACGAAAA', 'CCGAAAA', 'AATAAAA']],
      dtype='|S8')
```

Code 23-12

The second function is the **Counter** (Code 23-13), which counts the occurrence of each of the words in the DNA string. It receives the *aps* from **AllPossStrings** and the DNA string. Basically, it has a double-nested loop that considers each cell in the matrix. This function returns a matrix that can then be converted to an image for the purposes of display. The results from this example are shown in Figure 23-5.

FIGURE 23-5 The chaos representation from Code 23-13.

```
# chaos.py
def Counter( aps, data ):
    # aps from all possible strings
    N = len( aps )
    ctr = zeros( (N,N), int )
    for i in range( N ):
        for j in range( N ):
            ctr[i,j] = data.count( aps[i,j] )
    return ctr

>>> data = genbank.ReadGenbank( 'datagenbank/nc_00918.gb.txt')
>>> dna = genbank.ParseDNA( data )
>>> len( dna )
1551335
>>> aps = AllPossStrings( 'cgat',7 )
>>> ctr = Counter( aps, dna )
>>> akando.a2i( ctr ).save('fig5.gif')
```

Code 23-13

23.4.2 A Simpler Method

A simpler method is provided by Jeffrey (1990) and it has similar results. A two-dimensional grid is created and a point is placed at the center. As each letter in the DNA string is considered, the point is moved from its current location halfway to the appropriate corner. Figure 23-6 shows the movement of the point for the sequence TGAT. At each location a 1.0 is added to the appropriate pixel. As the system progresses, an image is formed, as shown in Figure 23-7.

The function **ChaosJeffrey** in Code 23-14 computes the image for a given sequence. The *me* is a two-dimensional data point that is moved about. At each location, the matrix *A* is incremented. This function can replace **AllPossStrings** and **Counter**.

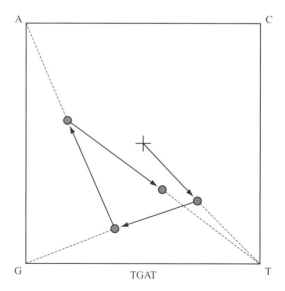

FIGURE 23-6. The first few movements in an approach offered by Jeffrey (1990).

23.4 Chaos Representation

FIGURE 23-7. An image created by the method discussed in Jeffrey (1990).

```
# chaos.py
def ChaosJeffrey( seq, D=256 ):
    # D is the dimension of the picture space
    N = len( seq )
    A = zeros( (D,D) )
    abet = 'acgt'
    tg = array( [[0,0], [0,D], [D,0], [D,D]] )
    me = array([D/2., D/2.])
    for i in range( N ):
        ndx = abet.find( seq[i] )
        half = (me+tg[ndx])/2.
        v,h = half.astype(int)
        A[v,h ] += 1
        me = half + 0
    return A
```

Code 23-14

23.4.3 Comparing Chaos Images of Different Species

Chaos images create an interesting pattern, but are they useful? Let's look at an example in which chaos images are used to delineate species and analyze genomes from differing crops. Some of the genomes contained very large sequences, and these were separated into 1 Mb strings. Three of the 1 Mb segments selected were extracted from genomes that came from maize (file 0), cress (file 1), rice (file 2), pine (file 3), and multiple strains of wheat (files 4, 5, 6).

Code 23-15 shows **Run2**, which reads each of the files and determines if they have multiple genes or are a single gene. Multiple genes are concatenated to make a long string. The string is considered to be a 1 MB segment, and currently there is a limit of *NP* = 3 strings for each file. This facilitates the creation of a program that will run in a reasonable amount of time, and the restriction of *NP* will be lifted later. For each run, the image is created and stored as a GIF image.

```
# chaos.py
def Run2( flist ):
    aps = AllPossStrings( 'ACGT',7)
    NF = len( flist ) # the number of files
    for i in range( NF ):
        print 'Working on ',flist[i]
        # get the data
        dna = fasta.Fasta( 'bacteria/' + flist[i] )
        # count the number of genes
        NG = len( dna )
        # if there are multiple genes then combine
        if NG >1:
            t = []
            for j in range( NG ):
                t.append( dna[j][1] )
            st = ''.join(t)
            del t
        else:
            st = dna[0][1]
        del dna
        # for every million bases make a plot
        NP = int(len( st )/1000000)
        if NP > 3:
            NP = 3
        for j in range( NP ):
            print '\tPortion',j,'of',NP
            ctr = Counter( aps, st[j*1000000:j*1000000+1000000] )
            akando.a2i( ctr ).save('data/work/chaos'+str(i)+'c'+str(j)+'.gif' )
>>> Run2( flist )
```

Code 23-15

This particular test generated 21 images, and Figure 23-8 shows results from five of the six genomes. Each column is a set of three images from three different 1 MB segments of the genome. The first four columns are maize, cress, rice, and pine. The last two columns are different strains of wheat, and as the sixth file was similar to these two it is not shown. There are strong similarities in each column and differences between the rows, which indicates that it is possible to sort the images according to their species.

23.4.4 Organizing the Data

The images are sufficient for organization according to species. Again following Deschavanne et al. (1999), the images are organized by principal component analysis (see again Section 19.3). Figure 23-8 shows only the first three images produced by each file. These genomes are quite large and can produce hundreds of images each created from 1 million bases. For the files used here, there were 454 images generated (see the directory *data/chaos* on the website associated with this book). The files in this directory are named *chaosXcN.gif* where *X* indicates the file number and *N* indi-

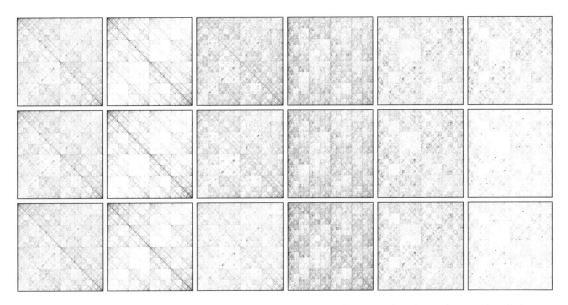

FIGURE 23-8 Three samples from different files. The columns represent: maize, cress, rice, pine, and two strains of wheat.

cates the segment of the genome. An organized presentation should thus be able to sort the images according to their X number.

To create an organized presentation, the eigenimages of these chaos images are computed. Code 19-20 in Chapter 19 provides the **LoadImages** function that will read in all of the image data, and Code 23-16 gathers the names of the files. In a Windows system, a file named *Thumbs.db* is created to contain thumbnail images of the file contents, and so this file is removed from the list.

The list *pics* becomes the list of image data. For this example there are 454 arrays in *pics*. The average is subtracted from the images by **SubAvg**. The covariance matrix is computed through **Lcov**, and the eigenimages are created through **Eimage**. Since the display used here will be two-dimensional the best two eigenimages are kept and stored in the list *pca*. Conversion of the images to two-dimensional data points follows Code 19-25 and uses **Coeffs**. The results are shown in Figure 23-9.

```
>>> mglist = os.listdir( 'data/chaos/')
>>> if 'Thumbs.db' in mglist:
       mglist.remove( 'Thumbs.db' )
>>> pics = eigimage.LoadImages( 'data/chaso/', mglist )
>>> avg, data = eigimage.SubAvg( pics )
>>> lcov = eigimage.Lcov( data )
>>> emg, evls = eigimage.Eimage( lcov, data )
>>> pca = [ emg[0], emg[1] ]
>>> cffs = eigimage.Coeffs( pca, data )
>>> akando.PlotMultiple( 'fig9.txt', cffs)
```

Code 23-16

The final step is to determine if these points are actually organized. To accomplish this, the data points are encoded according to the genome they represent. The function **PlotColors** in Code 23-17 receives the list of image names, and the 454 points are

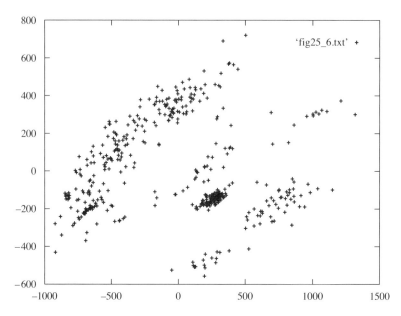

FIGURE 23-9 The PCA plot of the 454 images.

plotted in Figure 23-9. In this case it is known that there are seven different files, and thus a list of seven empty lists is created. Each file name is considered, and the number following the letters "chaos" is converted to an integer. This is used as an index and the data point associated with this file is stored in the appropriate list in *pts*. After all of the data points have been sorted, separate files for each are stored. The plotting command plots all of the files in the same window assigning a different marker for each file, as shown in Figure 23-10.

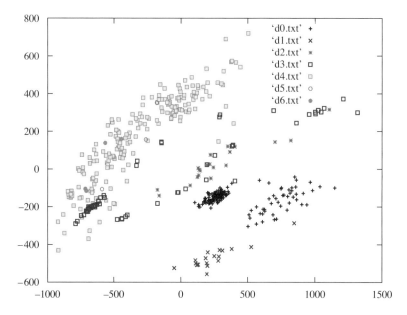

FIGURE 23-10 Each marker represents chaos images from different genomes.

23.4 Chaos Representation   361

```
# chaos.py
def PlotColors( mglist, cffs ):
    N = 7
    pts = []
    for i in range( 7 ):
        pts.append( [] )
    for i in range( len( cffs )):
        d = int( mglist[i][5] )
        pts[d].append( cffs[i] )
    for i in range( 7 ):
        akando.PlotMultiple( 'd'+str(i)+'.txt' , array(pts[i]))

gnuplot> plot 'd0.txt', 'd1.txt', 'd2.txt', 'd3.txt', 'd4.txt', 'd5.txt',
'd6.txt'
```

Code 23-17

As seen in this file, there are definite regions that contain all of the markers for a particular genome. It should be noted that *d4, d5*, and *d6* are different types of the same plant—wheat. The *d3* is a bit scattered through the space and is not well organized. It should be remembered that originally there were 454 eigenvectors and only two are used for this plot. The first ten eigenvalues are printed in Code 23-18, and it is clear that more than two are nontrivial values. Thus there is information in the third eigenimage that is not represented in Figure 23-10. The data in Code 23-18 creates Figure 23-11, which illustrates the same data points plotted using the second and third eigenimages.

In this case the *d3* is well separated from the rest. The point to be made is that these two-dimensional images do not depict the entire story. In cases such as this, more than two dimensions are necessary to separate the data. Commonly, the PCA analysis is reduced to two dimensions purely for the convenience of display. In this experiment each 1 MB segment from the large genomes was converted to a chaos

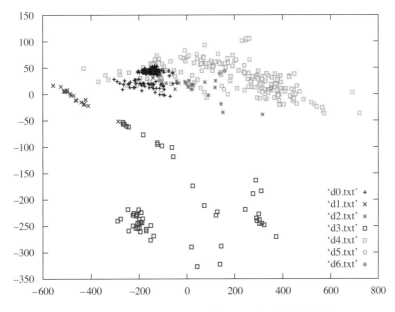

FIGURE 23-11 The same data points plotted according to the second and third eigenimages.

image. PCA was used to organize the images, which were demonstrated by two different views of three-dimensional PCA space.

```
>>> evls[:10].astype(int)
array([10443,  5240,  1842,   786,   512,   354,   279,   241,   202,   179])
>>> pca = [ emg[1], emg[2] ]
>>> cffs = eigimage.Coeffs( pca, data )
>>> PlotColors( mglist, cffs )
```

Code 23-18

23.5 Summary

DNA information is commonly expressed as a string of characters, but it is also possible to represent DNA as a vector or matrix of numerical values. Several different encodings for doing this were considered in this chapter. Some were based on physical properties of the DNA, and others were simply conversions of the string to a numerical format. Numerical formats allow for the use of a new set of tools for the purposes of analysis and alignment. In this case the frequencies of the numerical values were used to find repetitive tendencies in coding regions and to align sequences.

Bibliography

Chang. (2006). *Chang*. Retrieved from http://www.hku.hk/facmed/images/document/04research/database/presentations/cq_chang.ppt.

Deschavanne, Patrick J. et al. (1999). Genomic Signature: Characterization and Classification of Species Assessed by Chaos Game Representation of Sequences. *Molecular Biology and Evolution*, 16 (10), 1391–1399.

Down, T. A., and T. J. P. Hubbard. (2002). Computational Detection and Location of Transcription Start Sites in Mammalian Genomic DNA. *Genome Research*, 12, 458–461.

Jeffrey, H. J. (1990). Chaos game representation of gene structure. *Nucleic Acids Research*, 18 (8), 2163–2170.

Taylor, T., M. Rivera, G. Wilson, and I. I. Vaisman. (2005). New method for protein secondary structure assignment based on a simple topological descriptor. *Proteins*, 60 (3), 513–524.

Tempel, S., et al. (2006). Domain organization within repeated DNA sequences: Application to the study of a family of transposable elements. *Bioinformatics*, 22 (16), 1948–1954.

Wang, Z., and W. R. Atchley. (2006). Spectral Analysis of Sequence Variability in Basic-Helix-Loop-Helix (bHLH) Protein Domains. *Evolutionary Bioinformatics*, 2, 199–208.

Yu, Z-G., and G-Y. Chen. (2000). Rescaled range and transition matrix analysis of DNA sequences. *Communications in Theoretical Physics*, 33, 673.

Problems

1. Write a program to convert amino acid strings to "residue interface propensities." A web search may be used to find tables of these values for each amino acid.

2. Convert similar DNA strings to a numerical format using Chang's method. Align these strings using a correlation.

3. Repeat Problem 2 using binary encoding. Correlate each channel independently and sum the four correlations. Does this encoding work better than that of Chang's?

4. Download Genbank files for two mammals (human, mouse, bovine, etc.) and one plant. These should be large files (at least 1 MB). Create the images similar to Figure 23-8 for these files.

5. Using the results of Problem 4, create the PCA organization of the data (see Figure 23-10).

24 Gene Expression Array Files

Gene expression data (or microarrays) have quickly become an important component in modern bioinformatics. Microarray data is available on many websites in many formats. This chapter will consider the images generated by some microarray systems and develop Python codes to read them.

24.1 Raw Data

The gene expression array is a slide that contains several thousand spots of genes. A typical experiment will use two different sources, each with its own dye (commonly Cy3 and Cy5). The detection process creates an image for each dye that consists of circular spots of varying intensities. The intensity of a single spot indicates the amount of source that is attached to that particular spot on the slide. Since there are two dyes, the experiment creates two images that are combined to form a single color image. The convention is that Cy5 is shown as red and Cy3 is shown as green.

Consider two simple cases: (1) a sample from a sick patient, with a red dye attached, and (2) a sample from a healthy patient, with a green dye. If a particular gene is responsible for the patient's illness, it should be present for the sick patient and absent for the healthy patient. Thus there is much more red dye at that spot than green. That is what would happen in a perfect world. In reality the difference in fluorescence of the dyes can be attributed to gene regulation or anomalies that occur while performing the experiment. Furthermore, some genes may be responsible for the phenotype (the patient's sickness or other factors) but are not necessarily differentially expressed in the experiment.

The image data can be stored in different formats, including raw data and specific image formats. This chapter will consider two common formats: raw data and GEL files. Both of these files are quite easy to read.

24.1.1 Reading Raw Data in Python

The **file.read** function in Python reads the data as a string, and the **fromstring** command converts data to integers or whichever format the user defines. Code 24-1 shows the case of reading in binary data and converting it to 16-bit numbers. The variable *bindata* is a string of bytes read from the file that represents the data in a binary format. The **fromstring** command will convert the string into 32-bit integers. If *bindata* contains 4 bytes, a single integer is returned. If *bindata* contains 8 bytes, two integers are returned. The *int16* in this case changes the conversion to 16-bit integers.

Desktop computers have developed using two basic architectures from Intel and Motorola. Macintoshes have used the Motorola technology (although not exclusively), and PCs have used the Intel technology. One of the differences between these two is the way data is stored. Numbers requiring more than one byte are stored in different orders by the two different architectures. Thus data stored in one architecture will appear to be garbage if read by the other. For example, a 16-bit number requires 2 bytes and the Motorola system stores them with the high byte first (*big endian*) whereas the Intel systems store the low byte first (*little endian*). The **byteswap** command will convert the data from one *endian* type to the other. So the last line shows this optional case.

```
>>> from numpy import fromstring, int16
>>> bindata = file( binaryDataFile ).read()
>>> data = fromstring( bindata, int16 )
# option
>>> data = fromstring( bindata, int16 ).byteswap()
```

Code 24-1

In the directory *marray* on the website associated with this book there are two raw data files—*cy3.raw* and *cy5.raw*—each with a small header of 13 bytes. Next there are 4 bytes that indicate the size of the data. The rest of the file is the binary, 16-bit, little endian data. The function **ReadRawFile** in Code 24-2 opens the file for reading in a binary format according to the option '*rb*'. The file opens with a pointer at the first location in the file. This pointer is moved 13 bytes to the beginning of the dimensions. Four bytes are read and converted to *int16* integers. The dimensions of this file happen to be $V=7000, H=2200$. The data is read and converted to an unsigned 16-bit integer, *uint16*. The final step converts the long vector of data into a matrix of the appropriate size.

```
# marray.py
def ReadRawFile( fname ):
    # read as binary data
    fp = file( fname, 'rb' )
    # move to dimensions
    fp.seek( 13L )
    # read dimensions
    a = fp.read( 4 )
    H,V = fromstring( a, int16 )
    # get data
    a = fp.read( V*H )
    data = fromstring( a, uint16 )
    data = data.reshape( (V,H) )
    return data
>>> cy3 = ReadRawFile( 'marray/cy3.raw' )
>>> cy5 = ReadRawFile( 'marray/cy5.raw' )
```

Code 24-2

These data arrays are 16-bit, but computer screens can generally show only 8 bits of resolution. Thus, if *cy3* were converted to an image, very little of the data could be seen.

24.1.2 Dealing with 16-Bit Data

Data that is 16-bits can still be manipulated, but it cannot be shown on the computer monitor. Code 24-3 computes the histogram of one of the arrays. The data in this array is shifted 8 bits to the right with the >>8 command, which throws away the 8 least significant bits. Thus the 16-bit numbers are converted to 8-bit numbers by keeping only the 8 most significant bits. This histogram is plotted using the log scale in Figure 24-1, which demonstrates that the majority of the information has very small pixel values. Thus, if the 16 bit image were converted to an image without any processing, only a few pixels would be bright enough to be seen. In these images these bright pixels are due to dirt or imperfections on the slides. In fact, the brightest pixels are exactly the pixels that need to be discarded.

To isolate a select group of bits, the binary AND operation is used. The ampersand selects the bits to display in Code 24-4. The hexadecimal number 0x03FC is 0000 0011 1111 1100, which is the mask of which bits to keep in the 16-bit number. A small portion of this array is shown in Figure 24-2.

```
>>> hst = akando.RangeHistogram( ravel(cy3)>>8, 65536 )
>>> akando.PlotSave('fig1.txt', hst )

gnuplot> set logscale y
gnuplot> plot 'fig1.txt'
```

Code 24-3

```
>>> a = cy3 & 0x02FC
>>> akando.a2i( a[1000:1500,1000:1500] ).show()
```

Code 24-4

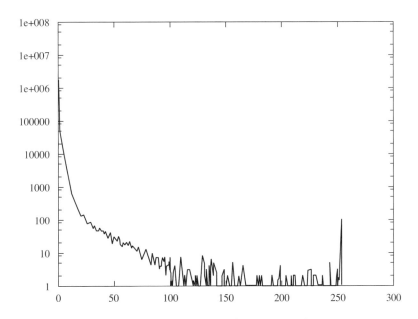

FIGURE 24-1 A histogram of the raw values of pixel values for a microarray image.

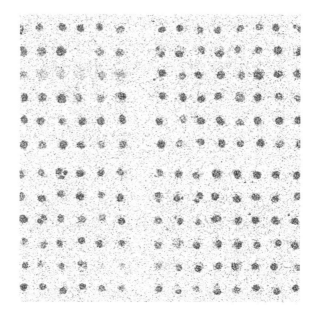

FIGURE 24-2 A small portion of the expression array from Code 24-4.

Commonly, the Cy3 dye is shown as green and the Cy5 dye is shown as red in an RGB image. The image in Figure 24-2 shows too much background, and so the mask is shifted a little in Code 24-5. The data is masked and converted to arrays named *red* and *green*. Again only a subset of the data is used for this example. A new image *blue* is also created since the last **Image.merge** command will require three channels of color data. The RGB image *mg* that has been created is shown in Figure 24-3. When

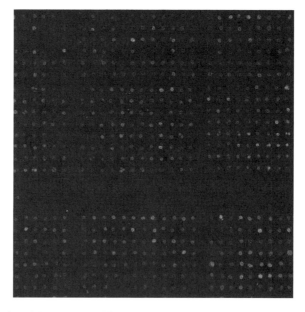

FIGURE 24-3 A portion of the array as an RGB image from Code 24-5.

displayed in color, some of the dots will be reddish, some greenish, and some yellow (which is a mix of red and green).

```
>>> from PIL import Image
>>> a = cy3 & 0x0FF0
>>> b = cy5 & 0x0FF0
>>> red = akando.a2i( b[1000:2000,1000:2000] )
>>> green = akando.a2i( a[1000:2000,1000:2000] )
>>> blue = Image.new( 'L',(1000,1000) )
>>> mg = Image.merge( 'RGB', (red, green, blue) )
>>> mg.show()
```

Code 24-5

24.2 GEL Files

Web-based databases contain images of microarrays. For example, the National Library of Medicine (NLM) stores images in a GEL format. These are uncompressed TIFF images with additional headers. This means that some TIFF image readers can display these images, but it also means that it is easy to write a program to extract the data. The files used in this example are from an NLM website found at http://www.ncbi.nlm.nih.gov/geo/query/acc.cgi?acc=GSM121436. This site contains two zip files for the two different dyes. For this example, the file GSM121436_a.gel will be considered.

The Python Image Library (PIL) has a problem reading these images, but it is easy to write a program to read them. The TIFF file format has numerous variations, and it is not uncommon for readers to have problems with some TIFF images. The TIFF format contains a header, TIFF tags, and image data (Murray and van Ryper, 1995). These files can be quite complicated, and although this section is by no means a complete study of the TIFF variations it will provide enough to read the GEL images from the NLM.

The basic difference between TIFF and GEL is that the TIFF file format allows the user to add extra tag information to the file. The image information is, however, not altered in the GEL files. These files have three sections, although TIFF images can have many more. Figure 24-4 shows the three sections in the order in which they are found in the file. The header is first, then the image data, and finally the tags stored in

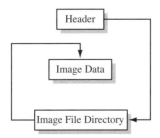

FIGURE 24-4 The arrangement of data in a GEL file.

the image file directory (IFD). The arrows, however, indicate the manner in which the data is read.

24.2.1 TIFF Headers

The header for a TIFF file contains 8 bytes. Code 24-6 opens the file for reading as a binary file and reads in the first 8 bytes. These are printed to the console but in a confusing manner. The *data* is actually a string but each character in the string represents 1 byte from the file. Because the file is not now storing text, some of the characters in *data* do not have an alphabet character equivalent. The first byte in the file is *0x49* (the *0x* indicates the hexadecimal representation instead of the decimal representation). As shown in Code 24-6, the ASCII character for *0x49* is the letter *I*. Thus, the first character printed in *data* is the *I*. Actually, the first two bytes in the file are *0x49* and so two *I*s are printed. The third byte is *0x2a*, which has the asterisk (*) as its ASCII equivalent. The fourth byte is *0x00*, which has no character assigned to it, and so Python prints \x00 to represent this byte. While printing *data* does provide information about each byte, some are shown as characters and others as their byte values. The last line uses **fromstring** to convert each character in the string to an unsigned 8-bit number (a byte), and it uses the **map** function to repeat the operation for all of the values. This command shows the byte value of all of the characters.

```
>>> fp = file( 'gsm121436_a.gel','rb' )
>>> data = fp.read(8 )
>>> data
'II*\x00\xbe\xbbs\x00'
>>> chr( 0x49 )
'I'
>>> map( hex, fromstring(data,'uint8') )
['0x49', '0x49', '0x2a', '0x0', '0xbe', '0xbb', '0x73', '0x0']
```

Code 24-6

The first two bytes in a TIFF represent the byte order format. The *II* represents Intel based technology (little endian), and a TIFF file starting with *MM* would represents Motorola technology (big endian). The next two bytes represent the version, and for a TIFF image these two bytes are *0x2a 0x00*. The final four bytes are the IFDOffset—or basically the location of the tag. In this case the four bytes are *[be bb 73 00]*. This is a 4-byte integer and the equivalent decimal value is 7,584,702. Code 24-7 shows this conversion.

```
>>> fromstring( data[4:], int )
array([7584702])
```

Code 24-7

24.2.2 The Image File Directory

The image file directory (IFD) contains information pertinent to the image. Because a TIFF file can contain multiple images, each image will need its own header. In this case the file contains a single image. The location of the header is *[be bb 73 00]*, and so the file pointer is moved to this location, as shown in Code 24-8. The result from **fromstring** is the array shown in Code 24-7, where the *[0]* is therefore added to the command to extract the integer. The **seek** command moves the file pointer to this location.

```
>>> tag = fromstring( data[4:], int )[0]
>>> fp.seek( tag )
```
Code 24-8

The beginning of the IFD is a list of tags that are associated with this image. Some tags are required, some are optional, and in this case some are unique to the GEL format. The first two bytes indicate the number of tags. Code 24-9 reads these two bytes and indicates that this file contains 27 different tags.

```
>>> a = fp.read( 2 )
>>> ntags = fromstring( a, int16 )[0]
>>> ntags
27
```
Code 24-9

Every tag contains 12 bytes. The first two bytes are an identifier, the next two bytes indicate the data type, the next four bytes indicate the number of items in the tag data, and the last four bytes is the data or an offset. The first tag is read and displayed in Code 24-10.

```
>>> a = fp.read( 12 )
>>> map( hex, fromstring(a,'uint8') )
['0xfe', '0x0', '0x4', '0x0', '0x1', '0x0', '0x0', '0x0', '0x0', '0x0',
 '0x0', '0x0']
```
Code 24-10

There are many types of tags that can be found in a TIFF file, and each one has an identifier. In this case the two bytes for the identifier are *[fe 00]*, which corresponds to describing that *NewSubFileType*. This will not be explored further. Tables of TIFF tag identifiers are found in Murray and van Ryper (1995) or on a multitude of websites. There are several tags that are crucial to viewing this file:

ImageWidth : [00 01] (256)
ImageHeight : [01 01] (257)
BitsPerSample: [02 01] (258)
Compression: [03 01] (259)
StripOffsets: [11 01] (273)

These include the image width, height, type of data, and the location of the data.

The numbers in the square brackets are the byte value, and the numbers in the parentheses are the decimal equivalents. The difference between a TIFF file and a GEL file can now be elucidated. The GEL files will add tags that are designated for

Molecular Dynamics Inc. These identifiers are of the type *[82 an]* where *n* ranges from 5 to c. A full list of TIFF tags including these specialty tags can be seen at Digital Preservation (2007).

There are 27 tags, and five are of interest here. Code 24-11 shows **FindImportantTags**, which is a simple program that will locate only the tags of interest. This function assumes that the file pointer *fp* is already located at the beginning of the first tag. The list *important* contains the IDs of the five tags of interest. These tags are all of the same type, and the last four bytes in each tag is a 32-bit integer of the value. Inside the *for* loop each tag is read to determine if it belongs to the list of important tags. If it does, the *ndx* becomes the location in *important* that matches the ID. The data is then read, converted to an integer, and stored in the appropriate location of *data*. It should be noted that other tags do not contain data that is a 32-bit integer, and so this program will not work for all types of tags. In this case the five numbers represent the image width (2,297 pixels), height (1,651 pixels), number of bits per pixel (16), the compression (1 = no compression), and the beginning location of the image data.

```
# marray.py
def FindImportantTags( fp, ntags ):
    important = [ 256, 257, 258, 259, 273]
    data = zeros( 5, int )
    for i in range( ntags ):
        tag = fp.read(12)
        idf = fromstring( tag[:2], int16 )[0]
        if idf in important:
            ndx = important.index(idf)
            d = fromstring( tag[8:], int)[0]
            data[ndx] = d
    return data

>>> FindImportantTags( fp, 27 )
array([2297, 1651,   16,    1,    8])
```

Code 24-11

24.2.3 Reading the Data

The programs that follow assume that the data is 16-bit and uncompressed. Additional assumptions are that each file contains only a single, grayscale image. Alterations to these assumptions will require alterations to the codes in this section.

The image size and the location of the data are known from **FindImportantTags**. In Code 24-12 the function **ReadGEL** will read the data with the above assumptions. It gets the location of the tags and moves to location *tag*. The number of tags is *ntags*, and the five important data values are read. Two *if* statements are used to catch cases in which the assumptions are violated. Finally, the file pointer is moved to the beginning of the data. The data is read, converted to unsigned 16-bit integers, and then reshaped to become a matrix.

```
# marray.py
def ReadGEL( fname ):
    fp = file( fname, 'rb' )
    # read header
    head = fp.read(8 )
    if head[:2] != 'II':
        print "Not an Intel based image."
        return -1
    # read location of tag
    tag = fromstring( head[4:], int )[0]
    # go there
    fp.seek( tag )
    # number of tags
    a = fp.read(2)
    ntags = fromstring( a, int16 )[0]
    width, height, nbits, comp, begin = FindImportantTags( fp, ntags )
    # confirm
    if nbits!=16:
        print "Not a 16-bit image"
        return -2
    if comp != 1:
        print "Image Compressed"
        return -3
    # read
    fp.seek( begin )
    data = fp.read( width*height*2)
    data = fromstring( data, uint16 )
    data = reshape( data, (height,width) )
    return data
>>> fname = 'data/gel/gsm121436_a.gel'
>>> data = ReadGEL( fname )
>>> akando.a2i( data ).show()
```

Code 24-12

The reason that these codes may be needed is that many TIFF readers cannot handle 16-bit data. Other readers may convert the 16-bit data to 8-bit data, which is undesirable. The **ReadGEL** function will extract the original data and bypass TIFF readers.

24.3 Summary

Gene expression (or microarray) files contain the image created by the scanning machine. This chapter explored two types of file and how to read these files. The popular GEL file is a TIFF file with additional headers. As long as this is an uncompressed file, it is quite easy to read. It should be noted that different labs do store data in different formats. The codes in this chapter are thus a good start in building programs to read multiple formats of gene expression images.

Bibliography

Digital Preservation. (2007). *Digital Preservation*. Retrieved from http://www.digitalpreservation.gov/formats/content/tiff_tags.shtml#table_1.

Murray, J. D., and W. van Ryper. (1995). *Encyclopedia of Graphics File Formats*. Sebastopol, CA: O'Reilly.

Problems

1. Write a program to read all of the Molecular Dynamics Inc. headers from a GEL file.

2. Write a program that automatically selects which eight bits are best for viewing a 16-bit image.

3. Write a program that encodes 16 bits of grayscale data into a three-color image (in which there are 24 bits that can be used).

4. Write a program that creates a GEL file. It should be readable by a third-party program that is capable of reading and displaying TIFF images.

5. Read a GEL file and store it as a JPG image. You will need to encode the 16 bits of data into two-color channels (8 bits each). Read the JPG image and compare it with the original data. Does JPG introduce small errors?

6. Read a GEL file. Determine how many bits are actually necessary. For example, if the first bit is completely random for pixels on a spot, it could be considered as unnecessary.

7. Create your own version of a GEL file in which you can insert headers that are not used by other vendors. Create headers for your name.

25 Spot Finding and Measurement

Expression array images contain thousands of spots, and the intensity of each of these needs to be measured. Before this can occur, it is necessary to find the spots. This chapter presents a method for finding the center of spots in a slide.

25.1 Spot Finding

Although microarray slides are printed by a computerized robot, the user defines the number of spots and the geometry of the spots on the slide. For example, the slides used here as examples contain 16×24 blocks, and each block contains 5×5 spots. Often individual spots are referenced by their block location as well as their location within the block.

Figure 25-1 shows just a small portion with six blocks each containing 5×5 spots. These spots are mostly in a regular geometry, but the geometry is not exact. It is difficult to see here that these GEL images also tend to be skewed. The number of spots in the top row and the bottom are the same. The distance from the first spot to the last spot in a row should be the same, but they are not. There is a slightly different magnification for different regions in these large images. The result is that a column of spots will not be exactly vertical and the angle at which it is tilted changes depending on the location of the column in the image. The same applies to rows.

One approach is to coarsely estimate where the spots would be using information about the original geometry and then to use a fine approach to actually isolate the individual spots. This, of course, is not the only plausible approach for isolating the spots.

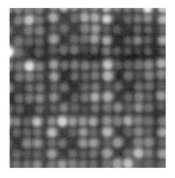

FIGURE 25-1 A small portion of a slide from the National Library of Medicine, with six blocks, each containing 5×5 spots. Please see http://www.ncbi.nlm.nih.gov/projects/geo/info/disclaimer.html

25.1.1 Intensity Variations

The intensities of the spots vary widely, which is important when measuring the DNA reactions. However, for spot finding this wide variation is a nuisance. Consequently, the **log** of the intensities is computed to give the spots similar intensities. Code 25-1 reads the GEL file and computes the **log**. There is also a bias subtraction to make the pixels with the lowest intensity close to 0. In actuality Figure 25.1 is the log of the intensities.

```
>>> from numpy import greater, log
>>> fname = 'data/gel/gsm121421_a.gel'
>>> data = ReadGEL( fname )
>>> lata = log( data )
>>> lata -= 7.2
>>> lata *= greater( lata, 0 )
>>> akando.a2i( lata ).show()
```

Code 25-1

25.1.2 Block Location

The first step is to find the beginning of the blocks. Each block is a 5×5 array of spots that are spaced roughly 17 pixels apart, and the blocks are about 90 pixels apart, leaving 22 pixels between the last spot of one block and the first spot of another block.

Since the spots are mostly vertical, it is assumed that over a small range (500 rows) they are vertical. This makes the detection of the columns quite easy. Code 25-2 sums 500 rows of the image and creates the plot in Figure 25-2. There are spikes at the location of the spot columns. However, the graph is not perfect. There are spuri-

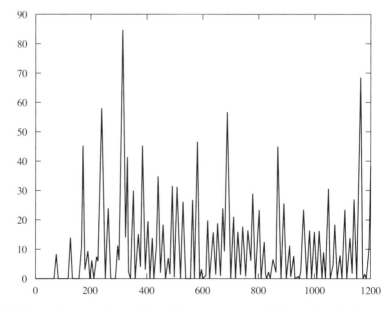

FIGURE 25-2 The vertical sum of rows 500 to 1,000 of the processed GEL image. Only the first part of the vector is shown due to its long length.

ous spikes due to the background (for example, $x = 90$) and very weak spikes in the columns ($x = 600$).

```
>>> vsum = lata[500:1000].sum(0)
>>> akando.PlotSave('fig2.txt', vsum )
```
Code 25-2

The pattern being sought is a set of five peaks spaced 17 pixels apart and 22 pixels to the sixth spot. The pattern search begins with the identification of the location of the peaks, including those peaks that are weak. A peak is defined as the case where $vsum[k] > vsum[k-1]$ and $vsum[k] > vsum[k+1]$ are both true. The function **MarkPeaks** in Code 25-3 creates a matrix M that has three rows. The top row is the vector and the next two are the same vector shifted one element to the left and right. Thus, the kth column contains $vsum[k]$, $vsum[k+1]$, and $vsum[k-1]$. The mx indicates which row is a maximum for each column. If $mx[k] = n$, then for column k the nth row contains the largest value. Columns in which the first row contains the maximum are returned as nz. This is the location of the peaks in vec.

```
# marray.py
def MarkPeaks( vec ):
    N = len( vec )
    M = zeros( (3,N), float )
    M[0] = vec + 0
    M[1,:-1] = vec[1:] + 0
    M[2,1:] = vec[:-1] + 0
    mx = M.argmax(0)
    nz = nonzero( equal(mx,0) )[0]
    return nz
```
Code 25-3

The peak locations are already sorted from lowest to highest in terms of their locations, not their heights. Code 25-4 shows the subtraction of consecutive points. The pattern 17, 17, 17, 17, 22 is evident. This segment shows the presence of two columns.

```
>>> mp = MarkPeaks(vsum )
>>> mp[100:110] - mp[99:109]
array([17, 18, 16, 17, 23, 17, 17, 17, 17, 21])
```
Code 25-4

The function **Scorefor5** in Code 25-5 considers every peak in mp and computes the difference of five consecutive peaks. It searches for the pattern 17, 17, 17, 17, 22. The $scores[i]$ is the squared error between the target pattern and the actual differences between consecutive peaks starting with the ith peak.

```
# marray.py
def Scorefor5( nz ):
    vec = array( nz )
    diff = vec[1:] - vec[:-1]
```

```
            mask = array( [17,17,17,17,22] )
            N = len( diff )
            scores = zeros( N-5, float )
            for i in range( N-5 ):
                a = abs(mask-diff[i:i+5] )
                scores[i] = (a*a).sum()
            return scores

>>> sc5 = Scorefor5( mp )
```

Code 25-5

Every peak in *mp* will get a score, but it is only the best scores that are sought. In this case the best score is the lowest value. With 24 blocks in the horizontal direction, the top 30 scores are shown in Code 25-6, and there is a definitive set of 24 scores that are far below all other scores. The location of these scores (and consequently the horizontal location of the beginning of each block) is given by *mp[ag[:24]}*.

```
>>> ag = argsort( sc5 )
>>> sc5[ ag[:30]])
array([  1.,   1.,   1.,   1.,   1.,   1.,   2.,   2.,   2.,   2.,   2.,   2.,
         3.,   3.,   4.,   4.,   4.,   4.,   4.,   5.,   5.,   6.,       25.,  25.,  33.,
        34.,  35.,  41.,  41.,  41.])
>>> mp[ag[:24]]
array([ 869, 1321, 1500, 1682,  418,  778, 1139,  958,  507, 1951,  237, 1861,
        327, 1048, 2133,  688, 2041, 1410, 1771,  597, 1591, 1229, 2167,  142])
```

Code 25-6

Code 25-6 computed the location of the first spot in each block for 500 rows of the image. This process needs to be repeated for all rows of the image. In Code 25-7 the function **VertMarks** repeats these computations for all 500 row segments of the image. The *gamma* is the maximum score allowed, and from viewing the results in Code 25-6 a threshold of 20 is set as the default. The value of *K* is 500 until the last segment of the image is considered, which may not have 500 rows. The *mark* is the array that indicates the location of the beginning of each block. Code 25-8 repeats those computations searching for the vertical location of the blocks.

```
# marray.py
def VertMarks( lata , gamma=20 ):
    # allocate
    V,H = lata.shape
    mark = zeros( (V,H) )
    # for each 500 row chunk
    for i in range( 0, V, 500 ):
        # cut up in to 500 row chunks: vsum
        vsum = lata[i:i+500].sum(0)
        # find the 120 best peaks
```

25.1 Spot Finding

```
        mp = MarkPeaks( vsum )
        sc5 = Scorefor5( mp )
        # mark
        ag = argsort( sc5 )
        vec = zeros( H )
        for j in ag:
            if sc5[ag[j]] < gamma:
                vec[mp[ag[j]]] = 1
        K = 500
        if i+500>V:
            K = V-i
        for j in range( K ):
            mark[i+j] = vec + 0
    return mark
```

Code 25-7

```
def HorzMarks( lata, gamma=15 ) :
    # allocate
    V,H = lata.shape
    mark = zeros( (V,H) )
    # for each 500 row chunk
    for i in range( 0, H, 500 ):
        print i
        # cut up in to 500 row chunks; vsum
        hsum = lata[:,i:i+500].sum(1)
        # find the best peaks
        mp = MarkPeaks( hsum )
        sc5 = Scorefor5( mp )
        # mark
        ag = argsort( sc5 )
        vec = zeros( V )
        for j in ag:
            if sc5[ag[j]] < gamma:
                vec[mp[ag[j]]] = 1
        K = 500
        if i+500>H:
            K = H-i
        for j in range( K ):
            mark[:,i+j] = vec + 0
    return mark
```

Code 25-8

The intersections of the masks generated from **VertMarks** and **HorzMarks** is the location of the beginning of the blocks. These intersections are computed in **FinalMarks** in Code 25-9. The two masks contain binary values (0 or 1) with 1's marking the vertical and horizontal locations of the blocks. The intersection of these two blocks is computed by adding the two arrays and searching for values above 1.9. Technically, the values in *mark* should be only 0, 1, or 2, but since this is an array of floats the numbers may not be exact. The **greater** function is used to find pixels close to the value of 2. The **nonzero** function returns two lists with the *x* and *y* values respectively. It is much preferred to have a list of points (*x, y*) instead. The functions **array**, **transpose**, and **list** are used to make the conversion.

```
# marray.py
def FinalMarks( vmark, hmark ):
    mark = vmark + hmark
    mark = greater( mark, 1.9 )
    nz = mark.nonzero( )
    a = array( nz )
    a = a.transpose()
    b = list( a )
    return b

>>> vmark = VertMarks( lata )
>>> hmark = HorzMarks( lata )
>>> fm = FinalMarks( vmark, hmark )
>>> fm[0]
array([  67, 1588])
```

Code 25-9

The final step is to create a function that will mark locations in the image for a given set of points. The function **MarkCenters** in Code 25-10 makes 5 × 5 crosses at each location specified in the list *pts*. These are overlaid on top of the image *lata* and are guaranteed to be 10% brighter than any other pixel in the matrix, as shown in Figure 25-3.

```
# marray.py
def MarkCenters( lata, pts ):
    mx = lata.max()*1.1
    N = len( pts )
    marked = lata + 0
    for i in range( N ):
        v,h = pts[i]
        marked[v-2:v+3,h] = ones( 5 )*mx
        marked[v,h-2:h+3] = ones( 5 )*mx
    return marked

>>> marked = MarkCenters( lata, fm )
```

Code 25-10

25.1.3 The Coarse Grid

The coarse grid marks the estimated location of all of the spots in an array. With the beginning of the blocks known, the associated 5 × 5 array of spots is easy to estimate.

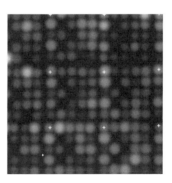

FIGURE 25-3 The plus signs mark the beginning of each block.

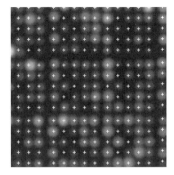

FIGURE 25-4 The plus signs mark the initial estimate of the center of each spot.

In this case the spots are 17 pixels apart and the function **InGrid** in Code 25-11 fills out the 5 × 5 array for each point in *fm*. A portion of the results is shown in Figure 25-4.

```
# marray.py
def InGrid( fm ):
    pts = []
    for p in fm:
        v,h = p
        for i in range( 5 ):
            for j in range( 5 ):
                pts.append( (v+i*17,h+j*17) )
    return pts

>>> pts = InGrid( fm )
>>> marked = MarkCenters( lata, pts )
```

Code 25-11

25.1.4 Fine-Tuning the Spot Locations

The spots are not exactly situated in the 5 × 5 array. Recall that these spots are on the order of a few microns in size. It is quite difficult to arrange and detect them in a perfect grid. Thus the next step is to consider each point in *pt* and to attempt to center it on its spot. Figure 25-5 shows the profile of one of the spots. It is not centered in the frame since its mark from *pts* is not on the center of the spot. However, the spot (like all good spots) has a smooth humped profile. The peak is easily detected and the location of the center mark has moved.

The function **FindTop** in Code 25-12 finds the peak in a vector and the function **AllSpots** in Code 25-13 repeats this for all spots. This function is set up to receive the original data, but in the example the *lata* is used. The **FindTop** function is called twice to fine-tune the location in the vertical and horizontal directions. The array *npts* is the array of newly estimated locations of the centers of the spots, and the results are shown in Figure 25-6. The *dist* is the distance that a point moved, and it is limited to 10 pixels. The reason is that a very dark pixel causes problems in **FindTop** since the neighboring spot will have the closest peak to the center. Thus any pixel that is scheduled to move more than 10 pixels is not moved at all.

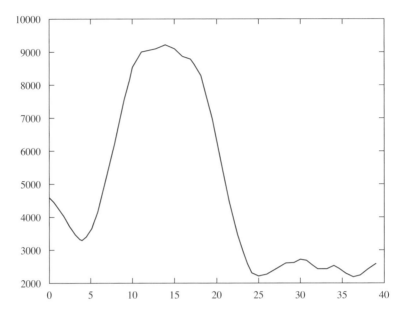

FIGURE 25-5 A plot of pixel values through an estimated center. In this case the estimated center is not located at the center of the circle.

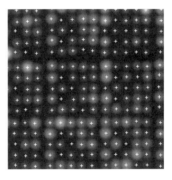

FIGURE 25-6 The plus signs mark the final estimation of the center of each spot.

```
# marray.py
def FindTop( vec ):
    D = len( vec )
    svec = akando.Smooth( vec, 3 )
    k = D/2
    drc = 1
    if svec[k-1] > svec[k]: drc = -1
    while svec[k+drc] > svec[k] and 5<=k<=35:
        k += drc
    return k
```

Code 25-12

```
# marray.py
def AllSpots( pts, data, Gamma=2000 ):
    N = len( pts )
    npts = []
    for i in range( N ):
```

```
        #if i%1000==0: print i
        v,h = pts[i]
        hsamp = data[v,h-20:h+20]
        nh = FindTop( hsamp )
        vsamp = data[v-20:v+20,h]
        nv = FindTop( vsamp )
        nv, nh = nv+v-20, nh+h-20
        dist = sqrt( (nv-v)*(nv-v) +(nh-h)*(nh-h) )
        if data[nv,nh] > Gamma and dist < 10:
            npts.append( (nv,nh) )
        else:
            npts.append( (v,h) )
    npts = array( npts )
    return npts
>>> npts = AllSpots( pts, lata, 0.1 )
>>> marked = MarkCenters( lata, npts )
```

Code 25-13

The final set of marks is in the list *npts*. These programs were not exactly perfect in that some of the blocks were not detected. Correction of these faults can be found in the problems at the end of this chapter.

25.2 Spot Measurements

Once the spots are located, the next step is to measure the qualities of the spot. This includes spot intensity, background intensity, roundness, and smoothness. The spots shown in Figure 25-1 are high quality, and quality measures are probably not too necessary. In some experiments, such as those shown in Figure 24-3, the spot qualities are poor and quality measures can assist in determining if a spot is valid.

Measuring the spot intensity requires that the on-target and off-target pixels be separated from each other. This means that the pixels inside the spot need to be identified. It is not required that the absolute border of the spot be identified, and pixels around the border of the spot can actually be ignored.

A typical spot centered in the frame is shown in Figure 25-7 and was obtained via Code 25-14. This is a 20 × 20 frame, and it does capture some of the neighboring spots, some of which can be brighter than the target spot.

FIGURE 25-7 A typical spot centered in a 20 × 20 frame.

```
>>> npts[1049]
array([213, 809])
>>> cut = data[213-20:213+20, 809-20:809+20]
>>> akando.a2i(cut).show()
```

Code 25-14

There are different philosophies about measuring spot and background intensities. In this simple approach the intensity is defined as the average of the pixels near the center of the spot. These pixels are contained within **Isolate** in Code 25-15. The pixels near the center are averaged and values deviating too far from the average are considered off target. The results of the example are shown in Figure 25-8.

```
from numpy import less
# marray.py
def Isolate( cut ):
    V,H = cut.shape
    V2, H2 = V/2, H/2
    avg = cut[V2-3:V2+4,H2-2:H2+4].mean()
    df = abs( avg - cut )
    targ = less( df, 1000 ).astype(int)
    return targ

>>> targ = Isolate( cut )
```

Code 25-15

A box that encompassed the target spot is determined by finding the extent of the white pixels in the vertical and horizontal directions. The box is expanded just slightly so that the border of the pixel contains no white pixels. To calculate the on-target value, the box is shrunk and values are collected from the *cut* only where *targ* is white.

The **FindLimits** function in Code 25-16 finds the borders and expands them 1 pixel. The **Measure** function in Code 25-17 cuts out the region for the single event *snip* and isolates those pixels that are white as *ontarg*. The values from these on-target pixels are then gathered in *valson* and the *intensity* is calculated. This function

FIGURE 25-8 A single spot is isolated in a smaller frame to delineate on-target pixels from other pixels.

returns the average target pixel intensity (*intensity*) and the average background pixel intensity (*backg*).

```
# marray.py
def FindLimits( targ ):
    V, H = targ.shape
    V2, H2 = V/2, H/2
    a = 1-targ[V2:0:-1, H2]
    up = V2 - nonzero(a)[0][0] - 1
    a = 1-targ[V2:, H2]
    down = nonzero(a)[0][0] + 1 + V2
    a = 1-targ[V2, H2:0:-1]
    left = H2 - nonzero(a)[0][0] - 1
    a = 1-targ[V2, H2:]
    right = nonzero(a)[0][0] + 1 + H2
    return up, down, left, right
```
Code 25-16

```
# marray.py
def Measure( targ, limits, cut ):
    up, down, left, right = limits
    snip = cut[up+2:down-2, left+2:right-2]
    ontarg = nonzero( snip )
    valson = snip[ontarg]
    intensity = valson.mean()
    border = concatenate( (cut[up,left:right],cut[down,left:right],\
                          cut[up:down,left], cut[up:down,right]))
    backg = border.mean()
    return intensity, backg
```
Code 25-17

These are only two of the measures that can be made for a spot. Quite often the standard deviations of the target and background pixels are also computed. Several quality measures have also been proposed. Wang, Ghosh, and Guo (2001) proposed the quality measure of the following:

$$q_{com} = (q_{size} \times q_{snr} \times q_{bkg1} \times q_{bkg2})^{1/4} \times q_{sat} \qquad (25\text{-}1)$$

Each of the q terms measures a different feature of the spot. The q_{size} measured the size of spot A and compared it to spot A_0:

$$q_{size} = \exp\left(-\frac{|A - A_0|}{A_0}\right) \qquad (25\text{-}2)$$

The q_{snr} measures the signal-to-noise ratio as

$$q_{snr} = 1 - \left[\frac{bkg_1}{sig + bkg_1}\right] = \frac{sig}{sig + bkg_1} \qquad (25\text{-}3)$$

The *bkg* is the standard deviation of the background, and the *sig* is the standard deviation of the spot. The saturation q_{sat} was 1 if 10% of the pixels were close to the maximum value. Two background measurements, q_{bkg1} and q_{bkg2}, were made to measure the local and excessive backgrounds as

$$q_{bkg1} = \frac{f_1}{CV_{bkg}} \qquad (25\text{-}4)$$

and

$$q_{bkg2} = f_2 \left\{ 1 - \frac{bkg_1}{bkg_1 + bkg_0} \right\}. \qquad (25\text{-}5)$$

Here the bkg_0 is the average of the background of all of the spots, and CV_{bkg} is the coefficient of variation. The *f* terms represent normalization, so that that spot with the largest measures is set to 1.

25.3 Summary

A microarray image contains thousands of spots laid out in a geometry specified in the lab. The spots are not exactly centered on the lattice, and the entire image is not centered in the frame. Furthermore, there is a bit of warping so that the scale at the top of the image differs slightly from those at the bottom.

Spot finding is the task of identifying the location of all of the spots in an array. This task is complicated by the fact that some spots are not visible. Usually, the spots are laid out in a two-tier lattice. In one of the cases shown here, there are 5 × 5 groups of spots that are arranged in the shape of a lattice. This knowledge is useful when attempting to find the spots.

The approach used here first performs a coarse finding in which the lattice is arranged. The next step is fine tuning, which considers each spot individually and finds the location that is a small deviation from the original lattice structure.

Bibliography

Wang, X., S. Ghosh, and S. W. Guo. (2001). Quantitative quality control in microarray image processing and data acquisition. *Nucleic Acids Research*, 29 (15), e75.

Problems

1. For a given GEL file, measure the vertical sums (Figure 25.2) for different window sizes (number of rows). Is there a window size that is too large?

2. Compute the amount of rotation within a GEL image. Basically, what is the angle that will make the columns of spots exactly vertical?

3. Compute the amount of skew within a GEL image. In other words, what is the scaling factor necessary to make the first and last rows exactly the same length? Is this skew linear?

4. Comments at the end of Section 25.1.4 indicate that the spot finding algorithm missed blocks around the perimeter. Identify the reasons for these exceptions.

5. Write a program that measures the circularity of a single spot. The circularity is a measure of how close the perimeter of the spot is to a perfect circle.

6. Write a program that measures the standard deviation of pixels on a single spot.

7. Write a program that performs Problem 6 for all spots. Identify the spots that have a high texture measure (the standard deviation divided by the mean).

8. Write a program that performs Equation (25-1) for a single spot.

9. Write a program that repeats Problem 8 for all spots. Identify the poor quality spots.

26 Spreadsheet Arrays and Data Displays

The previous chapter demonstrated methods by which array images can be read. Array scanners are accompanied by spot-reading software, and each array image is converted into a spreadsheet that contains information on spot locations, intensities, the gene represented at each spot, quality measures, and normalizations. Analysis of the data often begins with the spreadsheet information rather than the image.

A study usually has multiple array experiments and thus multiple array images. The platform (array architecture) used in the experiments is the same, but each trial produces a different set of spot intensities. Therefore a study will consist of a single platform file and multiple experiment files. These files are stored as tab delimited spreadsheet files so they can be read by any spreadsheet program or text editor. It is very easy to create Python programs to read these files.

26.1 Reading Spreadsheets

Reading a tab delimited spreadsheet is quite simple. The file is opened, read, and closed. The data is then parsed to extract the information that is required. The functions developed here will demonstrate the extraction of some of the data from the file. Extending the programs to extract all of the data is not difficult.

26.1.1 The Platform File

The platform file describes the architecture of the gene chip. Usually, each experiment runs several trials using the same architecture, and so each experiment has just one platform file. For example, an experiment presented at the National Library of Medicine (NLM) considers a COX-1 or COX-2 knockout effect in the brain. This page presents among other items a file for the expression array platform (GPL4006) and the files for the individual experiments (of which there are currently 30). The platform file can be read by a spreadsheet, and the first portion is shown in Figure 26-1.

The first ten lines are comments beginning with a pound sign (#). Line 11 is a header line that describes the columns, and lines 12 through 16907 contain data entries. It is easy to read part or all of this information. The first step is to save the file as a tab-delimited text file (which is the format that is downloaded from the website). The **ReadPlatform** function in Code 26-1 reads the file and splits the text on newline characters. This creates a list *rows* in which each item is a row from the spreadsheet. The index k is incremented until it finds the first line that does not begin with a pound sign. The next line is the header (line 11 in Figure 26-1), and it is stored in *header*. A

	A	B	C	D	E	F	G	H	I	J
1	#ID =									
2	#Grid Position = Position of spot on microarray									
3	#Name = Gene Name									
4	#Clone_description = Description of source of cDNA for microarray									
5	#Gene Names = alternative gene names									
6	#GB_ACC = GenBank Accesssion number									
7	#Symbol = Gene Symbol									
8	#Mouse Unigene = Unigene number									
9	#LocusLink ID = Locus link ID number									
10	#SPOT_ID =									
11	"ID**	Grid Position**	Name**	Clone_description	Gene Names**	GB_ACC**	Symbol**	Mouse Unigene**	LocusLink ID**	SPOT_ID*
12	1	GRID- 1-R1C1				no value	AA407331	<http://www.ncb	204508	106839
13	2	GRID- 1-R1C2	coiled-coil do	Homo sapiens cDN	RIKEN cDNA 1	BG063259	<Ccdc2	24494	67694
14	3	GRID- 1-R1C3	RIKEN cDNA	Homo sapiens U5	RIKEN cDNA 1	AW538380	2610031L17R	7039	68879
15	4	GRID- 1-R1C4	solute carrier	C. griseus mRNA fo	solute carrier fa	BG077242	<Slc35a1	22684	24060
16	5	GRID- 1-R1C5	expressed se	Homo sapiens unk	expressed sequ	BG077569	<AI839920	41440	103836
17	6	GRID- 1-R2C1	proteasome	Mus musculus prot	proteasome (pro	BG064542	<Psmb1	42197	19170

FIGURE 26-1 A section from the spreadsheet for the expression array platform GPL4006. The triangles indicate that the text is wider than the cell size.

dictionary *dct* is created, and each line of data will be stored in the dictionary. The dictionary key is the first column of the spreadsheet—the ID. The value of the dictionary entry is the rest of the data. Of course, it is possible to modify this program to store information in a list or in arrays as the user deems necessary.

This function returns the *header* and the *dct*. Since it is possible to have more than one experiment considered in an analysis, the *header* and *dct* are received with variable names that reflect the platform ID (which is GPL4006).

```
# marray.py
def ReadPlatform( fname ):
    d = file( fname ).read()
    rows = d.split('\n')
    # find title line
    k = 0
    while rows [k][0] == '#':
        k +=1
    header = rows [k]
    k +=1
    # parse title
    header = header.split('\t')
    # read in data
    NRows = len( d1) - k
    dct = {}
    for i in range( NRows ):
        items = rows[k+i].split('\t')
        if len(items)>1:
            n = int( items[0] )
            dct[n] = items[1:]
    return header, dct
>>> t4006, p4006 = ReadPlatform( 'marray/gpl4006.txt')
```

Code 26-1

26.1.2 The Z-Ratio File

The data from this experiment is the Z-ratio for each gene, which is the difference between two expressions. These simple files have only two columns: the gene ID and the Z-ratio value. In Code 26-2 the **Read2Col** function follows the same protocol as **ReadPlatform**. The exception is that within the *for* loop the second column of data is read and converted to a float. The dictionary created by this function will have the same keys as the dictionary returned by **ReadPlatform**, but the value will be a single float.

```
# marray.py
def Read2Col( fname ):
    d = file( fname ).read()
    rows = d.split('\n')
    # find title line
    k = 0
    while rows[k][0] == '#':
        k +=1
    header = rows[k]
    k +=1
    # parse header
    header = header.split('\t')
    # read data
    NRows = len( rows )-k
    dct = {}
    for i in range( NRows ):
        items = rows[k+i].split('\t')
        if len( items ) > 1:
            n = int( items[0] )
            val = float( items[1] )
            dct[n] = val
    return header, dct

>>> h121421, d121421 = Read2Col( 'marray/gsm121421.txt')
```

Code 26-2

The data from the experiment and the platform are now linked through the ID, which in both cases is the key of the dictionary. Code 26-3 shows the value and platform entries for gene 59. Of course, the headers for both are available as *t4006* and *h124121* from Codes 26-1 and 26-2.

```
>>> p4006[59]
['GRID-  3-R4C3', 'Transcribed sequences', '', 'expressed sequence C87102',
 'BG067676 http://www.ncbi.nlm.nih.gov/entrez/query.fcgi?cmd=Search&db
=Nucleotide&term=BG067676>', '', '25465', '97327', '']
>>> d121421[59]
-0.98522820
```

Code 26-3

26.1.3 Reading Two Channel Files

Some experiments store the information for the Cy3 and Cy5 expression levels. For this section an experiment measuring variations of normal humans is used (GSE65532007). There are ten arrays in this experiment, and each one contains gene expressions from two individuals. Some of the slides have two males, some have two females, and some have one of each. One possible question of this data is to find genes that are expressed in males but not in females. This experiment uses the HU800 platform. This format differs from the GPL4671 platform, but the **ReadPlatform** function will still read it. Figure 26-2 shows the beginning of the GPL4671 information.

The data files are very different from those in the Z-ratio files as they contain four sections. The first is several lines of information about the experiment, and these lines do not begin with a single character such as the pound sign. The second section contains analyzed data presented in 18 columns. The third section contains the spot intensity data, which is then used to create the second section. Section 3 contains 34 columns of data. Section 4 is very small and contains information about the filter used.

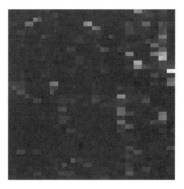

FIGURE 26-2 A heat map of the data. Because this is a color image, it will not distinguish the red and green portions well in print.

The third section contains information such as the intensity of the spot, the background intensity, standard deviations of both spots, and spot quality. There are two channels (Cy3 and Cy5), and there are several columns for each of these measurements. There are also columns for the gene ID, location on the slide, and gene name. The second section contains the gene ID, location, and name. There are also columns for the normalized values of the two channels.

In this chapter only a part of the data is used, and therefore only small portions of the data are read. Small modifications to the programs that follow will read other portions of the files. The function **ReadSection2** in Code 26-4 extracts only the information from the second section of the file. Basically, each row of data is stored without change in a dictionary where the key is the gene ID.

```
# marray.py
def ReadSection2( fname ):
    d = file( fname ).read()
    rows = d.split('\n')
    # find row that "Begin Measurements"
    k = 0
    while not("Begin Measurements" in rows[k]):
        k+=1
    k +=1
    # title
    header = rows[k]
    header = header.split('\t')
    k+=1
    # read data
    data = []
    while not ('End Measurements' in rows[k] ):
        items = rows[k].split('\t')
        data.append( items )
        k+=1
    return header, data

>>> h151667, d151667 = ReadSection2( 'hu800/GSM151667.txt' )
```

Code 26-4

There are 18 columns in this section, but for the following analysis only two of these columns are relevant. The headers for the columns can be seen either by examining the file in a spreadsheet or by printing *h151667*. The two columns of interest here are labeled *ch1 Normalized & BackSubtr* and *ch2 Normalized & BackSubtr*. These columns are the intensities of the two channels with the background intensities subtracted from the spot intensities. Furthermore, channel 2 has been normalized to channel 1 to eliminate biases present in the dyes Cy3 and Cy5. One such bias is that the dyes react differently to the laser illumination and detector.

The extraction of the pertinent information occurs in **GetMeasurements** in Code 26-5. This extracts only the gene ID (used as a key for the dictionary), the location of the spot, and the two aforementioned intensities. The spot location is stored in four columns to represent the block location and the location of the spot within the block. The variables *v* and *h* are computed to find the absolute location of the spot. These locations will be used later in displaying the information. The *ch1val* and *ch2val* are the values extracted from columns 14 and 15, which are the columns containing the desired data. In the example, gene 59 is located in row 2 (index starts at 0), column 18, and it has values of 1375.599 and 612.476 for the two channels.

```
# marray.py
def GetMeasurements( hudata ):
    dct = {}
    for d in hudata:
        k = int( d[0] )
        v = (int(d[1])-1)*20+int(d[3] )-1
        h = (int(d[2])-1)*20+int(d[4] )-1
        name = d[5]
        ch1val, ch2val = float( d[14]), float(d[15])
        dct[k] = (v,h,name,ch1val, ch2val )
    return dct

>>> dct151667 = GetMeasurements( d151667 )
>>> dct151667[59]
(2, 18, 'sperm associated antigen 1', 1375.599, 612.476)
```

Code 26-5

26.2 Displaying the Data

Microarrays contain thousands of data points, and visual analysis is one of the first steps in validating and understanding the information. This section reviews some of the popular methods of displaying microarray information.

26.2.1 The Heat Map

The *heat map* displays the intensities of the spots with 1 pixel per gene. One of the two channels is shown in red intensity and the other in green intensity. Thus any shade of yellow represents an equal mix of the two channels. The first step is to extract the information from the dictionary returned by **GetMeasurements**. In Code 26-6

HeatMapData collects all of the spot locations and puts them in *vs* and *hs*. The maximum values of these provide the size of the grid, *VxH*. The grids *heatg* and *heatr* are constructed for the two channels of data. The second *for* loop places the values from the dictionary into the proper location of the grid.

```
# marray.py
def HeatMapData( gmdct ):
    # gmdct from GetMeasurements
    N = len( gmdct )
    k = gmdct.keys()
    vs, hs = zeros( N, int ), zeros( N, int )
    for i in range( N ):
        vs[i], hs[i] = gmdct[k[i]][:2]
    V, H = vs.max(), hs.max()
    heatg = zeros( (V+1,H+1), int )
    heatr = zeros( (V+1,H+1), int )
    for i in k:
        v,h = gmdct[i][:2]
        heatg[v,h] = gmdct[i][3]
        heatr[v,h] = gmdct[i][4]
    return heatg, heart
```

Code 26-6

The second step is to convert this information into an image. In Code 26-7 the function **HeatMapPix** receives two arrays and assigns them to the red and green portions of an RGB image. However, the conversion process for these images uses two different scales. The **akando.a2i** function converts an array that has integer values ranging from 0 to 255 to a grayscale image. It differs from **akando.a2if** in that it does not rescale the data. In this case the minimum and maximum values in *heatg* and *heatr* are very different. The two arrays would thus be scaled differently, and the resultant image would be meaningless.

An alternative is to manually scale the two images to be on the same scale. In **HeatMapPix** the data is scaled according to the average and standard deviations of the arrays. The idea is that values of 0 and 255 should align with the 3-sigma limits of the distribution of values in the arrays. There are, of course, other methods to use to perform this scaling. The masks are used to convert all numbers less than the lower 3-sigma limit to 0 and all of those greater than the upper 3-sigma limit to 1. Figure 26-2 shows the heat map. Because this is a color image, it will not distinguish the red and green portions well in print. The color version is stored as *marray/heat.png* on the website that is associated with this book.

26.2 Displaying the Data

```
# marray.py
from PIL import Image
def HeatMapPix( heatg, heatr ):
    V,H = heatg.shape
    avgr, avgg = heatg.mean(), heatr.mean()
    devr, devg = heatg.std(), heatr.std()
    mnr, mxr = avgr-3*devr, avgr+3*devr
    r = (heatr-mnr)/(mxr-mnr)
    rmask = greater( r, 0 ).astype(int)
    r = r*rmask
    rmask = greater( r, 1 ).astype(int)
    r = rmask*1 + (1-rmask)*r
    mng, mxg = avgg-3*devg, avgg+3*devg
    g = (heatg-mng)/(mxg-mng)
    gmask = greater( g, 0 ).astype(int)
    g = g*gmask
    gmask = greater( g, 1 ).astype(int)
    g = gmask*1 + (1-gmask)*g
    r = akando.a2if( r*255 )
    g = akando.a2if( g*255 )
    b = Image.new( 'L', (H,V) )
    mg = Image.merge( 'RGB', (r,g,b) )
    return mg
>>> heatg, heatr = HeatMapData( dct151667 )
>>> mg = HeatMapPix( heatg, heatr )
>>> mg.show()
```

Code 26-7

26.2.2 The R Versus G Graph

The heat map is good for finding global biases, but it does not do a good job in displaying the distribution of data. A simple and only slightly more effective method is an R versus G graph. This and the ensuing graphs work better if the intensity values are removed from the dictionary and placed into a matrix. For this case, a 1600 × 2 matrix is created in which the two columns are the values from channel-1 and channel-2. This conversion is performed in **Dct2Mat** in Code 26-8.

```
# marray.py
def Dct2Mat( dct ):
    N = len( dct )
    mat = zeros( (N,2), float )
    matkeys = dct.keys()
    for i in range( N ):
        mat[i] = dct[matkeys[i]][3:]
    return matkeys, mat
```

Code 26-8

The *mat* is the matrix with the two columns. The R versus G graph is merely the plot of one column versus the other. Figure 26-3 shows this plot derived from Code 26-9. If the values of the two channels were identical for all genes, these marks would create a 45-degree line. Deviations from this line show cases in which a gene in one channel is expressed more than the other.

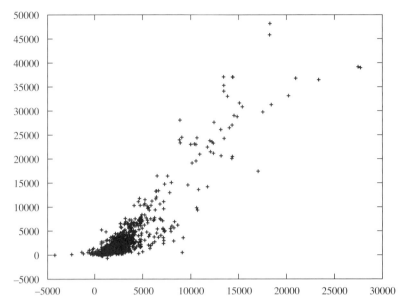

FIGURE 26-3 The plot of R versus G.

```
>>> mkeys, mat = Dct2Mat( dct151667 )
>>> akando.PlotMultiple( 'fig4.txt', mat )
```

Code 26-9

The R versus G graph does, however, have some problems. Commonly, a twofold expression is desired (one channel is expressed at least twice as much as the other). In this plot the vertical distance for a twofold expression varies for different locations on the x-axis. Furthermore, the data tends to fall along a 45-degree line that extends the scale of the graph. A graph with better resolution can be achieved by rotating the data 45 degrees in a clockwise direction, as with the R/G versus I graph.

26.2.3 The R/G Versus I Graph

Some of these problems are eliminated with the R/G versus I graph. In this case the x-axis is the intensity of both channels, and the y-axis is the ratio of the two channels. In Code 26-10 the function **RGvsI** creates a new matrix, with the ratio in the first column and the intensity in the second column. This data is plotted in Figure 26-4.

```
# marray.py
def RGvsI( data ):
    V,H = data.shape
    mat = zeros( (V,H) )
    mat[:,0] = (data[:,0]+data[:,1])/2 # intensity
    mat[:,1] = data[:,1]/data[:,0] # ratio
    return mat

>>> mat2 = RGvsI( mat )
```

Code 26-10

26.2 Displaying the Data

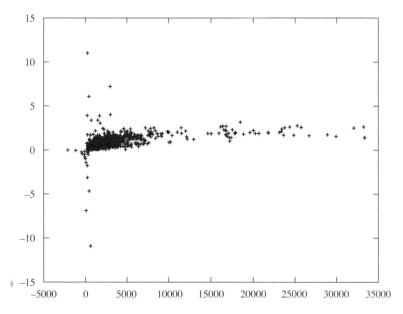

FIGURE 26-4 An R/G versus I plot illustrating the data in Code 26-10.

The horizontal nature of the data is basically the R versus G plot rotated. Deviations in the vertical direction are genes that are expressed differently in the two channels. The large vertical deviation near $x = 0$ is a mathematical problem with this approach. At low intensities, small variations in values create significant deviations in the ratios. Researchers debate whether this data is valid or not. Some say that the data variation is too great and large deviations are meaningless. Others argue that perhaps the R/G ratio is not the perfect measurement and that this plot does not justify discarding the data.

26.2.4 M Versus A Graph

The M versus A graph plots the log of the ratios to the log of the intensities, as shown in Code 26-11. This plot has several advantages. The first is that it compresses the range of the previous graph, which makes it easier to see trends in the data. The second is that a gene that is expressed or suppressed appears on the same scale. In the R/G versus I plot, a gene that is expressed in the red channel but not the green channel will have a value of 2 or greater. A gene expressed in the opposite manner will have a value between 0 and 1/2, and this range is quite small in the graph. In the M versus A plot, the average of the data is at $y = 0$, and the red expressed genes are $y > 2$ while the green expressed genes are $y < -2$. In this example **log2** is used, but any **log** function will work. Users of older versions of Python may not have a **log2** function and should use

$$\log_2(x) = \frac{\log(x)}{\log(2)}. \tag{26-1}$$

Figure 26-5 shows this plot.

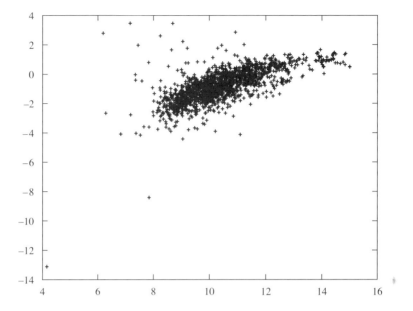

FIGURE 26-5 An M versus A plot illustrating the data in Code 26-11.

```
# marray.py
def MvsA( data ):
    V,H = data.shape
    mat = zeros( (V,H) )
    mat[:,1] = log2( abs(data[:,1]/data[:,0])) #M
    mat[:,0] = log2( sqrt( abs(data[:,1]*data[:,0]))) #A
    return mat

>>> mat3 = MvsA( mat )
```

Code 26-11

This plot has the same ideal as the R/G versus I graph. The x-axis is related to the intensity of the spot, and the y-axis is related to the R/G ratio. The advantages are that genes expressed in one channel but not the other appear at the same distance $y = 0$. This plot also reveals a bias because the cloud has an upward slant. This requires normalization, a topic that is discussed is in Chapter 27.

26.3 Summary

After a microarray image has been analyzed, the data is usually stored in a spreadsheet. Some data repositories simply store the spreadsheets. This chapter uses Python code to read a few different types of spreadsheets. This chapter also considers four methods that are used to display the values from a large number of spots: (1) heat maps, (2) R versus G graphs, (3) R/G versus I graphs, and (4) M versus A graphs.

Bibliography

GSE6553. (2007). *GSE6553*. Retrieved from http://www.ncbi.nlm.nih.gov/geo/query/acc.cgi?acc=GSE6553.

National Library of Medicine. (2007). *NLM*. Retrieved from http://www.ncbi.nlm.nih.gov/geo/gds/gds_browse.cgi?gds=2602.

Problems

1. Extract from a platform file (such as GPL4006) the names of the genes and alphabetize them.

2. Extract the gene names from a platform file and a data file (associated with that platform) and show that both files have exactly the same list of gene names.

3. Read a platform file and color code each gene. Create a map with 1 pixel per spot (similar to the heat map) that displays the color-coded genes.

4. Read a Z-ratio file (such as GSM121241.txt). Create a heat map based on the Z-ratio values.

5. Read a two-channel file (such as GMS151667) that can create a heat map image based on circularity measures. Is there a region in the image in which the circularity is less?

6. Read a two-channel file (such as GMS151667) that can create a heat map image based on background intensities.

7. Create a plot of the signal to noise ratio (SNR) versus intensity from a two-channel file.

8. In Code 26-4, the program reads two columns from Section 2 of the spreadsheet. These columns are *ch1 Normalized & BackSubtr* and *ch2 Normalized & BackSubtr*. Write a new program that reads *ch1 Intensity* and *ch2 Intensity* from Section 3 of the spreadsheet.

27

Applications with Expression Arrays

The M versus A plot from the previous chapter displays additional problems in the data. At lower intensities, there is a tendency for the second channel to be more expressed than the red channel. This is seen in the data, which has a general slant rather than a horizontal distribution.

27.1 LOESS Normalization

The LOESS normalization procedure, which was discussed in Section 5.3, is a simple method of removing this bias. The argument as to the validity of removing the bias is beyond the scope of this text. This method basically separates the data along the horizontal axis. For example, the data can be separated into five groups, with 20% of the data points in each group. Obviously, the groups will not have the same width along the x-axis. The average of each group is computed and subtracted. This simple method has the problem of having data points that are neighbors but are in two groups having different values subtracted.

Fortunately, the bias is linear. Code 27-1 determines the sort order of the first column of data. Each column is thus sorted according to this order and stored in *nmat*. The data is then separated into 20 groups, each with the same number of data points. The average for each group is computed and plotted in Figure 27-1. The line indicates that the bias is linear, which means that it is possible to use a linear normalization.

```
>>> from numpy import transpose
>>> ag = argsort( mat3[:,0] ) # mat3 from Code 26-11
>>> N = 1600; step = N/20
>>> nmat = zeros( mat.shape, float )
>>> nmat[:,1] = mat3[:,1][ag]
>>> nmat[:,0] = mat3[:,0][ag]
>>> avgs = zeros((2,20), float )
>>> for i in range( 20 ):
        avgs[:,i] = nmat[i*step:(i+1)*step].mean(0)
>>> akando.PlotMultiple('fig1.txt', transpose( avgs ) )
```

Code 27-1

Figure 27-2 demonstrates this normalization. The graph is separated into K number of groups, and each group has the same number of data points. The boundaries of three groups are shown as vertical lines. The boundaries are farther apart in the diagram than they will be in practice. The local average is computed for each group. The average for the middle boundary shown is the average of all of the data points in the

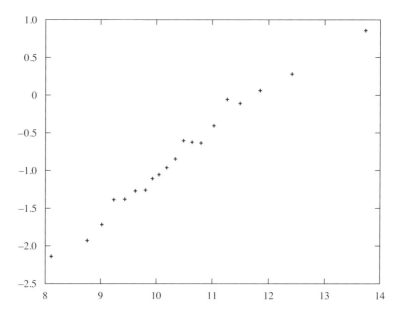

FIGURE 27-1 The plot of averages for each group versus the group location.

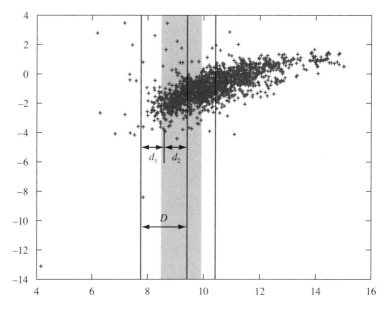

FIGURE 27-2 A graph showing three of the many boundaries (vertical lines), the averaging window for the middle boundary, and the distances from a selected data point to the encompassing boundaries.

shaded box. The extents of the box are halfway to the neighboring boundaries. A data point is modified by removing the weighted average of its encompassing boundaries. In the figure a data point is shown as being a distance d_1 and d_2 from the boundaries. A data point will be modified by

$$y'_i = y - \left(\frac{D - d_1}{D}a_1 + \frac{D - d_2}{D}a_2\right). \tag{27-1}$$

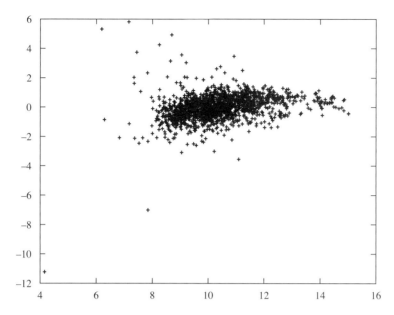

FIGURE 27-3. The data after the *Loess* function in Code 27-2.

The function **Loess** in Code 27-2 first sorts the data to make it easier to manipulate. The data points are rearranged so that the data for the *x*-axis is sequential. The boundaries are then located, and the first and last one are moved slightly to extend just beyond the first and last data points. The average for each boundary is computed, and then the *for* loop computes the weighted average for each data point and subtracts this average. In this case the number of boundaries is 21, making 20 groups with an equal number of data points. The results are plotted in Figure 27-3. The final two lines in this function sort the data back to the original order. This is necessary to align the data points after LOESS with the dictionaries that contain the gene names.

```
# marray.py
def Loess( mat ):
    N = len( mat )
    step = N/20
    # sort the data
    agg = mat[:,0].argsort( )
    smat = zeros( mat.shape, float )
    smat[:,0] = mat[:,0][agg]
    smat[:,1] = mat[:,1][agg]
    # find the boundaries
    ag = range( 20 )
    boundaries = smat[:,0][ag]
    boundaries = concatenate( (boundaries, [smat[-1,0]] ))
    boundaries[0] *= 0.999
    boundaries[-1] *= 1.001
    # find averages at the boundaries
    avgs = zeros( 21, float )
    avgs[0] = smat[:10,1].mean()
    for i in range( 1,21 ):
        avgs[i] = smat[i*20-10:i*20+10,1].mean()
    avgs[20] = smat[-10:,1].mean()
    # move points
    for i in range( N ):
```

```
            x = smat[i,0]
            nz = ( less( boundaries, x )).nonzero()[0]
            k = nz[-1]
            # weighted avg
            a1, a2 = avgs[k], avgs[k+1]
            d1, d2 = x-boundaries[k], boundaries[k+1]-x
            D = d1+d2
            wavg = (D-d1)/float(D)*a1 + (D-d2)/float(D)*a2
            smat[i,1] -= wavg
        # unsort
        umat = mat + 0
        umat[agg,1] = smat[:,1]
        return smat, agg
>>> umat = Loess( mat3 )
```

Code 27-2

27.2 Expressed Genes

The average of the data is now close to 0 for each group. Since this is the M versus A style of data, outliers are considered to be points above or below a certain threshold. Often a twofold threshold is used, and so data above $y = 2$ and below $y = -2$ are considered as genes that are significantly expressed between the two samples on the slide.

Code 27-3 plots the genes that have an M value above 2 (the first *for* loop) or below -2 (the second *for* loop). In Figure 27-4 the boxed points are those that rated as a twofold increase. As seen, most of the data points are in the low-intensity range where there is an argument as to their validity.

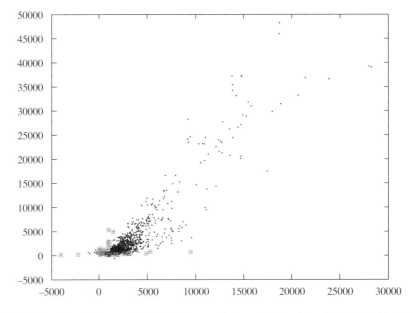

FIGURE 27-4 The data points are plotted in the R versus G format but the data points that had large expression are marked with boxes.

```
>>> nz2 = nonzero( less( umat[:,1], -2 ))[0]
>>> nz1 = nonzero( greater( umat[:,1], 2 ))[0]
>>> nz = concatenate( (nz1,nz2) )
>>> len( nz )
42
>>> temp = zeros( (42,2) )
>>> for i in range( 42 ):
      temp[i] = mat[nz[i]]+0

>>> akando.PlotMultiple('fig4a.txt', mat )
>>> akando.PlotMultiple('fig4b.txt', temp )
gnuplot> plot 'fig1.txt', 'fig2.txt' with points pt 3
```

Code 27-3

27.3 Multiple Slides

This experiment (GSE6553) uses the following multiple slides:

GSM151667	Sample ID F51_M58
GSM151668	Sample ID M58_M57
GSM151669	Sample ID M57_M56
GSM151670	Sample ID: M56_M55
GSM151671	Sample ID: M55_F53
GSM151672	Sample ID: F53_F52
GSM151673	Sample ID F52_F51
GSM151674	Sample ID: F53_F51
GSM151675	Sample ID: F52_M57
GSM151676	Sample ID: M55_M58

There were three females (F51, F52, F53) and four males (M55, M56, M57, and M58). Each slide contained gene probes for signal transduction (322 genes), inflammatory responses (50 genes), chemokines and their receptors (12 genes), and Th1 and Th2 cytokines (20 genes). Except for M56, each individual is tested on three slides.

The LOESS process normalized the data in a single slide but does nothing to normalize slides to each other. There are many parameters (some not very controllable) that affect the result of a gene expression experiment, and these vary from slide to slide. Methods of normalizing across multiple slides have therefore been proposed. In a recent study by Yang et al., (2002) simple statistics have been used to normalize the slides to each other. Python code for these methods is presented here.

27.3.1 Normalization

There are two major steps in normalization that were presented in Section 5.3 along with GnuPlot commands that create box plots or candlestick plots (Yang et al., 2002). The first is to make all of the averages the same, and the second is to make all of the standard deviations the same.

Code 27-4 shows **ReadManyFiles**, which reads the different files and extracts the matrix information. Code 27-5 shows **ManyMVA**, which performs the conversion to M versus A and performs the LOESS normalization. The **ManyStats** function

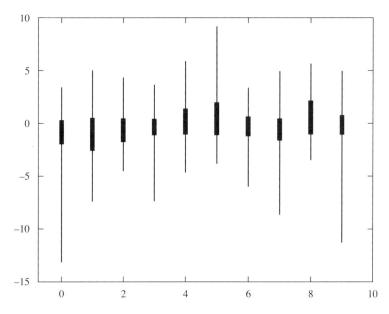

FIGURE 27-5 The boxplots representing the average and standard deviations of the ten slides.

in Code 27-6 converts the data to a format that GnuPlot can read. The data is shown in Figure 27-5.

```
# marray.py
def ReadManyFiles( dr, nms ):
    N = len( nms )
    data = []
    for i in range( N ):
        header, hudata = ReadSection2( dr +'/' + nms[i] )
        dct = GetMeasurements( hudata )
        data.append( Dct2Mat( dct ))
    return data

>>> nms = ['GSM151667.txt', 'GSM151668.txt', 'GSM151669.txt', 'GSM151670.txt',
'GSM151671.txt', 'GSM151672.txt', 'GSM151673.txt', 'GSM151674.txt',
'GSM151675.txt', 'GSM151676.txt']
>>> data = ReadManyFiles( 'hu800', nms )
```

Code 27-4

```
# marray.py
def ManyMVA( data ):
    N = len( data )
    D = len( data[0][1] )
    M = zeros( (D,N), float )
    A = zeros( (D,N), float )
    for i in range( N ):
        mva = MvsA( data[i][1] )
        A[:,i] = mva[:,0] + 0
        M[:,i] = mva[:,1] + 0
        mat = zeros( (D,2), float )
        mat[:,0] = A[:,i]+0
        mat[:,1] = M[:,i]+0
        smat, agg = Loess( mat )
        M[:,i] = smat[:,1] + 0
```

```
        return M, A
>>> M,A = ManyMVA( data )
```

Code 27-5

```
# marray.py
def ManyStats( M ):
    D,N = M.shape
    mat = zeros( (N,5) )
    for i in range( N ):
        mat[i,0] = i
        avg = M[:,i].mean()
        dev = M[:,i].std()
        mat[i,1] = avg - dev
        mat[i,2] = M[:,i].min()
        mat[i,3] = M[:,i].max()
        mat[i,4] = avg + dev
    akando.PlotMultiple( 'file.txt', mat )
>>> ManyStats( M )
```

Code 27-6

The first normalization step is to move the box plots up or down to make all of the averages the same. This is a very simple step. The function **SameAverage** in Code 27-7 computes the mean of all of the arrays and then subtracts the mean from them. Thus the average of each slide is now 0. The output of **ManyMVA** includes the matrix *M*. In this case there are ten columns in this matrix representing the ten slides. Therefore the *avgs* in the function will be the average of each column. Thus all ten columns are managed in a single command. Figure 27-6 shows the results with all boxes now centered at *y = 0*.

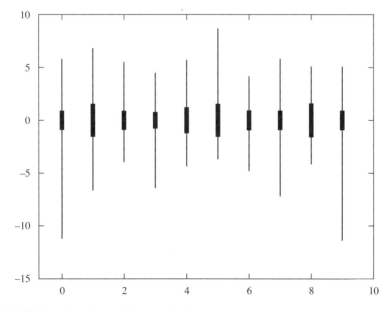

FIGURE 27-6 The box plots after the first normalization.

```
# marray.py
def SameAverage( M ):
    avgs = M.mean(0)
    nM = M - avgs
    return nM

>>> nM1 = SameAverage( M )
>>> ManyStats( nM1 )
```

Code 27-7

The second normalization is to make all of the standard deviations the same. These are made to be 1 by dividing the previously normalized data by the individual standard deviations. The function **SameStdev** in Code 27-8 performs this, and the results are shown in Figure 27-7, where all of the boxes are now the same.

```
# marray.py
def SameStdev( M ):
    dev = M.std(0)
    nM = M / dev
    return nM

>>> nM2 = SameStdev( nM1 )
>>> ManyStats( nM2 )
```

Code 27-8

Now the slides have been normalized and they can be compared to each other. An outlier in one slide with a value of $y = 4$ should have the same meaning as a point with $y = 4$ in another slide.

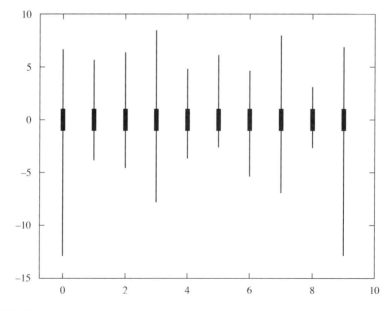

FIGURE 27-7 The box plots after the second normalization.

27.3.2 Extracting Outliers

Finally, the biological questions can begin to be answered. Depending on the question, the task is to find genes that are outliers in some slides but not others. Consider the question of finding genes expressed in men but not women. In the female-female (FF) slides, it is expected that these genes will not be strongly expressed and that the ratio M should be close to 1. The ratio M is also close to 1 for male-male (MM) slides, where the genes are strongly expressed. In the mixed slides (MF or FM) the male genes should be expressed in one channel but not the others. Thus one possible way of finding these genes is to find genes differentially expressed in the MF and FM slides but not in the MM or FF slides.

The function **Outliers** in Code 27-9 examines all of the data files and extracts genes that are above or below the *sigma* threshold. In Code 27-10 the program **SpecificOutliers** also receives a list of binary values that code which slides are desired to have a differentially expressed gene. The genes are scored according to the number of slides that obey the rules. In this case the search is thus for a gene that is differentially expressed in three of the slides but not in the other seven. The experiment determines which is the code for *yes* and therefore in which slides the genes are differentially expressed. In this test there are 26 genes that obtained a score of 9, which indicates that the gene was differentially expressed in the MF files but not in the FF or MM files except for only one file. Code 27-11 prints these genes. The next step would be to consider the function of the genes and the pathways that connect them.

```
# marray.py
def Outliers( M, sigma=3 ):
    D,N = M.shape
    outies = []
    for i in range( N ):
        a = greater( M[:,i], sigma )
        hit = list(nonzero( a )[0])
        a = less( M[:,i], -sigma )
        hit += list(nonzero( a )[0])
        outies.append( hit )
    return outies
```
Code 27-9

```
# marray.py
from numpy import take
def SpecificOutliers( outies, yes ):
    # collect outies in yes
    couts = []
    for i in range( len( outies )):
        if yes[i]==1:
            couts.extend( outies[i] )
    group = list( set( couts )) # list of yes expressed
    N = len( group )
    scores = zeros( N, int )
    for i in range( N ):
        for j in range( len( outies )):
            if group[i] in outies[j] and yes[j]==1:
                scores[i] += 1
            if group[i] not in outies[j] and yes[j]==0:
                scores[i] += 1
```

```
        ag = argsort( scores )[::-1]
        hits = take( group, ag )
        cnts = scores[ag]
        return hits, cnts
>>> couts = Outliers( M,2 )
>>> yes = [1,0,0,0,1,0,0,0,1,0]
>>> hits, cnts = SpecificOutliers( couts, yes )
```

Code 27-10

```
>>> for i in range( 26 ):
        print i, dct[ hits[i] ][5]

0 protein phosphatase 1, regulatory (inhibitor) subunit 8
1 retinoic acid- and interferon-inducible protein (58kD)
2 differentially expressed in hematopoietic lineages
3 integrin-linked kinase
4 major histocompatibility complex, class II, DR beta 1
5 tumor necrosis factor receptor superfamily, member 1B
6 SIAH1
7 EST
8 karyopherin (importin) beta 3
9 homeo box B2
10 ESTs
11 EST
12 Mucosal addressin cell adhesion molecule-1
13 protein kinase, AMP-activated, beta 2 non-catalytic subunit
14 NME4
15 HNRPDL
16 DKFZP586D1519
17 Hs.150555
18 KIAA0689
19 cAMP responsive element binding protein 3 (luman)
20 DKFZP586M1523
21 protease inhibitor 8 (ovalbumin type)
22 PDE6B
23 interleukin 13 receptor, alpha 1
24 Hs.150555
25 Human beta-1D integrin mRNA, cytoplasmic domain, partial cds
```

Code 27-11

This is a very simple test and certainly many more questions can be asked of this data. The purpose of this book is to present the use of Python in bioinformatics. While these biological questions are extremely important, it is not within our scope here to pursue these questions. Rather it is the intent of this book to provide readers with Python tools to pursue the questions that interest them most.

27.4 Summary

The data from a microarray will generally have biases from several sources. These could be dependent upon location in the plate, the nonlinearity of the detector response, etc. Thus normalization of the data is applied. A common normalization is LOESS, which considers subsets of the data for normalization. Another type of normalization compares multiple microarray slides so that different experiments can be compared with each other. The literature is still active with arguments as to the necessity and types of normalization that need to be used. The purpose of this chapter was to simply create programs to perform some of the methods.

Bibliography

National Library of Medicine. (2007). *GSE6553*. Retrieved from http://www.ncbi.nlm.nih.gov/geo/query/acc.cgi?acc=GSE6553

Yang, Y. H., et al. (2002). Normalization for cNDA microarray data: A robust composite method addressing single and multiple slide systematic variation. *Nucleic Acids Research*, 30 (4), 1–10.

Problems

1. Repeat LOESS normalizing using different window sizes. Is there a big difference?

2. For the example in Section 27.3, find the genes that are expressed in the MM slides but not in the MF, FM, or FF slides.

3. For the data in Section 27.3, find the genes that are differentially expressed in all slides.

4. For the data in Section 27.3, find the genes with the widest variation in differential expression.

5. Refer to Problem 26.8. Find the genes in the files in Section 27.3 in which the male had a high intensity and the female had a low intensity.

Index

A

affine gap, 114
akando, 37
alignment, 2, 12, 333, 337
    dynamic programming, 101
    gapped alignments, 179
    multiple sequence alignment, 157
    numerical sequence alignment, 345
    sequence alignment, 89
anchor, 51
and (logical), 367
antipersistent, 353
append, 5, 9, 11, 37
arange, 21, 27
argsort, 21, 39, 203, 256
array
    function, 13
    numerical, 8, 13, 16
arrow matrix, 102–113
ASN.1, 71, 84
assembly, 157, 160, 171, 179, 249
astype, 19
atof, 7
atoi, 7
average, 53
axis, 24, 259

B

backtrace, 105, 111–114, 182–184, 192
big endian, 26, 366
binary, 5, 25, 85, 323
binary tree, 198, 199
biopython, 12, 87, 287
BLOSUM matrix, 93, 94, 102, 178, 195
BlosumScore, 95, 96
BMU, 275, 314
box plots, 405, 407, 408
brain, 389
byteswap, 26, 366

C

C++, 1, 3–7, 10
candlesticks (GnuPlot), 59, 405
catsequence, 170
chaos, 354–360
chdir. *See* os.chdir
class, 3, 6, 51
ClustalW, 179, 194
cluster variance, 258, 269, 271
clustering, 255, 275, 311
coding DNA, 76, 80–83, 309, 328
coding region, 3, 76, 89, 254, 327, 362
codon, 82, 309, 327
codon frequency, 309, 315
command line, 49, 52
complex number, 2, 4, 13
complexity, 241
concatenate, 22
conditional probability, 66, 67
consensus sequence, 170
constituents, 255
constructor, 6
contig, 158
convert (Image), 30
convolution, 333
copy.copy, 18, 150
copy.deepcopy, 150
correlation, 37, 47, 333, 350
cos, 17, 38, 326
cost function, 144, 152, 251
count. *See* string.count
covariance matrix, 2, 63, 290, 359
crop (Image), 33

D

dancer, 37
DC term, 320
debugger, 11
deep copy. *See* copy.deepcopy
def, 6

413

default, 6
deletion, 90
destructor, 6
determinant, 64
dictionary, 9, 82, 127, 198, 228, 390
dictionary tree, 207
digital Fourier transform, 320
dimensionality reduction, 292, 312
divmod, 4, 113, 201
document analysis, 232
dot, 2, 25
dot product, 2, 16, 141, 291
dtype, 15
dynamic programming, 12, 89, 101, 159, 188

E

eigenvalue, 290
eigenvector, 290, 361
elif, 96
else, 4
emission, 125
energy surface, 143–144
equal, 17–19
error handling, 2, 6
execfile, 7
exp, 17
extend (list), 409

F

fast Fourier transform, 322
FASTA, 71
FFT. *See* fast Fourier transform
fft. *See* scipy.fftpack.fft
fft2. *See* scipy.fftpack.fft2
fftpack. *See* scipy.fftpack
file, 5
find. *See* string.find
float, 4
Floyd-Warshall, 266
fly embryo, 341–342
for, 2
Fortran, 1, 3, 25
Fourier transform, 27, 319
from - import, 2
fromstring, 26, 365
function, 6

G

GA. *See* genetic algorithm
gap penalty, 95, 103, 114, 195

gapped alignments, 179
gaps, 75, 91, 114, 159, 179
Gaussian distribution, 23, 55, 336
GC content, 347
GEL image, 369, 375,
Genbank, 71, 72, 74
gene expression arrays, 61
genetic algorithm, 12, 139, 174
getdata (Image), 35
getpixel (Image), 30
GIF, 31–32, 39, 358
global alignment, 91, 112
Gnuplot, 37–38
greater, 18
greater_equal, 19
greedy, 157

H

has_key (dictionary), 10
heat map, 392
Hidden Markov Model, 125
hierarchical clustering, 273
HMM. *See* Hidden Markov Model
Hurst exponent, 351–353
HWHM, 56
Hydrophobicity, 345

I

Idle, 11
If, 2
ifft. *See* scipy.fftpack.ifft
ifft2. *See* scipy.fftpack.ifft2
Image, 29
Image.merge, 31, 35, 368
Image.new, 31, 35
Image.open, 29
Image.save, 36
Image.split, 31, 35
imaginary part, 292
import, 7
inbreeding, 147
indels, 90
index, 9
indices, 8, 10, 23
inherit, 6
inner product, 16
insert, 9
insertion, 90
instance, 49

integer, 4, 6, 13
inverse Fourier transform, 319

J
Java, 1, 3, 14
jets, 115
join. *See* string.join
JPEG, 30–32

K
k-means clustering, 255, 259
key (dictionary), 9, 198, 228
knockout, 389
kurtosis, 56

L
less, 19
less_equal, 19
linalg, 64, 291
linear regression, 300, 352
linguistic complexity, 241
linked lists, 197
list, 8, 9
listdir. *See* os.listdir
little endian, 26, 366, 370
local alignment, 91, 112
logical_and, 19, 285, 312
logical_not, 20
logical_or, 19
logical_xor, 19
LOESS, 56, 61, 401
log, 17, 352, 367
log2, 17, 323
log odds, 68, 93, 131

M
map, 25, 92
matrix, 2, 13
mean, 23, 53, 55
merge. *See* Image.merge
microarrays, 365, 369, 393
mode (Image), 29, 30
module, 3, 7
moments, 56
multadd, 323, 334
multiple sequence alignment, 157
multivariate statistics, 63

N
Needleman-Wunsch, 112, 113, 117
negative documents, 232, 236, 238
nested repeats, 119
noncoding DNA, 327–328
nongreedy, 157, 169, 249
nonzero, 13, 175, 257, 350, 379
normal distribution, 55, 60, 63, 221
normalization, 56, 134, 401
not, 20
not_equal, 19
numpy, 2, 13, 56, 63, 92, 108

O
object oriented programming, 3
odds, 68, 131
ones, 13
os module, 11
os.chdir, 11, 49, 350
os.listdir, 232, 359
outer, 16, 255
outer product, 16
overloading, 6

P
PAM matrix, 94, 102, 117, 178, 195
Parseval's theorem, 321, 325
PCA. *See* principal component analysis
percolation tree, 209
periodic signals, 319, 331
persistent, 353
PIL. *See* Python Image Library
polar coordinates, 264–265
pop (list), 9, 216, 257
Porter stemming, 230
positive documents, 232
power spectrum, 322
principal component analysis, 12, 289, 312
print, 4
protein, 68, 75, 82–83, 89, 139, 298, 345
putdata (Image), 35
putpixel (Image), 30
Python Image Library, 29, 52

R
random, 3, 17, 23,
random.normal, 23, 52, 56
random.rand, 19, 23
random.ranf, 2, 17, 23

random.seed, 3, 53
random.shuffle, 23
range, 10
ravel, 22, 35
readlines, 5
real part, 293, 349
recurrent HMM, 130
reload, 7
remove (list), 9
repeats, 119
replace. *See* string.replace
reshape, 22
resize (Image), 33
return, 9
reverse [::-1] (string), 9, 81
rooted tree, 198
rotate (Image), 33
running average, 54

S

scipy, 13, 27, 324, 338
scipy.fftpack, 324, 338
scipy.fftpack.fft, 47
scipy.fftpack.fft2, 47, 325, 338
scipy.fftpack.ifft, 47, 325
scipy.fftpack.ifft2, 47, 325, 338
scoring matrix, 95, 102–117
seed. *See* random.seed
seek, 5
self, 6
self-organizing map, 12, 275, 314
sensitivity analysis, 302
Shakespeare, 209, 254
shallow copy, 150, 151
shape, 15
show, 30
shuffle, 23, 149, 250
simulated annealing, 139–143
sin, 17
singular valued decomposition, 290, 297, 306
skewness, 56,
slicing, 8, 14, 20, 39, 108, 231
Smith-Waterman, 112–117, 182–183
Smooth, 41–43, 328, 349, 381
SOM. *See* self-organizing map

sort (list), 21, 189, 218
species identification, 309, 354
splices, 76–81
split. *See* Image.split
split. *See* string.split
spot finding, 375
sqrt, 17
standard deviation, 23, 54, 221, 253, 286, 392, 405
steady state, 300
str, 6, 25, 37
string, 4–5
string.count, 76, 154
string.find, 74, 76
string.index, 95
string.join, 72, 121, 151
string.lstrip, 182
string.replace, 72, 78
string.rstrip, 182
string.split, 31
suffix trees, 217–220, 230
sum, 24
superstring, 246–254
SVD. *See* singular valued decomposition
Swiss roll, 262

T

Take, 21, 121, 259
tandem repeats, 119, 123
tell, 5
tensor, 13–14, 23
text mining, 12, 227, 232
third party software, 2, 27, 29, 36, 52,
TIFF image, 31, 369–374
tolist, 25
tostring, 25
transition, 134–137
transpose (array), 2, 34
tree, 197
tuple, 8–9, 13, 46, 108, 184, 327

U

unrooted tree, 198
UPGMA, 199–202
upper (string), 353

V
value (dictionary), 9
variance, 54
vector, 13–18, 21

W
wavelength, 43, 319, 329
while, 5

word count matrix, 232
word frequency matrix, 232, 234, 238
write, 5

Z
zero sum, 59, 63, 298
zeros, 13